From Turbulence to Climate

Springer-Verlag Berlin Heidelberg GmbH

Martin Beniston

From Turbulence to Climate

Numerical Investigations of the Atmosphere with a Hierarchy of Models

With 168 Figures and 28 Tables

Springer

Prof. Dr. Martin Beniston
University of Fribourg
Institute of Geography
Pérolles
1700 Fribourg
Switzerland

ISBN 978-3-642-63731-5

Cip data applied for
Die Deutsche Bibliothek - CIP-Einheitsaufnahme
Beniston, Martin: From turbulence to climate : numerical investigations of the atmosphere with a hierarchy of models ; with 28 tables / Martin Beniston.-
Berlin ; Heidelberg ; New York ; Barcelona ; Budapest ; Hong Kong ; Milan ; Paris ; Santa Clara ; Singapore ; Tokyo : Springer, 1998

ISBN 978-3-642-63731-5 ISBN 978-3-642-58787-0 (eBook)
DOI 10.1007/978-3-642-58787-0

Production: ProduServ GmbH Verlagsservice, Berlin
Typesetting: Camera-ready by author; Cover design: Springer-Verlag, E. Kirchner
SPIN:10629296 32/3020-5 4 3 2 1 0 - Printed on acid-free paper

Preface

This volume is a summary of work undertaken since 1978 when I completed my doctoral dissertation at the Laboratoire de Météorologie Dynamique (LMD) of the Ecole Normale Supérieure in Paris. Since that time, my career has taken me through different countries and institutions, many of which have strong links to climate research in one form or another. It would be difficult to thank the numerous persons whom I have met in the course of my professional life and who contributed directly (for example as co-authors), or indirectly, to the papers which I have published in various journals.

I am particularly indebted to Dr. Gilles Sommeria, who at the time was my thesis supervisor, for his advice during the preparation of my dissertation, and for his enthusiasm which encouraged me to take up a career in scientific research. My thanks also go to Prof. Ernst Augstein, Prof. Hans Hinzpeter, and Prof. Klaus Hasselmann for giving me the opportunity to carry out advanced research at the Max-Planck-Institute for Meteorology, from the early to mid-1980s.

In Switzerland, the support of the Swiss National Science Foundation is acknowledged for the continued development of modeling activities in both the mesoscale and the global climate fields since my arrival in this country in 1985. I would like to express particular thanks to Prof. Atsumu Ohmura of the Geography Institute of ETH-Zürich, who has supported my efforts at establishing the first global climate modeling group in Switzerland. This has involved establishing collaborations with a number of Swiss and international institutes. Prof. Ohmura was prepared to take the risk, in the early 1990s, that climate modeling could be a viable avenue of research in Switzerland, at a time when it was unclear whether the project would succeed due to problems related to funding, computer time allocation, and the engagement of our international partners.

I would like to thank the reviewers of this text, Prof. Huw C. Davies (Atmospheric Physics, ETH-Zürich), Prof. Hartmut Grassl (World Climate Research Program, WMO, Geneva), and Prof. Michel M. Verstraete (Joint Research Center - CEC, Ispra, Italy), whose comments led to substantial improvements to the original manuscript.

I am indebted to Mrs. Sylvie Bovel-Yerly, who edited and formatted the manuscript in an extremely competent and timely manner.

Finally, my thoughts go to my wife, Martine, and my son, Daniel, whose encouragement allowed this volume to materialize. During the intensive drafting phases of this volume, it was not always easy to reconcile the real world with the virtual, numerical world of atmospheric models!

Fribourg, Switzerland
July 1997

Martin Beniston

Table of Contents

1. Introduction

1.1 Issues related to numerical modeling

Mathematical models represent the manner in which we synthesize our understanding of a particular problem of physics, chemistry, biology, or economics. In the field of meteorological and climate research, atmospheric models are those which attempt to take into account various elements of the system which may be important to its evolution, whatever their scale. The rationale for any form of modeling is that it is usually impossible to conduct experiments with the atmosphere - except inadvertently - so that the understanding of the fundamental underlying mechanisms governing atmospheric flows, climate, and climate change can come about only through observations or modeling. Observational data provides much information on current or past atmospheric conditions, but does not provide any picture of the future evolution of the system; it is assumed here that extrapolation to the future based on past observations cannot be considered as a valid technique for most purposes. Models therefore constitute the *only predictive tool* available for investigating a particular system or process, whether it is the behavior of air pollution in a mountain valley or the response of the global atmosphere to enhanced concentrations of radiatively-active gases. Advanced models also enable investigations of certain processes which may be ultimately relevant, such as feedback mechanisms between different elements of the system. Sensitivity analyses can be attempted which observation and the real world does not allow.

Models are central to our concepts of environmental physics; they cannot be dissociated from observational data, however, although it is debatable whether the link between data and models is currently treated in the most optimal manner. It is often erroneously considered that there is a "natural" separation between models and experimentation or observation, and indeed much of the scientific community today is split into two distinct groups, those working with data and those working with models. There is, however, no intrinsic reason to separate the two approaches to investigations of the natural environment: data and models are by necessity complementary. Used in isolation, each approach is often meaningless.

In order to interpret observational data, one needs to have *a priori* formulated some form of conceptual model to explain the information the data is providing. A very simple example is the measurement of temperature with a mercury or alcohol thermometer; the thermometer is providing the observer with a measurement of the height of fluid in the tube, which is converted to a value of temperature. However, the fact remains that the thermometer is not measuring temperature directly, but the length of the column of mercury or alcohol, which according to molecular physics and thermodynamic principles expands or contracts with changes in temperature. The observer has in this case already a theoretical idea of the processes involved which lead to the desired data, i. e, temperature. This is also the case with more sophisticated observational techniques: satellite images of climatological variables (temperature and moisture) are not direct measurements of those variables but measurements of radiation values which need to be converted to the required physical variables based on the theory of radiative transfer. Here again the observer needs to have a concept, or model, in mind which allows a coherent interpretation of the information received by the satellite. In climate research, proxy data are those which infer a particular climate variable from other sources of data, such as temperature from isotopes contained in lakes or ocean records, or from pollen types contained in sediments which point to past environmental conditions. As for remote-sensing techniques, these are not *direct* measurement of the required climatological variable, and as a consequence there is the need for some form of model to understand the processes linking the proxy record to a particular value of temperature or precipitation.

In summary, therefore, a model is a tool which allows us to test our understanding of a particular system in physico-mathematical terms. The processes predicted by the model need to be upheld by observational evidence; if a particular parameterization incorporated in a model were to be disproved by systematic analysis of data, then the model would need to be modified to accommodate the conclusions drawn from observations. As in most areas associated with basic physics, data is used here to test the limits of the model concept, and therefore the type of data selected for this task should not chosen at random.

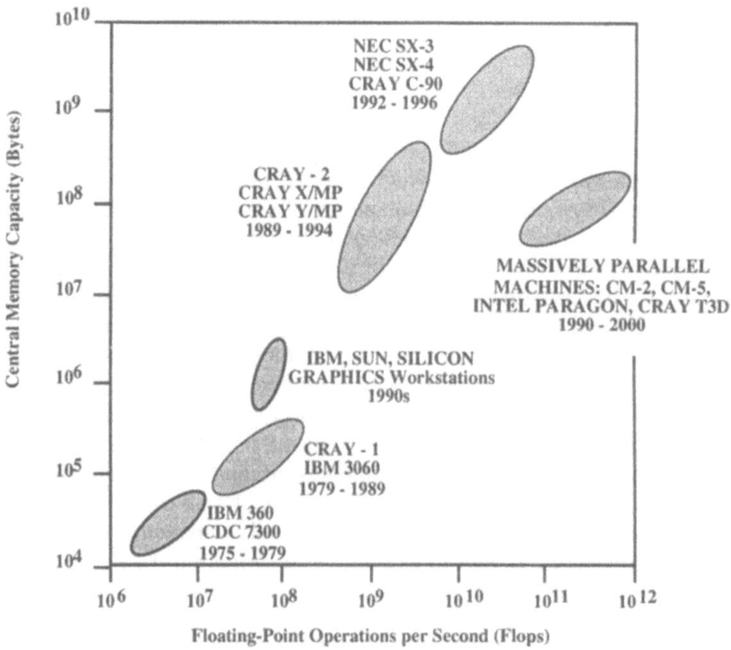

Figure 1.1: Evolution of computer speed and memory size since the 1970s

The interest in numerical simulation of atmospheric physics began already in the late 1940s with the advent of computers capable of treating some of the complexity inherent to meteorology, such as the ENIAC machine built in 1946, although probably the first attempt to compute the future state of the atmosphere from a given initial state by the application of the laws of motion was that of Richardson (1922). Computers have since the 1960s undergone an extraordinary technological revolution unparalleled in any other industry; speed and capacity of modern machines double about every 18 months. Figure 1.1 provides a schematic illustration of the evolution of computational power from the early 1970s through to the end of this century; both set of axes are logarithmic.

In a first phase, atmospheric models were limited to rather simple representations of processes in one or two spatial dimensions; advances were also made in the numerical treatment of primitive equation sets (e.g., Phillips, 1957; 1959; Hinkelmann, 1959). Various models were developed in the early sixties to study phenomena acting on vastly different time and space scales. In particular, a major breakthrough in General Circulation Models (GCM) can be attributed amongst others to Smagorinsky (1962; 1967) whose pioneering work at GFDL (the Geophysical Fluid Dynamics Laboratory at Princeton University, USA) paved the way for the much more complex GCMs to come. Numerical models were developed widely from the 1970s onwards, and followed closely progress made in computing speed and capacity. Models became more sophisticated and contained more detailed physics. For example, the Deardorff (1972) ABL model was subsequently modified by Sommeria (1976) to take into account phase changes of water. Pielke's (1974) mesoscale model has undergone constant evolution to become one of the most widely used models of its type globally, both for air pollution studies and, more recently, to investigate small-scale processes which could be relevant in a climatological context.

Numerical methods have also considerably evolved. Finite difference schemes, upgraded finite difference with spline interpolation (Pielke, 1984), spectral models such as the ECMWF (European Center for Medium-range Weather Forecasts, Reading, UK) weather forecast model, finite volumes, finite element techniques are all used in numerical simulation models. Advanced programming software are increasingly being developed to allow complex models to be implemented on vector supercomputers and, more recently, on massively parallel machines (see for example Gruber et al., 1990).

This work reflects a summary of over 15 years of research in numerical modeling by the author, with supporting material from a number of other sources, taken either from close collaborators or from the wider international literature, which are intended to help clarify certain issues. Its focus is on the treatement of small-scale processes in a hierarchy of numerical models of the atmosphere. While one of the more obvious small-scale features in atmospheric models is turbulence, a major part of this work will not be devoted to turbulence as a stand-alone process, but rather with turbulent-related phenomena such as those associated with cumulus-type clouds in boundary-layer models, pollution dispersion in regional-scale models, cloud-radiation interactions in mesoscale models, and boundary-layer physics in continental-scale limited area models (LAM) and global-scale general circulation models (GCMs). The objective is to underline the importance of the small scales on the larger-scales and the implications of interactions between scales. It will be seen that these are fundamental in the time evolution of atmospheric dynamics and thermodynamics, whatever the scale considered; as a consequence, it will be shown that a coherent mathematical description of small-scale processes in atmospheric models and, indeed, in most geophysical fluid dynamics applications, is essential if one wishes to accurately simulate atmospheric physics and climate. In this manuscript, small-scale processes which have a physical meaning as illustrated in Figure 1.2 will often be addressed in models through a variety of subgrid-scale parameterizations. The subgrid-scale processes are generally grid-scale dependent, in that some form of diagnosis based on model solutions at the resolved scales is required, but are independent of the physical scale of the process considered.

In order to address many of the above-mentionned issues, the present work has been divided into three principal sections, which include:

- a brief overview of the problem of scales and scale-separation,
- a description of microscale, mesoscale, and global-scale models, with examples of subgrid-scale physical parameterizations and results from model simulations,
- analyses of couplings between models of different scales, and links between models and observational data.

1.2 Issues of scale

Figure 1.2 illustrates the range of possible atmospheric scales (adapted from Orlanski, 1975); these may be broadly subdivided into the following three scale categories:

- The microscale, characteristic of turbulent processes
- The mesoscale, where regional-scale atmospheric circulations dominate
- The macroscale, associated with synoptic features and climate

It is seen that key processes are located along a diagonal in the space-time diagram. It is to be noted that processes taking place at large scales are associated with long time scales, while those operating at small scales are generally short-lived features. Large-scale phenomena such as planetary waves do not act over very short periods, and turbulent processes would not be expected to persist for more than a few minutes at a time. This diagonality of atmospheric processes essentially allows particular combinations of spatial and temporal scales to be modeled independently of the other scales; for example, when investigating regional-scale flows in a mountainous region (valley flows) or at the boundary between a land surface and a lake or an ocean (sea breeze), simulations would typically make use of a mesoscale model and not a climate model nor a very detailed turbulence-scale model. Certain phenomena indicated in Figure 1.2, such as fronts or hurricanes, are included in the mesoscale segment of the diagram, even though they are related to the larger-scale. Fronts and hurricanes travel across entire oceans and generally affect regions much greater than their characteristic size. However, if one follows a front or a hurricane as it translates across a part of the globe, its spatial and temporal dimensions enter into the mesoscale range.

In an overview paper, O'Neill (1988) makes the point that, in many environmental studies, the scale of observation determines the level of interest, and that a subdivision of processes into different scales is useful for a particular class of problems. Furthermore, '*no single division stands out as fundamental. There certainly seems no good reason to force all problems into a single framework*' (op. cit., p. 32). It is additionally emphasized (op. cit., p. 33) that '*the theory recommends focusing on a single level but the appropriate level should be based on the problem at hand. It is not useful to force a new problem area into the mold that was appropriate for other problems at other levels*'. Partitioning of scales is common to investigate particular sets of problems, such as turbulence, mesoscale processes, synoptic and climatological-scale flows.

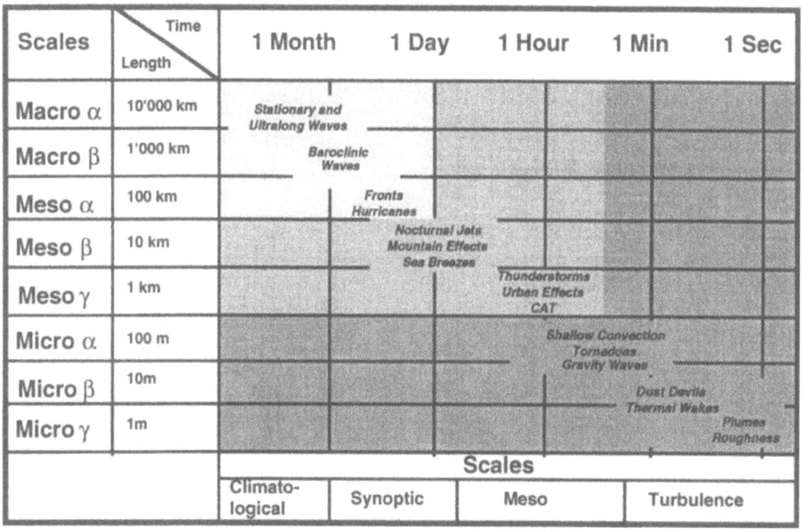

Figure 1.2: Characteristic scales of the atmosphere (adapted from Orlanski, 1975)

When modeling a particular scale of interest, however, processes acting at a smaller scale will play a role in the evolution of the dynamic and thermodynamic state of the atmosphere, i.e., it is not because there is an arbitrary scale separation which determines the type of model which can be applied to a particular problem that the model is completely independent of the other scales. In the example of a mesoscale model applied to the sea breeze, turbulent fluxes of momentum, heat, and moisture in the atmospheric boundary layer and in the surface layer, while much smaller than the scale of the sea-breeze circulation itself, are nonetheless key features which determine the nature, strength, and evolution of the breeze. Interactions between scales smaller than that of the process under investigation, i.e., generally the grid-scale of a model, therefore need to be addressed in a physically coherent, and numerically efficient manner. The interaction between scales, however, becomes non-trivial when applied to non-linear systems. Fluid mechanics, which are used to describe atmospheric flows, are highly non-linear, essentially because the rate of change of a particular property depends on the transport of that property by the fluid flow, whose velocity **V** is one of the variables of the problem. Thus the conservation of momentum equation (which will be addressed in the next chapters) involves the manifestly non-linear term

$\mathbf{V}.\nabla\mathbf{V}$. Furthermore, thermal problems are crucial to atmospheric flows, so that temperature T and density ρ are among the unknown variables. Additional non-linearities arise through the coupling of these variables by the convection velocity, represented in the mass and internal energy balance equations by the terms $\nabla(\rho\mathbf{V})$ and $\mathbf{V}.\nabla T$. In a linear system, the ultimate effect of a forcing from the smaller scales to the forcing at the scale of interest is simply the superposition of the effects of each forcing taken individually. In a non-linear system, adding a small cause to one that is already present can induce dramatic effects that may be completely disproportionate to the amplitude of the forcing. This concept may be illustrated in the following simple example; if variable X is a unique state variable, k is a parameter, and λ an applied constraint, then X may be described as evolving according to:

$$dX/dt = \lambda - kX \quad 1.1$$

A stationary state value is given by:

$$\lambda - kX_s = 0, \text{ or } X_s = \lambda/k \quad 1.2$$

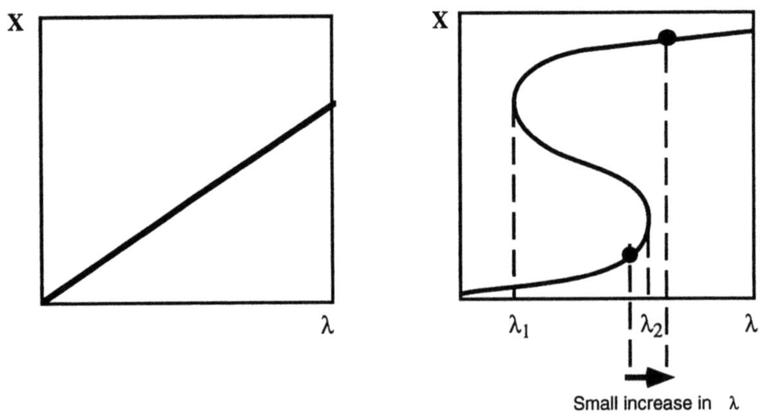

Figure 1.3: Response of linear system to changes in forcing function λ (left); Response of a non-linear system to small changes in λ (right)

Figure 1.3a represents this solution; by following the way X_s varies in terms of λ, it is seen that the solution is a straight line passing through the origin, i.e., a typical linear system. If the relationship between X_s and λ is non-linear, however, then one possible solution could be that illustrated in Figure 1.3b. In this curve, when $\lambda < \lambda_1$ or $\lambda > \lambda_2$, then the solutions are linear as in Figure 1.3b. For the interval $\lambda_1 < \lambda < \lambda_2$, however, the system exhibits several distinct solutions. A small change in the forcing function λ within this interval range, from zero to a small value just to the left of λ_2 on the lower branch will lead to a finite jump to the upper curve. This response is large compared to the small variation of λ which led to it. It also has consequences for the predictability of the system, which breaks down from the linear to the non-linear system.

A further illustrative example to stress the point of the complex behavior of non-linear equations is given for the difference equation of quadratic form:

$$X(i+1) = cX(i) \,.\, [1 - X(i)] \qquad 1.3$$

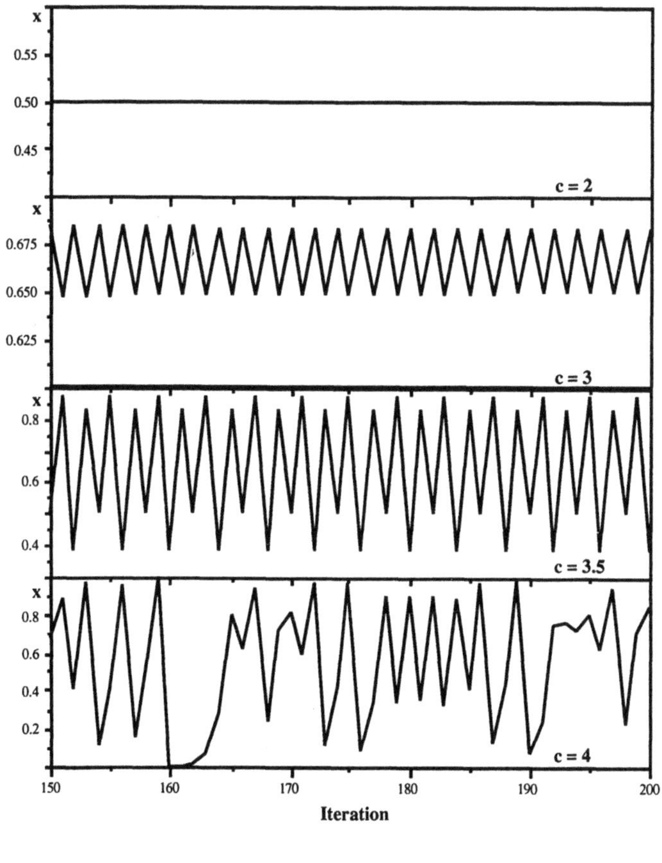

Figure 1.4: Solutions of X(i+1) = cX(i) [1 - X(i)] for different values of c, showing breakdown to chaos

where X is a particular variable, c is a forcing parameter, and i is the iteration step. Figure 1.4 illustrates the solution for different values of c at different iterations. For c=2, the solution is unique and the system is predictable in an unequivocal manner. For c=3, an oscillatory behavior becomes apparent; these oscillations amplify as c is increased, then cascade into four-period oscillations for c=3.5; as c approaches 4, the cascading process increases until the oscillations become more and more complex and, ultimately, tend towards chaos, i.e., the system becomes inherently unpredictable. This example again emphasizes the fact that small changes in a particular forcing function can lead to a wide range of responses, from the linear steady state to chaos. Another manner of representing this process is given in Figure 1.5, where values of X are plotted against c for 200 iterations of Equation 1.3 at each value of c. The left-hand part of the diagram represents the unique solution of Equation 1.3, and with higher values of c, periodic solutions increase until chaos is attained. At each period doubling, *bifurcations* in the solution are to be observed; unless it may be determined precisely at which iteration the solutions will bifurcate to a lower or an upper branch of the diagram, the predictability of the system breaks down with increasing values of c. A very minor change in c at the first bifurcation point will determine the subsequent evolution of the system as a function of c. Many systems in nature, including those associated with atmospheric flows, reveal the existence of period-doubling cascades which eventually lead to chaotic behavior.

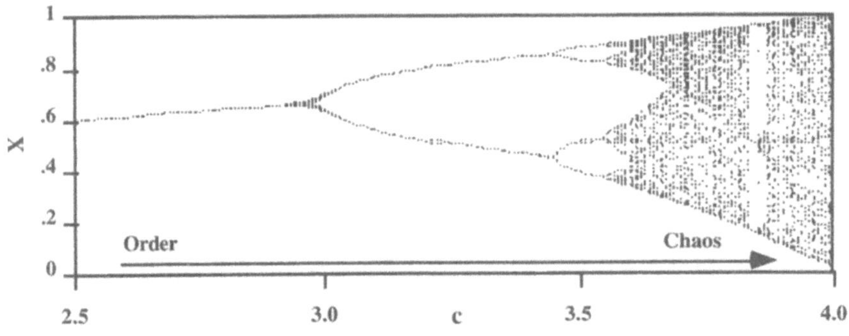

Figure 1.5: Bifurcation diagram for the solutions to Equation 1.3

An interesting case in point to illustrate the notion of bifurcations in a physical sense is related to Bénard convection, which will be returned to in Chapter 4, where Rayleigh-Bénard type convection in the atmosphere is simulated by a mesoscale model. The properties of a fluid at rest, where there is no bulk motion nor thermal forcing, will be everywhere uniform, i.e., the fluid is in a state of equilibrium. A very small perturbation of temperature, for example a small increase in temperature at the base of the fluid, would generate a momentary perturbation, but viscous forces would quickly restore the fluid to its original equilibrium. However, increasing the temperature above some critical threshold level would result in a bulk movement, or overturning, of the fluid whose motion is far from random. A structure of the motion in the form of cells, known as Bénard cells, appears when the critical temperature is exceeded, i.e., the flow enters a regime of thermal convection. In terms of the fluid at rest, the cells generate what may be referred to as symmetry breaking, because the properties of the fluid are no longer homogeneous. Moving in the horizontal plane, the symmetry breaking is identified by upward and downward motions of cells; a plot of vertical velocity across the domain would reveal, in a well-established set of cells, alternating regions of upward and downward motion, i.e., probably a two-period oscillation. A bifurcation has therefore occured in this sense, between the fluid at rest and the well-established cellular motion. Increasing the temperature perturbation in the fluid will ultimately result in a breakdown of the cellular patterns towards a state close to chaos - for example the motion of water when it begins to boil. The fluid therefore undergoes an increasing number of bifurcations, ultimately reaching its chaotic state for sufficiently elevated temperature forcings, where the motion of the fluid is by definition unpredictable. It should be noted that, interestingly, while the inception of Bénard cells can be predicted for a particular threshold of temperature for a given fluid, the direction of rotation of the cells is unpredictable and uncontrollable; once the direction of rotation is established in a particular cell, it will not change. In the Bénard cellular structure, which is one of the easier convective systems to investigate experimentally or numerically, there is a duality between randomness and determinism. In other words, several solutions may be possible for the same parameter; in the case of the Bénard cells, only chance will decide which of the solutions, i.e., a left-handed or a right-handed cell, will be realized.

While it is not the intention of this work to enter into the intricacies of chaos theory and complexity, the examples above have served to make the point that small-scale forcings within a non-linear system may not be set aside. In some instances, a long time may elapse before responses to a particular small-scale forcing manifest themselves; when they do, however, the system may enter a totally new state. Lorenz (1963) was one of the first to discover the implications of such considerations for the atmospheric sciences. He developed a nominally simple model of the climate system in which only three modes, each associated with a particular wavelength, were considered. The Lorenz model suggests that weather and climate are inherently unpredictable (Lorenz, 1963; 1968; 1969; 1982; Nicolis and Prigogine, 1989), since they exhibit one of the basic properties of chaotic dynamics, namely the sensitive dependency of the model solutions on initial conditions. Very minor changes in initial conditions can lead to very major differences in the evolution of the atmospheric state after some time; it is on the basis of this consideration that the well-known "butterfly effect" entered the popular jargon, i.e., the notion that the flutter of a butterfly wing in one location of the globe could trigger major climate phenomena, such as tropical cyclones and other extreme events, in another part of the world. Nicolis and Prigogine (1989) have extensively reviewed systems which exhibit chaotic behavior; chaos in nature reveals the fact that disorder at a certain scale is not incompatible with order at different scale. The apparent randomness of, for example, turbulent flow is not the result of imperfections in an experimental setup or of a complex environment which cannot be controled, but is indeed embedded in the dynamics of perfectly deterministic systems which may involve only a few variables. In these systems, all scales are important, and not simply the largest, most obvious dominant ones. If one accepts that chaos, and therefore unpredictability, are inherent to most systems requiring the solution of non-linear partial differential equations, the question may be posed as to the usefulness of attempting *any* form of atmospheric or climate simulations; experience with numerical weather prediction have demonstrated that the forecast is sensitive to the initial conditions, i.e., the model will in time generate very different solutions in response to small changes in the initial fields. The value in modeling complex environmental systems therefore resides in the possibility of undertaking a large number of sensitivity experiments, in which studies of the response of the system to a particular forcing may be analyzed. Interactions between different elements of the system may also be envisaged in order to further our understanding of the fundamental processes and controls involved. In this sense, models are a powerful analysis tool which not only complement, but often go beyond, the interpretation of observational data.

2. Fundamental Concepts Related to Atmospheric Modeling

2.1 Introduction

In order to adequately model atmospheric dynamics and thermodynamics, whatever the scale considered, the fundamental physical and mathematical laws governing atmospheric circulations need to be implemented in a form which is physically meaningful, mathematically rigorous, and computationally efficient. The scale of the model chosen will largely determine the simplifications which can be applied to a particular set of equations and the type of physical parameterization schemes to be incorporated. In essence, however, the underlying fundamental equations will be the same for all models; these are basically conservation equations which describe the characteristics of mass, motion, heat, moisture and the behavior of pollutant species in the atmosphere. These principles lead to a set of coupled equations which must be satisfied simultaneously and which include the sources and sinks of individual quantities.

In any atmospheric model, numerous physical factors linked to the dynamic and thermal characteristics of the atmosphere need to be addressed. Complex interactions and feedbacks occur between fundamental meteorological processes and elements of the terrestrial system; within the atmosphere itself, a multitude of complicating phenomena, ranging from turbulence and electromagnetic radiation to clouds and precipitation, require attention in models at all scales. A non-exhaustive list of factors which need to be incorporated in an atmospheric model include the following:

- *The nature of the underlying surface*. The earth's surface acts as a physical boundary to atmospheric dynamics and as a source of complex sources and sinks through spatial and temporal heterogeneities in roughness elements, heat, and moisture. These are reflected for example through land-use characteristics where surface-induced friction effects are a function of the type of land cover (urban areas, forests, water surfaces, grasslands, etc.). The thermodynamic structure of the atmospheric boundary layer (ABL) is also a function of surface temperature and moisture heterogeneities, related to vegetation type and the presence of rivers, lakes or oceans. In certain regions, mountain and valley systems exert a particularly strong influence on air flow, through the channeling effects of the orography, and by the generation of mountain and valley breezes according to local thermal conditions.

- *Atmospheric turbulence*. Many flows occuring in nature and in engineering applications are turbulent; except possibly in very stable conditions, the atmospheric boundary layer is inherently turbulent. It should be noted that turbulence is not a feature of *fluids* but of *fluid flows*. One major characteristics of such flows is their irregularity, or randomness, which makes a deterministic approach to the problem intractable; instead, turbulence is treated through statistical methods. Turbulent flows are always dissipative, since viscous shear stresses perform deformation work which increases the internal energy of the fluid at the expense of the kinetic energy of the turbulence. These characteristics have profound consequences for atmospheric flows and on features embedded within them, in particular diffusion - or dispersion - of heat , moisture, momentum, and air pollution. The intensity of turbulence may depend on purely dynamic factors, such as roughness of the surface or wind shear (the vertical deformation of the flow), or on thermal factors (principally atmospheric stability), or on a combination of both factors.
- *Solar, terrestrial, and atmospheric radiation.* Although the atmosphere depends for its heat on radiant energy from the sun, it absorbs directly only about 15% of the available solar energy. More than 33% of the energy is lost by reflection and back-scattering to space. The remainder, less than 50%, is absorbed at the Earth's surface and is then available for heating the atmosphere. Much of this energy is emitted back to space in the form of infrared (IR) radiation. However, different trace gases in the atmosphere, such as carbon dioxide (CO_2), methane (NH_4), nitrous oxides (N_2O), chloro-fluoro-carbons (CFC), and particularly water vapor absorb and re-emit some of this IR radiation back towards the earth's surface, a phenomenon often referred to as the "Greenhouse Effect". Since the beginning of the industrial era, greenhouse-gas (GHG) concentrations have risen at an unprecedented rate as a result of human activities. This has led in recent years to the concern by many climate specialists that the atmosphere will warm in coming decades at a rate far greater than that resulting from natural climatic fluctuations. The thermodynamic role played by radiation is fundamental to climate, but also to atmospheric processes at small and medium time and space scales. Clouds play a dominant role in all aspects of radiation, as they are by far the most important absorbers of radiation at essentially all wavelengths.
- *Cloud activity*. Cloud formations, especially the turbulent cumulus type, are capable of generating significant dynamic and thermodynamic modifications in the atmosphere. The formation of cloud condensation droplets is accompanied by the release of latent heat; at the cloud edges and cloud top, evaporation of liquid droplets leads to latent heat absorption and corresponding cooling of the neighboring cloud-free air. In dynamic terms, a cumulus-type cloud is an unstable and turbulent phenomenon; exchange of air at the cloud boundaries can result in secondary circulations which perturb atmospheric flows from the surface through to the upper troposphere. Stratiform clouds exert an influence on the atmosphere less through turbulent feedback mechanisms than through interactions with radiative fluxes; the high reflectivity of the upper boundary of these clouds will mitigate the incoming solar radiation, while the base of the stratiform clouds may be relatively opaque to outgoing terrestrial radiation. The

interaction between clouds and radiation in atmospheric models is still relatively poorly understood and remains a rather crudely parameterized feature, so that it is uncertain whether clouds, particularly stratiform clouds, are exerting a globally negative or positive feedback on the climate system.

- *Precipitation processes.* The triggering of precipitation is of course closely linked to cloud formation. In a cumulus-type cloud, precipitation is the sign of a stabilization or decay of cloud growth, and therefore an attenuation of the cloud dynamic and thermodynamic influence on its immediate environment. As rainwater exits its saturated cloud environment, it begins to evaporate and therefore cools the sub-cloud air layers. In the case of moderate to heavy rain, the combined effects of evaporational cooling and rainwater fallout lead to a reversal of vertical motion, cutting off the cloud from its low-level dynamic and moisture sources and leading to its eventual decay. These features of precipitation can significantly modify the ABL structure.

In order to adequately simulate a representative range of atmospheric processes, a numerical model should address as many of the above processes as possible. This is generally achieved through *parameterization*, where such phenomena are treated through simplified but physically-coherent mathematical representations. This section will provide a brief overview of the governing equation sets and a succinct overview of some of the notions which form the essence of numerical modeling: parameterization, methods of numerical solution, and initial and boundary conditions.

2.2 Governing equations

There exists an abundant literature describing fluid flow and thermodynamic principles as applied to the atmosphere. Comprehensive textbooks dealing with such issues include for example Holton (1972), Tennekes and Lumley (1972), Haugen (1973), Iribarne and Godson (1973), Dutton (1976), and Pielke (1984), so that only a summary of the governing equations and a very brief overview of some commonly-used numerical schemes will be provided here. As already mentioned in the introduction, all models make use of conservation principles in their formulation; the key equation sets may be written in a Cartesian representation as follows:

Conservation of zonal momentum:

$$\frac{\partial u}{\partial t} = -u\frac{\partial u}{\partial x} - v\frac{\partial u}{\partial y} - w\frac{\partial u}{\partial z} + fv - \frac{1}{\rho}\frac{\partial p}{\partial x} + F_u \qquad 2.1$$

Conservation of meridional momentum:

$$\frac{\partial v}{\partial t} = -u\frac{\partial v}{\partial x} - v\frac{\partial v}{\partial y} - w\frac{\partial v}{\partial z} - fu - \frac{1}{\rho}\frac{\partial p}{\partial y} + F_v \qquad 2.2$$

Conservation of vertical momentum:

$$\frac{\partial w}{\partial t} = -u\frac{\partial w}{\partial x} - v\frac{\partial w}{\partial y} - w\frac{\partial w}{\partial z} - f^{*}u - \frac{1}{\rho}\frac{\partial p}{\partial z} - g + F_v \qquad 2.3$$

Conservation of heat

$$\frac{\partial\theta}{\partial t} = -u\frac{\partial\theta}{\partial x} - v\frac{\partial\theta}{\partial y} - w\frac{\partial\theta}{\partial z} + F_\theta + C_\theta + R_\theta \qquad 2.4$$

Conservation of specific humidity:

$$\frac{\partial q}{\partial t} = -u\frac{\partial q}{\partial x} - v\frac{\partial q}{\partial y} - w\frac{\partial q}{\partial z} + F_q + C_q + R_q \qquad 2.5$$

Conservation of cloud liquid water:

$$\frac{\partial q_L}{\partial t} = -u\frac{\partial q_L}{\partial x} - v\frac{\partial q_L}{\partial y} - w\frac{\partial q_L}{\partial z} + F_{q_L} + C_{q_L} \qquad 2.6$$

Conservation of rainwater:

$$\frac{\partial q_r}{\partial t} = -u\frac{\partial q_r}{\partial x} - v\frac{\partial q_r}{\partial y} - w\frac{\partial q_r}{\partial z} + F_{q_r} + C_{q_r} \qquad 2.7$$

These equations, where the symbols have their usual meteorological meaning (a list of symbols is provided in Appendix A), consist of the local time rate of change on the left-hand side of the equations, and are made up on the right-hand side of the advective (grid-resolved) terms, pressure-gradient and Coriolis terms (Equations 2.1, 2.2 and 2.3, also referred to as the *Navier-Stokes Equations*), and additional quantities which represent the contributions to the tendency terms of atmospheric boundary and surface layer turbulence (F), cloud activity including thermodynamic effects of phase changes of water and precipitation (C), and radiative flux exchanges (R).

The governing equation set is completed by the equation of state for a perfect gas, which relates pressure, temperature, and density (and thereby links atmospheric dynamics and thermodynamics) through:

$$p = \rho R T_v \qquad 2.8$$

Here, T_V is the virtual temperature ($T_V = 1 + 0.61q$) which takes into account the influence of water vapor in atmosphere on pressure. In many models designed to simulated processes on the regional-scale or for the larger climatic scales, dynamic processes are assumed to be hydrostatic; the pressure term is then computed from the hydrostatic equation, one form of which is given below (making use of the equation of state and replacing temperature T by potential temperature θ):

$$\frac{\partial p^{R/cp}}{\partial z} = \frac{-g p_0^{R/cp}}{C_p \theta} \qquad 2.9$$

Conservation of mass is written in the form of a predictive equation for density, which reads:

$$\frac{\partial\rho}{\partial t} = -\frac{\partial\rho u}{\partial x} - \frac{\partial\rho v}{\partial y} - \frac{\partial\rho w}{\partial z} \qquad 2.10$$

For most models where incompressibility is assumed, however (i.e., $\partial\rho/\partial t = 0$), Equation 2.10 may be simplified to a familiar form of the continuity equation, which for shallow atmospheric systems reads:

$$\frac{\partial u}{\partial x} + \frac{\partial v}{\partial y} + \frac{\partial w}{\partial z} = 0 \qquad 2.11$$

Equation 2.11 takes on an additional importance when the hydrostatic equation (Equation 2.9) is applied to the system; the hydrostatic approximation is a consequence of the simplification of the vertical momentum equation (Equation 2.3) where, for most atmospheric processes outside the range of microscale turbulence and cumulus-cloud processes, the dominant forces represent a balance between the gravitational acceleration g and the vertical pressure gradient $\partial p/\partial z$. This implies a loss of the predictive capability of Equation 2.3 (i.e., $\partial w/\partial t \sim 0$), such that the only manner in which the vertical velocity can be computed is through the solution of the continuity equation (2.11). A diagnostic equation for non-hydrostatic pressure will be derived in Section 2.3.

In the predictive equation set (2.1 - 2.7), the advective terms [e.g., $- u (\partial u/\partial x) - v (\partial u/\partial y) - w (\partial u/\partial z)$] are all grid-resolved, i.e., they may be solved directly by a particular combination of a spatial discretization scheme and time integration algorithm at each model grid-point and for each time-step; this is also the case, in Equations 2.1 - 2.3 for the pressure-gradient and Coriolis terms. The remaining terms on the right-hand-side of the governing equation set (F, C, R) all require *parameterization.* Parameterization is required in two instances:

- when the process under consideration is smaller than the characteristic grid-mesh of the model
- when the complete physics describing a particular phenomenon would, if computed explicitely at each model grid-point and for every time-step, overload the computer resources

For each of the numerical models to be discussed in subsequent chapters of this monograph, particular parametric schemes addressing possible solutions to the F, C, and R terms will be described. It is worthwhile, however, to briefly review the manner in which the flux form of the equations are formulated; this form allows the separation of the different components of the equations into a set of terms representative of the large-scale (grid-resolved) forcings and small-scale (subgrid-scale) terms representative of turbulent motion. The equations which will be described below are valid for the atmospheric boundary layer, where turbulence induced by friction with the surface is prevalent.

As already noted, atmospheric flows are inherently turbulent in nature; turbulence may be identified on an anemograph as a series of fluctuations of velocity and direction as eddies of various scales pass the measurement point. Figure 2.1 illustrates a typical time-series of velocity fluctuations, which are clearly distributed about some average velocity. Peaks of velocity greater than the mean velocity result from an increase in momentum advected by turbulent eddies, while troughs of velocity lower than the average velocity result from eddies which remove momentum from the flow at the measurement point. It is therefore possible to define an instantaneous velocity by its average component and its fluctuating (or turbulent) component. This can be written as:

$$V = \overline{V} + V' \qquad 2.12$$

with the overbar denoting the mean quantity and the prime indicating the turbulent quantity of the flow.

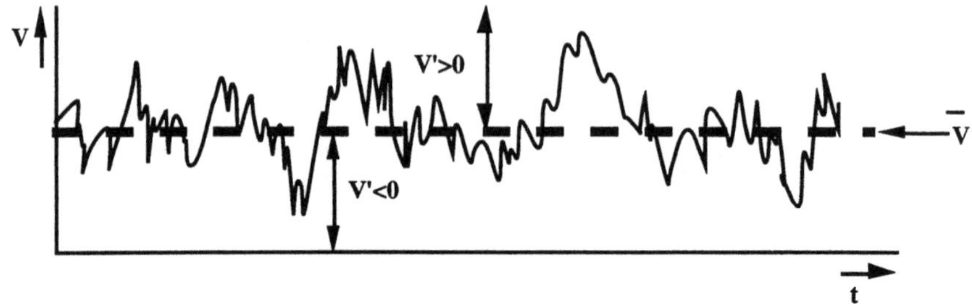

Figure 2.1: Example of an anemometer recording, illustrating the concepts of mean and fluctuating velocities

If this averaging method is now applied to the horizontal equations of motion, the flux form of the Navier-Stokes equation is obtained. The approximate form of the initial (non-transformed) momentum and continuity equations is given by:

$$\frac{\partial u}{\partial t} = -u\frac{\partial u}{\partial x} - v\frac{\partial u}{\partial y} - w\frac{\partial u}{\partial z} - \frac{1}{\rho}\frac{\partial p}{\partial x} + fv \qquad 2.13$$

$$\frac{\partial v}{\partial t} = -u\frac{\partial v}{\partial x} - v\frac{\partial v}{\partial y} - w\frac{\partial v}{\partial z} - \frac{1}{\rho}\frac{\partial p}{\partial y} - fu \qquad 2.14$$

$$\frac{\partial \rho}{\partial t} + \frac{\partial \rho u}{\partial x} + \frac{\partial \rho v}{\partial y} + \frac{\partial \rho w}{\partial z} = 0 \qquad 2.15$$

Multiplying Equation 2.13 by ρ and Equation 2.15 by u, and adding the result, the resulting equation reads:

$$\frac{\partial \rho u}{\partial t} = -\frac{\partial \rho u^2}{\partial x} - \frac{\partial \rho uv}{\partial y} - \frac{\partial \rho uw}{\partial z} - \frac{\partial p}{\partial x} + \rho fv \qquad 2.16$$

and similarly for the v-component:

$$\frac{\partial \rho v}{\partial t} = -\frac{\partial \rho uv}{\partial x} - \frac{\partial \rho v^2}{\partial y} - \frac{\partial \rho vw}{\partial z} - \frac{\partial p}{\partial y} - \rho fu \qquad 2.17$$

Substituting now for the instantaneous values as shown in equation 2.12, it is possible to separate values representative of the mean flow from those characteristic of turbulence. For example, applying equation 2.12 to the first term on the right-hand-side of equation 2.17 and averaging yields:

$$\overline{\rho uv} = (\overline{\rho} + \rho')(\overline{u} + u')(\overline{v} + v') = \overline{\rho uv} + \overline{\rho u'v'} \qquad 2.18$$

where ρ' has been neglected for a nearly-incompressible atmosphere, and by definition the average of a prime quantity is zero (i.e., the sum of all positive and negative fluctuations is zero), so that:

$$\overline{\overline{u}\;v'} = 0\;;\quad \overline{u'\;\overline{v}} = 0,\ \text{since}\ \overline{u'} = \overline{v'} = 0\ \text{and}\ \overline{\overline{u}} = \overline{u}\;;\quad \overline{\overline{v}} = \overline{v} \qquad 2.19$$

Applying this methodology to all terms in equation 2.15 and 2.17 results in the following form of the horizontal Navier-Stokes equation:

$$\frac{\partial \overline{u}}{\partial t} = -\overline{u}\frac{\partial \overline{u}}{\partial x} - \overline{v}\frac{\partial \overline{u}}{\partial y} - \overline{w}\frac{\partial \overline{u}}{\partial z} + f\overline{v} - \frac{1}{\overline{\rho}}\frac{\partial \overline{p}}{\partial x} - \frac{1}{\overline{\rho}}\left[\frac{\partial \overline{\rho u'^2}}{\partial x} + \frac{\partial \overline{\rho u'v'}}{\partial y} + \frac{\partial \overline{\rho u'w'}}{\partial z}\right] \tag{2.20}$$

$$\frac{\partial \overline{v}}{\partial t} = -\overline{u}\frac{\partial \overline{v}}{\partial x} - \overline{v}\frac{\partial \overline{v}}{\partial y} - \overline{w}\frac{\partial \overline{v}}{\partial z} - f\overline{u} - \frac{1}{\overline{\rho}}\frac{\partial \overline{p}}{\partial y} - \frac{1}{\overline{\rho}}\left[\frac{\partial \overline{\rho u'v'}}{\partial x} + \frac{\partial \overline{\rho v'^2}}{\partial y} + \frac{\partial \overline{\rho v'w'}}{\partial z}\right] \tag{2.21}$$

A **B** **C** **D** **E**

Equations 2.20 and 2.21 are made up of the acceleration term **A**, which consists of the advection by the mean flow **B**, Coriolis and pressure-gradient forces **C** and **D**, and transport of momentum by three dimensional turbulence **E**. This last term depends on turbulent fluctuations to transfer momentum from one location in space to another. The **E** terms are often referred to as the *eddy stress terms*. As the direct mesurement of these turbulent terms is not a straightforward matter compared to the mean flow variables (**B** terms), it is therefore customary to approximate (or *parameterize*) the eddy stress terms in what it technically known as *closure*, which for the Navier-Stokes equations is by no means trivial. As was perhaps observed in the preceding discussion, the averaging technique led to separation of mean and turbulent components of the flow, the latter being second-order (i.e. multiples of primed variables also known as *correlations*). If one wishes to solve the second-order terms through time dependant equations (i.e., relations involving ($\partial \overline{\rho u'^2} /\partial t$), ($\partial \overline{\rho u'w'} /\partial t$), etc.), it becomes apparent that for each time-dependent second-order equation, one is generating a third-order term. The form of the second-order equation for potential temperature θ, for example, is the following:

$$\frac{\partial \overline{u_i'\theta'}}{\partial t} = -\frac{\partial \overline{u_k u_i'\theta'}}{\partial x_k} - (\overline{u_i'u_k'} - \frac{2}{3}\delta_{ik}\overline{E})\frac{\partial \overline{\theta}}{\partial x_k} - \frac{2}{3}\overline{E}\frac{\partial \overline{\theta}}{\partial x_i} - \overline{u_k'\theta'}\frac{\partial \overline{u_i}}{\partial x_k}$$
$$- \frac{\partial \overline{u_i'u_j'\theta'}}{\partial x_j} + \frac{1}{\overline{\rho}}\overline{p'}\frac{\partial \overline{\theta}}{\partial x_i} - 2\varepsilon_{ikl}\Omega_k\overline{u_l'\theta'} - \frac{1}{\overline{\rho}}\frac{\partial \overline{p'\theta'}}{\partial x_i} + \delta_{i3}\frac{g}{\overline{\theta}}\overline{\theta'^2} \tag{2.22}$$

where E is the turbulent kinetic energy defined as $E = \sqrt{[u'^2 + v'^2 + w'^2]}$. If one were to derive time-dependent relations for the third-order terms, e.g., ($\partial \overline{\rho u'v'w'} /\partial t$), one would generate fourth-order terms, and so on. In other words, whatever the order of the tendency (time-dependent) term, a term of higher-order appears on the right-hand side of the momentum equation. It is then a matter of deciding to which order one wishes to close (or *truncate*) the equations, since in the flux form the equations cannot yield exact solutions. Particular methods must therefore be devised in order to parameterize the turbulent fluxes such as $\rho\,\overline{u'w'}$, $\rho\,\overline{v'w'}$, etc. These will be described for each of the atmospheric models incorporating ABL closure schemes in Chapter 3.

The smaller the grid of the model, i.e., the more one approaches the turbulent end of the atmospheric scale spectra illustrated in Figure 1.2 (lower-left portion of that figure), the less the likelihood that the hydrostatic assumption will be valid. This is the case for fine grid-mesh models, where turbulence and/or vertical motion in clouds or resulting from strong temperature heterogeneities may dominate over the advection terms. In this case a non-hydrostatic solution needs to be applied in order for the model solutions to retain a degree of coherence.

A non-hydrostatic form of the vertical pressure gradient may be derived by taking the divergence of the Navier-Stokes equations. This reads, in tensor notation:

$$\frac{\partial}{\partial t}\frac{\partial}{\partial x_j}\rho_0 \overline{u_i} = -\frac{\partial^2}{\partial x_i \partial x_j}(\rho_0 \overline{u_i}\;\overline{u_j}) - \frac{\partial^2}{\partial x_i \partial x_j}(\rho_0 \overline{u'_i\; u'_j}) - \frac{\partial^2 p'}{\partial x^2_i} - \frac{\partial}{\partial x_i}\left[\frac{\partial p_0}{\partial x}\delta_{i1} + \frac{\partial p_0}{\partial x}\delta_{i2}\right]$$

$$+ \delta_{i3}\, g \frac{\partial}{\partial x_i}\frac{\alpha'}{\alpha^2_0} - 2\,\varepsilon_{ijk}\,\Omega_j \frac{\partial}{\partial x}\rho_0 \overline{u_k} \qquad 2.23$$

The terms with subscript 0 represent the large-scale value of the variable, and the primed terms the deviations from the large-scale fields. α is the specific volume, which is the inverse of density ρ. From basic atmospheric thermodynamics, the ratio α' / α_0 may be approximated by the following relationship:

$$\frac{\alpha'}{\alpha_0} = \frac{\theta'}{\theta_0} - \frac{c_v}{c_p}\frac{p'}{p_0} \qquad 2.24$$

Use of the continuity equation and the approximation given in Equation 2.24 yields a diagnostic second-order differential equation for the pressure perturbation:

$$\frac{\partial^2 p'}{\partial x^2_i} + g\frac{c_v}{c_p}\frac{\partial}{\partial z}\left(\frac{\rho_0 p'}{p_0}\right) = -\frac{\partial^2}{\partial x_i \partial x_j}(\rho_0 \overline{u_i}\;\overline{u_j}) - \frac{\partial^2}{\partial x_i \partial x_j}(\rho_0 \overline{u'_i\; u'_j}) - \frac{\partial^2 p'}{\partial x^2_i} - \frac{\partial}{\partial x_i}\left[\frac{\partial p_0}{\partial x}\delta_{i1} + \frac{\partial p_0}{\partial x}\delta_{i2}\right]$$

$$+ \delta_{i3}\, g \frac{\partial}{\partial x_i}\frac{\alpha'}{\alpha^2_0} - 2\,\varepsilon_{ijk}\,\Omega_j \frac{\partial}{\partial x}\rho_0 \overline{u_k} \qquad 2.25$$

For relatively shallow systems, the second term on the left-hand-side does not appear; in this case Equation 2.25 is referred to as a Poisson partial differential equation (Hildebrand, 1962). There are a number of numerical schemes which allow the solution of the Poisson equation for pressure. One of the drawbacks of the Poisson formulation is the increase in computational time required; small discrepancies in the differentiation operation are also likely to amplify computational errors. Models which use this type of formulation to determine pressure are referred to as *non-hydrostatic models*. In the hierarchy of models whose results will be discussed in the following chapters, only the Deardorff-Sommeria model (Deardorff, 1973; Sommeria, 1976) applies this kind of formulation. The mesoscale models and GCMs in principle have sufficiently large grid meshes to justify the use of the simpler hydrostatic equation described by Equation 2.9.

2.3 Numerical schemes

It is not the intention here to provide an in-depth description of approaches to the numerical solution of the fundamental equations summarized in Section 2.2, since many advanced textbooks exist on the subject (e.g., Richtmyer and Morton, 1967; Mesinger and Arakawa, 1976; Haltiner and Williams, 1980). A brief overview will be given here of the two principle approaches which have been applied to the numerical model systems which will be discussed in subsequent chapters: these comprise essentially finite-difference algorithms, applied to the small and meso-scale models, and spectral schemes used for global climate models.

In the Deardorff-Sommeria model (Deardorff, 1972; Sommeria and Deardorff,1976), as well as in the DREAMS mesoscale modeling system (Beniston, 1984; 1985), a simple finite-difference scheme has been applied. Early versions of DREAMS used "forward-upstream" discretization, but this scheme has the undesirable effect of having strong numerical diffusivity associated with it. The most commonly used explicit scheme involves a centered-in-space, centered-in-time (the so-called "leapfrog scheme") discretization to solve the basic equation set on the model grid mesh. While reasonably simple, this scheme has the advantage of neither amplifying nor damping meteorological wave modes. As an example of centered finite differencing, the one-dimensional advection equation for the zonal wind vector u, i.e. $\partial u/\partial t = -u\,(\partial u/\partial x)$ may be discretized as:

$$\frac{u(x, t + \Delta t) - u(x, t - \Delta t)}{2\Delta t} = -u\,(x, t)\,\frac{u(x + \Delta x, t) - u(x - \Delta x, t)}{2\Delta x} \qquad 2.26$$

In Equation 2.26, x represents the position of variable u in space, and t its position in time. The unknown in this equation is $u(x, t + \Delta t)$, so that it may be found by simple arithmetic integration; note that the method requires three time-levels to compute forward in time. This type of integration scheme must satisfy the Courant-Friedrichs-Levy (CFL) criterion (see for example Richtmyer and Morton, 1967) which, if transgressed, leads to numerical instability and a rapid divergence of the model solutions. The CFL condition is written as:

$$c\,\frac{\Delta t}{\Delta x} \leq 1 \qquad 2.27$$

where c is the fastest-allowed propagation mode in the model domain, Δt is the computational time-step, and Δx is the grid interval. Equation 2.27 expresses the fact that the advection of a meteorological quantity cannot cross more than one grid interval in one time step.

In addition to the fulfilment of the CFL criterion, a further control is required on the time integrations using the leapfrog scheme. The centered-in-time technique has the property of decoupling solutions at even and odd time-steps; in order to overcome this problem, a four-point time filter is required, and the so-called Asselin (1971) technique is used. For example, in the case of variable u, this is given as:

$$u(x, t + \Delta t) = .25\,[\,u(x, t + \Delta t) + 2\,(u(x, t)) + u(x, t - \Delta t)\,] \qquad 2.28$$

Spectral General Circulation Models (GCMs), on the other hand, are formulated in a different manner from Cartesian models, in that they transform the atmospheric variables into wave form. One reason for this is that over a sphere, the horizontal dimensions represent a continuum with no lateral boundaries. An additional advantage of the spectral formulation is that it is less costly in computing time than a grid-point representation. GCMs generally revert to a grid-point representa-tion in the vertical, because the constraints imposed by lower and upper boundary conditions significantly complicate the application of spectral techniques.

Spectral methods are based on Fourier's theorem, which states that any periodic fucntion can be defined in terms of sine and cosine waves. Over a spherical surface such as the Earth, the variation of a climatic variable around any latitude or longitude band is by necessity periodic and therefore can be represented as a sum of sine and cosine functions. The Fourier transform, i.e., the wave representation of the original variable, allows mathematical operations to be performed more easily and more rapidly.

In a Cartesian grid, horizontal and vertical exchanges of momentum, temperature, or humidity are computed by considering local gradients between adjacent grid-points. In a spectral model, however, information needs to be transfered between Cartesian and Fourier space.

Any variable X can be represented on a sphere as a Fourier expansion, i.e.:

$$X = \sum_{m=-M}^{M} \sum_{n=|m|}^{|m|+j} X(m,n)\ Y(m,n) \qquad 2.29$$

where the X(m,n) term represents the complex spectral coefficient and Y(M,n) the spherical harmonics which are a function of longitude λ and latitude ϕ. Y can be defined by:

$$Y(m,n) = P(m,n)\ (\sin\phi)\ \exp(im\lambda) \qquad 2.30$$

with P(m,n) a Legendre polynomial of degree n and zonal wavenumber m (inverse of the wavelength). Spectral coefficients of dynamic and thermodynamic variables are transformed onto a latitude-longitude grid. A Legendre transform is evaluated for each spectral variable at latitude λ_j. The latitudes are related to the model resolution through the roots of the associated zero-order Legendre polynomial. For any variable X, this may be written as:

$$X(m,n) \rightarrow X_{(m)}\ (\phi_j\,;t) = \sum_{n=|m|}^{|m|+j} X(m,n)\,P(m,n) \qquad 2.31$$

The resulting Fourier harmonics $X_{(m)}$ at each latitude λ_j and time t are transformed through FFT algorithms to longitudes $\lambda_l = 2\pi\lambda/L$, $1<\lambda<L$, such that:

$$X(m) \rightarrow Z\,(\lambda_l, \phi_j\,;t) = \sum_{m=-M}^{M} X_{(m)} \exp\ (im\lambda_l) \qquad 2.32$$

These two steps enable a spatial distribution of values at a series of grid-points to be achieved. Physical processes, such as radiation transfer, convection, and surface energy budgets can be computed in a Cartesian grid. Once these computations have been achieved, the variables are "translated" into wave space using FFTs, and the inverse Legendre transformations are constructed for each latitude, using Gaussian quadrature defined as:

$$Z(m,n) = \sum_{j=1}^{k} w_{j,k} (\phi_j)\, Z_{(m)} (\phi_j)\, P(m,n) (\sin \phi_j) \qquad 2.33$$

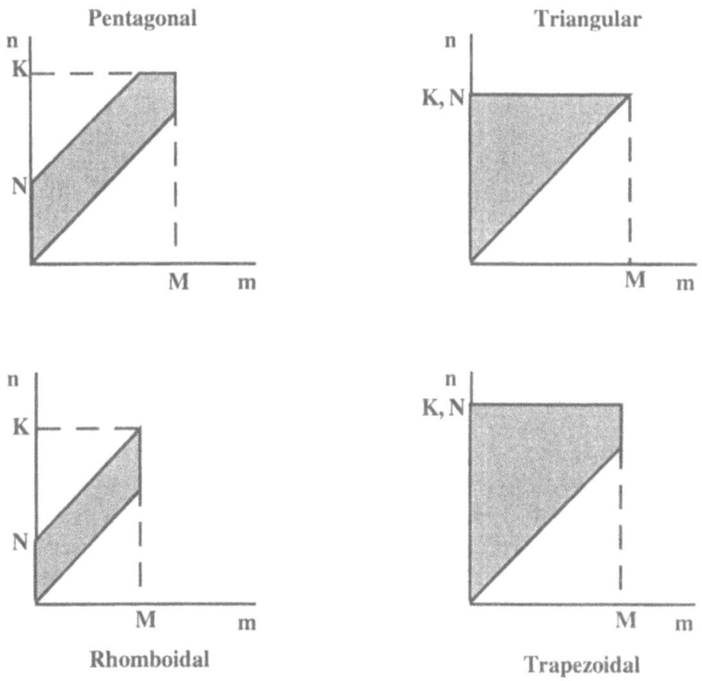

Figure 2.2: Truncation of spectra in GCMs

Once the model has been reassembled in spectral form, a time integration scheme is applied to establish a forecast of the meteorological fields.

Because of restrictions imposed by computational resources and the complexity of the physics being treated in a GCM, the number of wave numbers in a spectral model must be limited. The model is said to be *truncated* at a particular wave number; the truncation therefore represents the number of waves which are resolved around a latitude zone, and is the determining factor in the resolution of a GCM. Truncation is in fact the relationship between the largest Fourier wave number M, the highest degree of the associated Legendre polynomial K, and the highest degree of the Legendre polynomial of degree zero defined by the term N. Truncation types involving combinations of M, N, and K can take several forms, including pentagonal, triangular (for $M = N = K$), rhomboidal (for $K = N + M$) or trapezoidal (for $N = K > M$). These truncation relationships are illustrated in Figure 2.2; it should be noted that the most common truncations used in GCMs are triangular or rhomboidal.

2.4 Boundary and initial conditions

In any numerical model, appropriate initial and boundary conditions need to be applied so that the solutions are mathematically coherent. Model initialization is required to specify the state of the system at the beginning of the computation, by setting each of the dependent variables to some reference value obtained from observation or hypothesis. In most models, initialization is not a trivial process, because the information used to initialize a simulation must not only be processed in a manner suited to the internal model structure, but it must also be consistent with the laws and relations which define the model.

Boundary conditions are also a necessary part of a model simulation. These describe constraints on the evolution of the system, or interactions between the system and its environment which are not described explicitly by the model. Since lower and upper model boundaries are artificially enclosed, and additionally at the sides in small and mesoscale limited area models, it is necessary to specify the values of the dependent variables at these boundaries. Boundary conditions generally require updating at each time-step of the integration. For most models, it is convenient to define the top, lateral (for models other than global GCMs) and bottom boundary conditions separately, because the manner in which variables at these surfaces are specified is different for each boundary type; furthermore, only the bottom boundary condition has a physical meaning, since it generally coincides with the Earth's surface, whereas the other boundaries are incorporated for computational necessity.

The number of boundary conditions that can be applied to a particular model depends primarily on the form of the differential equations which has been adopted for the model. Model equations that have the correct number are described as *well-posed*, whereas they are *overspecified* if the number of conditions exceeds that actually used. Oliger and Sundström (1976) maintain that overspecification of model boundary conditions can, for certain solution algorithms, lead to physically-erroneous shortwave phenomena which travel across a model grid at the fastest wave speed allowed in the model; these are generated at the boundaries and, if unchecked through some mathematical filter function, are likely to severely contaminate the model physical variables.

3. Investigations with the Deardorff-Sommeria ABL Model

3.1 ABL turbulence

Deardorff (1972; 1973) was one of the pioneers of three-dimensional ABL modeling. He was one of the first to recognize that computers were in a rapidly developing phase and were, in the early 1970s, providing the type of resources which enable numerical solutions to primitive equations describing atmospheric turbulent flows. The advent of higher performance computers in the 1970s already made mathematical modeling competitive with other methods of investigation of the turbulent boundary-layer. Modeling attempts prior to full three-dimensional simulations essentially focused on two-dimensional problems for reasons of computer speed and memory.

However, key papers by Kraichnan (1967) and Lilly (1969) pointed to the fact that two-dimensional turbulence is substantially different from three-dimensional turbulence. Energy cascade in two dimensional turbulence is mainly directed towards low wavenumbers and becomes weak at higher wavenumbers. This cascading process leads to significant changes in flow behavior, so that the discrepancies between two- and three-dimensional turbulence are too great to be ignored; detailed turbulent modeling in the ABL must therefore be three-dimensional if it is to be realistic (Deardorff, 1973).

Deardorff justifies the use of models over, or as a complement to, other observational or experimental methods, in order to better understand some of the more subtle aspects of ABL turbulence. In his own words, "*One wishes to know how to relate the wind and temperature at anemometer level to externally-imposed conditions; to know at what rates the heat and momentum fluxes are decreasing with height within the "constant-flux" layer; to explain the presence of vanishing potential temperature gradients at heights as low as 50 m [above the surface]; to understand the cause of fluctuations of long period associated with unseen turbulent structures extending well up into the planetary boundary-layer; and to understand the manner in which the scale of the dominant eddies increases with height up past the top of the surface layer. For reasons such as these, the surface layer and the remaining bulk of the planetary boundary-layer should be studied as a single entity*".

Sommeria (1976) extended the ideas on turbulent modeling to the moist atmosphere, and incorporated in Deardorff's ABL model a set of closure equations which take into account turbulent processes generated by shallow cumulus convection.

This chapter will provide an overview of the highlights of the Deardorff-Sommeria model, in particular the detailed second-order closure equations derived to compute the contribution of turbulence to the kinematic and thermodynamic fields. It should be noted that, unlike the models which will be discussed in Chapters 4 and 5 of this work, turbulence in the ABL model is likely to be a major term in the governing equation set introduced in Chapter 2, in comparison to mesoscale and global models. The closure assumptions to be discussed below are designed for flow fields in which the turbulence in expected to account as much as 20% of the total momentum flux, except in the case of stable stratification.

3.1.1 The modeling of the ABL: second-order subgrid-scale closure

The derivation of second-moment equations is carried out as was discussed in Section 2.2 (Equations 2.12 - 2.21) in terms of large-scale or ensemble averaging of the governing equations. The Reynolds-averaged equations form the basis for a range of different closures. The equations presented below will all be in Cartesian notation; in addition, the following substitutions are made throughout:

$$\overline{u_i'u_j'} \rightarrow (\overline{u_i u_j} - \overline{u_i'}\ \overline{u_j'}) \qquad 3.1$$

Also, subgrid energy prior to Reynolds averaging is defined as:

$$\varepsilon' = 0.5\,(u'^2 + v'^2 + w'^2) \qquad 3.2$$

while after averaging, the generalized turbulence energy is given by:

$$\overline{E} = 0.5\,(\overline{u'^2} + \overline{v'^2} + \overline{w'^2}) \qquad 3.3$$

Following Lilly (1967), removing the factor $\frac{2}{3}\overline{E}$ in the case of the normal stresses yields the following expression for the Reynolds stresses:

$$\begin{aligned}
\frac{\partial}{\partial t}(\overline{u_i'u_j'} - \tfrac{2}{3}\delta_{ij}\overline{E}) = &- \frac{\partial}{\partial x_k}[\overline{u_k}\,(\overline{u_i'u_j'} - \tfrac{2}{3}\delta_{ij}\overline{E})] - (\overline{u_i'u_k'} - \tfrac{2}{3}\delta_{ik}\overline{E})\frac{\partial \overline{u_j}}{\partial x_k} \\
&- (\overline{u_j'u_k'} - \tfrac{2}{3}\delta_{jk}\overline{E})\frac{\partial \overline{u_i}}{\partial x_k} - \tfrac{2}{3}\overline{E}\,(\frac{\partial \overline{u_i}}{\partial x_j} + \frac{\partial \overline{u_j}}{\partial x_i}) \\
&+ \frac{\delta_{ij}}{3}(\overline{u_k'u_l'} - \tfrac{2}{3}\delta_{kl}\overline{E})(\frac{\partial \overline{u_k}}{\partial x_l} + \frac{\partial \overline{u_l}}{\partial x_k}) \\
&- \frac{\partial}{\partial x_k}[\overline{u'_k\,(\overline{u_i'u_j'} - \tfrac{2}{3}\delta_{ij}\overline{\varepsilon'})}] \\
&- \frac{1}{\rho}(\frac{\partial}{\partial x_j}\overline{u_i'p'} + \frac{\partial}{\partial x_i}\overline{u_j'p'}) + \frac{1}{\rho}\overline{p'(\frac{\partial u_i'}{\partial x_j} + \frac{\partial u_j'}{\partial x_i})} \\
&- 2\varepsilon_{ikl}\,\Omega_k((\overline{u_j'u_l'} - \tfrac{2}{3}\delta_{jl}\overline{E}) - 2\varepsilon_{jkl}\,\Omega_k((\overline{u_i'u_l'} - \tfrac{2}{3}\delta_{il}\overline{E}) \\
&+ \frac{g}{\theta_{v0}}(\delta_{i3}\overline{u_j'\theta'_v} + \delta_{j3}\overline{u_i'\theta'_v} - \tfrac{2}{3}\delta_{ij}\overline{u_3'\theta'_v})
\end{aligned} \qquad 3.4$$

The reason for removing the term $\frac{2}{3}\overline{E}$ from the normal stresses is because it allows an explicit separation of terms into isotropic and anisotropic components. The formal definition of anisotropy is here:

$$\left(\overline{u_i'u_j'} - \frac{2}{3}\delta_{ij}\overline{E} \right) \neq 0 \qquad 3.5$$

For potential temperature fluxes, the second-order closure equation for $\overline{u_i'\theta'_v}$ is:

$$\frac{\partial \overline{u_i'\theta_v'}}{\partial t} = -\frac{\partial}{\partial x_k}\left(\overline{u_k}\;\overline{u_i'\theta_v'} \right) - \left(\overline{u_i'u_k'} - \frac{2}{3}\delta_{ik}\overline{E} \right)\frac{\partial \overline{\theta_v}}{\partial x_k} - \frac{2}{3}\overline{E}\frac{\partial \overline{\theta_v}}{\partial x_i} - \overline{u_k'\theta_v'}\frac{\partial \overline{u_i}}{\partial x_k}$$
$$- \frac{\partial \overline{u_i'u_j'\theta_v'}}{\partial x_j} + \frac{1}{\overline{\rho}}\overline{p'}\frac{\partial \overline{\theta_v}}{\partial x_i} - 2\varepsilon_{ikl}\Omega_k\overline{u_l'\theta_v'} - \frac{1}{\overline{\rho}}\frac{\partial \overline{p'\theta_v'}}{\partial x_i} + \delta_{i3}\frac{g}{\overline{\theta_v}}\overline{\theta_v'^2} \qquad 3.6$$

In Equations 3.4 and 3.6, terms involving molecular transports of Reynolds stresses and fluxes have been omitted because their influence outside the laminar sub-layer are negligible. This is not the case for the molecular rate of dissipation ε, however, which appears in the turbulent kinetic energy equation given by:

$$\frac{\partial \overline{E}}{\partial t} = -\frac{\partial}{\partial x_k}\left(\overline{u_k}\;\overline{E} \right) - \left(\overline{u_i'u_j'} - \frac{2}{3}\delta_{ij}\overline{E} \right)\frac{\partial \overline{u_j}}{\partial x_j} - \frac{\partial}{\partial x_k}\left(\overline{u_k'\varepsilon'} + \frac{\overline{u_k'p'}}{\overline{\rho}} \right) + \frac{g}{\overline{\theta_v}}\overline{u_3'\theta_v'} - \varepsilon \qquad 3.7$$

In order to complete Equation 3.6, a formulation for the variance of potential temperature, $\overline{\theta_v'^2}$, is required. This takes the form:

$$\frac{\partial \overline{\theta_v'^2}}{\partial t} = -\frac{\partial}{\partial x_k}\left(\overline{u_k}\;\overline{\theta_v'^2} \right) - \frac{\partial}{\partial x_k}\left(\overline{u_k'\theta_v'^2} \right) - 2\,\overline{u_k'\theta_v'^2}\frac{\partial \overline{\theta_v}}{\partial x_i} - \chi \qquad 3.8$$

where:

$$\chi = 2\kappa\frac{\partial \overline{\theta'_v}}{\partial x_j}\frac{\partial \overline{\theta'_v}}{\partial x_j} \qquad 3.9$$

χ is a measure of the rate of smoothing of temperature variance by molecular thermal diffusivity.

Sommeria (1976) extended the above principles to include phase changes of water in the original Deardorff ABL model. The specific humidity, cloud liquid water, and rainwater specific humidities are thus written in analogy with Equation 3.6 as:

$$\frac{\partial \overline{u_i'q'}}{\partial t} = -\frac{\partial}{\partial x_k}\left(\overline{u_k}\;\overline{u_i'q'} \right) - \left(\overline{u_i'u_k'} - \frac{2}{3}\delta_{ik}\overline{E} \right)\frac{\partial \overline{q}}{\partial x_k} - \frac{2}{3}\overline{E}\frac{\partial \overline{q}}{\partial x_i} - \overline{u_k'q'}\frac{\partial \overline{u_l}}{\partial x_k}$$
$$- \frac{\partial \overline{u_i'u_j'q'}}{\partial x_j} + \frac{1}{\overline{\rho}}\overline{p'}\frac{\partial \overline{q}}{\partial x_i} - 2\varepsilon_{ikl}\Omega_k\overline{u_l'q'} - \frac{1}{\overline{\rho}}\frac{\partial \overline{p'q'}}{\partial x_i} \qquad 3.10$$

$$\frac{\partial \overline{u_i' q_L'}}{\partial t} = -\frac{\partial}{\partial x_k}(\overline{u_k}\ \overline{u_i' q_L'}) - (\overline{u_i' u_k'} - \frac{2}{3}\delta_{ik}\overline{E})\frac{\partial \overline{q_L}}{\partial x_k} - \frac{2}{3}\overline{E}\frac{\partial \overline{q_L}}{\partial x_i} - \overline{u_k' q_L'}\frac{\partial \overline{u_i}}{\partial x_k}$$

$$-\frac{\partial \overline{u_i' u_j' q_L'}}{\partial x_j} + \frac{1}{\overline{\rho}}\overline{p'}\frac{\partial \overline{q_L}}{\partial x_i} - 2\varepsilon_{ikl}\Omega_k \overline{u_l' q_L'} - \frac{1}{\overline{\rho}}\frac{\partial \overline{p' q_L'}}{\partial x_i} \qquad 3.11$$

$$\frac{\partial \overline{u_i' q_R'}}{\partial t} = -\frac{\partial}{\partial x_k}(\overline{u_k}\ \overline{u_i' q_R'}) - (\overline{u_i' u_k'} - \frac{2}{3}\delta_{ik}\overline{E})\frac{\partial \overline{q_R}}{\partial x_k} - \frac{2}{3}\overline{E}\frac{\partial \overline{q_R}}{\partial x_i} - \overline{u_k' q_R'}\frac{\partial \overline{u_i}}{\partial x_k}$$

$$-\frac{\partial \overline{u_i' u_j' q_R'}}{\partial x_j} + \frac{1}{\overline{\rho}}\overline{p'}\frac{\partial \overline{q_R}}{\partial x_i} - 2\varepsilon_{ikl}\Omega_k \overline{u_l' q_R'} - \frac{1}{\overline{\rho}}\frac{\partial \overline{p' q_R'}}{\partial x_i} \qquad 3.12$$

The variances of q, q_L, and q_R are computed through similar relationships to Equation 3.8.

In this form of second-order closure, a number of assumptions are involved; the first concerns the correlation of turbulent pressure terms, i.e.:

$$\frac{1}{\rho}\overline{p'\left(\frac{\partial u_i'}{\partial x_j} + \frac{\partial u_j'}{\partial x_i}\right)} = -c_m \frac{\overline{E}^{1/2}}{\Delta}(\overline{u_i' u_j'} - \frac{2}{3}\delta_{ij}\overline{E}) + C_v \qquad 3.13$$

$$\frac{1}{\overline{\rho}}\frac{\partial \overline{p'\theta_v'}}{\partial x_i} = -c_s \frac{\overline{E}^{1/2}}{\Delta}(\overline{u_i'\theta_v'}) + \delta_{i3} C_g \qquad 3.14$$

The humidity/pressure fluctuation terms have analogous expressions to Equation 3.14. c_m, c_s are constants of order 1, C_v is an effect of the compensation of velocity-deformation as derived by Lilly (1967) for quasi-isotropic turbulence, and C_g is a gravitational compensation term. Lilly (1967) writes these two terms as:

$$C_v = \frac{2}{5}\overline{E}\left(\frac{\partial \overline{u_i}}{\partial x_j} + \frac{\partial \overline{u_j}}{\partial x_i}\right) \qquad 3.15$$

$$C_g = -\frac{1}{3}\frac{g}{\overline{\theta_v}}\overline{\theta_v'^2} \qquad 3.16$$

Δ is the relevant length-scale entering into the determination of the magnitudes of the subgrid-scale variables. While this circumvents the problem of making mixing-length type assumptions, the c_m and c_s coefficients do have a grid dependency when the ratios of horizontal-to-vertical grid spacing are not close to unity.

The second major assumption incorporated into this closure scheme concerns the variables ε and κ. These quantities are defined as:

$$\varepsilon = c_E \frac{\overline{E}^{3/2}}{\Delta} \qquad 3.17$$

$$\kappa = c_\theta \frac{\overline{E}^{1/2}\,\overline{\theta_v'^2}}{\Delta} \qquad 3.18$$

where Δ replaces a mixing length. Constants c_E and c_θ are generally different from c_m and c_s because they are related to a damping process rather than an isotropizing process.

A third assumption required for the closure of the second-moment equations pertains to the triple correlations. For momentum, heat, and moisture, these are given as:

$$\overline{u'_i u'_j u'_k} = - c_{3m} \, \Delta \, \overline{E}^{\,1/2} \left(\frac{\partial}{\partial x_i} \overline{u_j' u_k'} + \frac{\partial}{\partial x_j} \overline{u_i' u_k'} + \frac{\partial}{\partial x_k} \overline{u_i' u_j'} \right) \qquad 3.19$$

$$\overline{u'_i u'_j \theta'_v} = - c_{3s} \, \Delta \, \overline{E}^{\,1/2} \left(\frac{\partial}{\partial x_i} \overline{u_j' \theta'_v} + \frac{\partial}{\partial x_j} \overline{u_i' \theta'_v} \right) \qquad 3.20$$

$$\overline{u'_i u'_j q'} = - c_{3q} \, \Delta \, \overline{E}^{\,1/2} \left(\frac{\partial}{\partial x_i} \overline{u_j' q'} + \frac{\partial}{\partial x_j} \overline{u_i' q'} \right) \qquad 3.21$$

$$\overline{u'_i u'_j q_L'} = - c_{3L} \, \Delta \, \overline{E}^{\,1/2} \left(\frac{\partial}{\partial x_i} \overline{u_j' q_L'} + \frac{\partial}{\partial x_j} \overline{u_i' q_L'} \right) \qquad 3.22$$

$$\overline{u'_i u'_j q_R'} = - c_{3R} \, \Delta \, \overline{E}^{\,1/2} \left(\frac{\partial}{\partial x_i} \overline{u_j' q_R'} + \frac{\partial}{\partial x_j} \overline{u_i' q_R'} \right) \qquad 3.23$$

with:

$$\overline{u'_i \theta'^2_v} = - c_{3\theta} \, \Delta \, \overline{E}^{\,1/2} \frac{\partial}{\partial x_i} \overline{\theta'^2_v} \qquad 3.24$$

and similar relationships for the moisture variances.

3.1.2 Simplifications to the full second-moment closure equations

Because in some instances it may not be justified to solve the full, and rather intractable, second-order turbulence closure equations introduced in Section 3.1.1, some simplifications can be applied in order to provide a more computationally-efficient framework for describing turbulence in a fine-scale ABL model. If only the isotropic source terms in the main equations are retained, as well as the damping and buoyancy terms, a set of approximate equations is obtained, i.e.:

$$\overline{u_i' u_j'} - \frac{2}{3}\delta_{ij}\,\overline{E} = - \frac{4}{15}\frac{\Delta}{c_m}\,\overline{E}^{\,1/2} \left(\frac{\partial \overline{u_i}}{\partial x_j} + \frac{\partial \overline{u_j}}{\partial x_i} \right) \qquad 3.25$$

$$\overline{u_i' \theta_v'} = - \frac{2}{3}\frac{\Delta}{c_s}\,\overline{E}^{\,1/2} \frac{\partial \overline{\theta_v}}{\partial x_i} + \frac{2}{3}\delta_{i3}\frac{g}{\theta_v}\frac{\Delta}{c_s\,E^{1/2}}\,\overline{\theta_v'^2} \qquad 3.26$$

$$\overline{u_i' q'} = - \frac{2}{3}\frac{\Delta}{c_q}\,\overline{E}^{\,1/2} \frac{\partial \overline{q}}{\partial x_i} \qquad 3.27$$

$$\overline{u_i' q_L'} = - \frac{2}{3}\frac{\Delta}{c_{qL}}\,\overline{E}^{\,1/2} \frac{\partial \overline{q_L}}{\partial x_i} \qquad 3.28$$

$$\overline{u_i'q_R'} = -\frac{2}{3}\frac{\Delta}{c_{qR}}\overline{E}^{1/2}\frac{\partial \overline{q_R}}{\partial x_i} \quad 3.29$$

$$\overline{\theta'^2_v} = -\frac{2}{3}\frac{\Delta^2}{c_s c_\theta}\frac{\partial \overline{\theta_v}}{\partial x_l}\frac{\partial \overline{\theta_v}}{\partial x_l} \Big/ \left(1 + \frac{2}{3}\frac{Ri}{c_s c_\theta}\right) \quad 3.30$$

$$\overline{E} = \frac{\Delta}{c_E}\left[\frac{4}{15}\frac{\Delta}{c_m}\frac{\partial \overline{u_i}}{\partial x_j}\left(\frac{\partial \overline{u_i}}{\partial x_j} + \frac{\partial \overline{u_j}}{\partial x_i}\right) + \frac{g}{\theta_v}\frac{\overline{u'_3\theta'_v}}{\overline{E}^{1/2}}\right] \quad 3.31$$

In Equation 3.30, Ri is the Richardson Number defined as:

$$Ri = \frac{g}{\theta_v}\frac{\Delta^2}{\overline{E}}\frac{\partial \overline{\theta_v}}{\partial x_3} \quad 3.32$$

3.1.3 Non-linear eddy-viscosity approach

The non-linear eddy-viscosity approach consists in further simplifying the equations discussed in Section 3.1.2 through the removal of buoyancy and mean isotropy effects. These simplifications yield:

$$\overline{u_i'u_j'} - \frac{2}{3}\delta_{ij}\overline{E} = -\frac{4}{15}\frac{\Delta}{c_m}\overline{E}^{1/2}\left(\frac{\partial \overline{u_i}}{\partial x_j} + \frac{\partial \overline{u_j}}{\partial x_i}\right) \quad 3.33$$

$$\overline{u_i'\theta_v'} = -\frac{2}{3}\frac{\Delta}{c_s}\overline{E}^{1/2}\frac{\partial \overline{\theta_v}}{\partial x_i} \quad 3.34$$

$$\overline{E} = E^* = \frac{4}{15}\frac{\Delta^2}{c_m c_E}\frac{\partial \overline{u_i}}{\partial x_j}\left(\frac{\partial \overline{u_i}}{\partial x_j} + \frac{\partial \overline{u_j}}{\partial x_i}\right) \quad 3.35$$

The humidity terms remain unchanged with respect to Equations 3.27 - 3.29, since they are not directly influencing buoyancy (phase changes of water would enter into the buoyancy term through the release of latent heat, implying a physical coupling between the moisture and temperature equations when clouds are generated in the model).

The above viscosity approach was used by Deardorff (1970; 1972) for both neutral and unstable stratifications. When attempting to relate this type of formulation to K-closure theory (as will be discussed in Chapter 4 for the DREAMS mesoscale model), the value of eddy coefficients may be inferred from the following set of relationships:

$$K'_m = \frac{4}{15}\frac{\Delta}{c_m}(E^{*1/2}) = (c\Delta^2)\, D / \sqrt{2} \quad 3.36$$

$$K'_s = \frac{2}{3}\frac{\Delta}{c_s}(E^{*1/2}) = 2.5\frac{c_m}{c_s}K'_m \quad 3.37$$

where:

$$c = [(\frac{4}{15c_m})^3 / c_E]^{1/4} \qquad 3.38$$

and:

$$D = \sqrt{[(\frac{\partial \overline{u_i}}{\partial x_j} + \frac{\partial \overline{u_j}}{\partial x_i})(\frac{\partial \overline{u_i}}{\partial x_j} + \frac{\partial \overline{u_j}}{\partial x_i})]} \qquad 3.39$$

is the magnitude of the local velocity deformation.

According to the type of problem considered, various combinations of approaches discussed in Sections 3.1.1, 3.1.2, and 3.1.3 are possible.

The Deardorff model was one of the pioneer models in the early 1970s for the investigation of shear and buoyancy-driven flows. The modifications by Sommeria (1976) and Sommeria and Deardorff (1977) to handle the phase changes of water were a decisive step towards detailed understanding of other important elements of ABL turbulence, particularly cloud fields. It is this aspect of the Deardorff-Sommeria model which will be investigated in the following section.

3.2 Use of the Deardorff-Sommeria model for parameterization purposes

3.2.1 Introduction

A large variety of mathematical models is now available for the study and representation of the planetary or atmospheric boundary layer (ABL). One may distinguish between layered parametric models which attempt to yield a simplified picture of its main features and more detailed models which attempt some representation of its turbulent structure. The second type contains ensemble-mean models with various degrees of sophistication in the treatment of higher order moments of the turbulent fluctuations, and subgrid turbulence models which attempt an explicit description of the turbulent features with some statistical treatment for unresolved small-scale turbulence. Each type of model has its own range of capabilities. Subgrid turbulence models are the most adequate if one looks for a reliable picture of the real world, but their size limits their use to a small number of study cases. They may adequately be replaced by ensemble-mean models for horizontally homogeneous cases but remain the best tool for the study of non-homogeneous cases, i.e., mainly those involving clouds. Conversely, layered parametric models can be used under a large range of meteorological conditions but need extensive validation against real data. As all the data needed for these validations are not generally available, it seems advisable to complement those with data from already validated subgrid turbulence models. This idea leads to the study conducted by Beniston and Sommeria (1981), where results from a three-dimensional ABL model are compared with hypotheses used in parameterization schemes and with results from parametric models.

The model used in this study is a slightly modified version of the one developed by Sommeria (1976) in collaboration with Deardorff, briefly described in Section 3.1. The data set is obtained for initial conditions similar to those used in the Puerto Rico simulation described in Sommeria and LeMone (1978). It is analyzed in view of possible applications to parameterization methods for non-precipitating convection and related turbulence. It is compared with one example of a cloud population model by Fraedrich (1976) and of a parametric layered model by Betts (1975, 1976). Some results may provide suggestions for further development of parametric models.

3.2.2 Description of the model and the data set

The main purpose of the model is to describe the detailed evolution of the turbulent meteorological field within a three-dimensional grid network covering a small domain of the atmosphere (a 2 km x 2 km x 2 km cube in the present case). The basic variables include the u, v, w components of the velocity, the virtual potential temperature θ_v, the specific humidity q, the specific content q_L of cloud water, and the deviation P' of the pressure from a prescribed large-scale field. The basic equations are the usual set of meteorological expressions for moist air with the anelastic approximation for density. Besides the equation of state and the continuity equation, it contains a Navier-Stokes equation with a Coriolis term and a buoyancy term including the buoyancy effects of water vapor and of cloud droplet loading, a thermodynamic equation which includes latent heat release, and conservation equations for vapor and liquid water, which include condensation and evaporation processes.

The set of equations is spatially averaged over the grid volumes. This implies a statistical treatment for the grid-averaged subgrid turbulent processes. The treatment is based on a simplified closure of the second-order moment equations, along with a representation of subgrid condensation as described in Sommeria and Deardorff (1977). In contrast to the ensemble mean models, the parameterized subgrid-scale processes here represent only a small fraction of the directly computed grid-scale turbulent processes.

The arbitrary or empirical assumptions made in such a model include mostly the value of the constants used in the closure method and the specified boundary conditions. The order of magnitude of these constants has been assessed by various studies (see, e.g., Deardorff, 1973) and their exact value does not seem critical for neutral or unstable thermal stratification. The case of stable stratification remains more delicate as the existence and the value of a critical Richardson Number is greatly dependent on the closure assumptions. The prescribed cyclic lateral houndary conditions prevent taking into account any detailed interaction between the computed field and a supposed large-scale forcing. However, large-scale effects coming from mean subsidence or advection can be included in the evolution of the horizontally-averaged variables. At the upper boundary, the vertical gradients of the variables are prescribed according to the observed values, and turbulent fluxes are set to zero. The lower boundary conditions assume a constant flux or surface layer extending from the Earth's surface to the first level where variables are computed. The surface fluxes may either be obtained experimentally or computed from the measured values of the variables at the earth's surface and the computed values at the first level of the model, with the use of a surface layer parameterization. This scheme itself relies on empirical formulas, the Businger et al. (1971) formulation being used here. In addition to these considerations, one should keep in mind the fact that the turbulent phenomena are computed on a 40 x 40 x 40 point grid, which limits the scale range of usable results to about 1-20. All these limitations, however, do not overcome the advantages of a three dimensional model as a kind of numerical laboratory for the validation of more synthetic or "compact" models.

Before a numerical experiment is made, a number of parameters must be specified from the observations. These includes external conditions such as the large-scale horizontal gradients of the components of velocity, temperature, moisture and pressure, as well as the lower boundary conditions for temperature and moisture, and some estimate of the radiational cooling for which a prescribed value is used here. Initial conditions are specified from the observations. The model, starting from horizontally homogeneous conditions, evolves toward a statistically steady state which is reached after approximately 1 hour of model time. This evolution of the turbulent field takes place independently of any adjustment of the mean profiles which generally do not correspond to a long-term equilibrium.

The results which have been selected here are related to specific properties of moist non-precipitating convection as observed in Trade-wind areas. These include properties of cloud populations used as hypotheses in various parametric models, tests of two types of parametric models, namely those of Betts (1975; 1976) and Fraedrich (1976), and statistical features which may appear useful for further developments of parameterization methods.

The main feature of the undisturbed Trade-wind boundary layer measured in the Puerto Rico experiment, and simulated by the model to steady state, is the presence of a well-mixed layer for moisture and potential temperature up to 550 m. It is surmounted by a conditionally unstable and moist layer up to 1100 m, and by a stable and dryer layer above.

The velocity profile shows a wind maximum near the middle of the mixed layer and a veering with height up to 1200 m. Other parameters prescribed as external conditions in the model include the latitude (19°N), the large-scale pressure gradient and a mean infrared cooling of 1.5×10^{-5} K/s. The large-scale pressure gradient is such that the geostrophic wind corresponds to the measured wind above 1200 m and obeys the thermal wind equations from there down to the ocean surface. The initial conditions, together with the large-scale forcing, have been chosen in order to obtain a quasi steady-state for the mean profiles, in particular for the moisture and temperature fields.

3.2.3 Tests of hypotheses used in parameterization schemes

Due to the restrictions imposed by too little and relatively imperfect atmospheric data, it has been necessary for workers in the field of tropical cumulus studies to make certain assumptions concerning the morphology and distribution of Trade-wind cumuli. Based on observations and numerical results, these hypotheses are then incorporated into cumulus parameterization schemes. The hypotheses, for the most part, have not been verified in the atmosphere to any great extent; it is in such cases that a numerical model may take over from scant meteorological data to validate certain assumptions with a reasonable degree of confidence.

The relatively fine grid length of 50 m in the present model allows the resolution of several classes of non-precipitating cumuli, both in the horizontal and in the vertical. The smallest cumuli observed in the model have an equivalent diameter of approximately 110 m, or an area of 9500 m^2, well resolved by the present grid; the largest clouds occupy about 25% of the horizontal grid. At any one time, there are up to 32 individual cumuli coexisting in the domain of integration at varying levels within the cloud layer. The model thus simulates clouds in sufficient number and of sufficiently different dimensions to give some confidence in the statistical features of the cloud population. The statistics, however, may be biased for small clouds when their size is of the order of the grid size, as well as for large clouds which cover a significant part of the domain.

Hypotheses concerning the morphology and distribution have been advanced which are common to most authors, and many of these have been succinctly included in Fraedrich's (1976) work on cumulus parameterization based on observable cloud characteristics. The first such hypothesis which has been tested using the simulation of cumuli by the present model is the relationship between cloud diameter and cloud thickness. Plank (1969) had already established the existence of such a relationship from aircraft observation of Florida cloud populations, and Fraedrich used data from the Atlantic Tropical Experiment (ATEX; cf. Fraedrich, 1976) to find a cloud radius to cloud thickness ratio γ. Hill (1974) used data from a numerical simulation to ascertain that a fairly constant ratio γ exists for a given category of cumuli. Using data on microfilms, over 100 individual cumuli were examined at varying points in time; it was possible to establish that despite a certain scatter of points, a link between cloud thickness and cloud radius does indeed exist. The average value for γ was found to be 0.4 ± 0.1. This value is well within the generally accepted limits (Fraedrich, 1976) of $0.2 \leq \gamma \leq 0.5$.

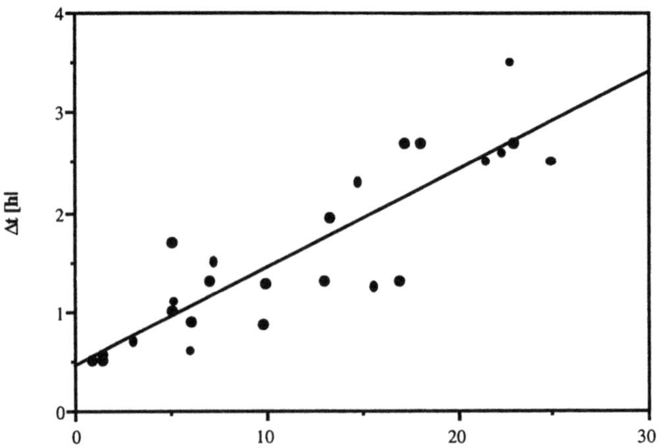

Figure 3.1: Relationship between maximum cloud area A_c and half-life Δt of model-simulated clouds

The smaller values of γ are generally associated with cumulonimbus formations, which characteristically have a large vertical extent and a relatively narrow base. The values found for the simulated clouds are in the range of the smaller non-precipitating cumuli; in the cases studied, the clouds rarely exceeded 1000 m in width or in height. One reason for the rather large scatter of the radius to thickness distribution is that the model is able to simulate clouds which have a large radius but which do not have great vertical extent. One probable cause for this is the strong stability which exists in the upper section of the model, which constrains vertical cloud development beyond the Trade-wind inversion. Such clouds do not have the "convective tower" configuration typical of tropical cumuli, and as such do not fit into the pattern of clouds having a relatively constant radius to thickness ratio. In general, however, Beniston and Sommeria (1981) have found that most of the clouds which do not reach as far as the Trades inversion in the model have a relatively constant value of the parameter γ.

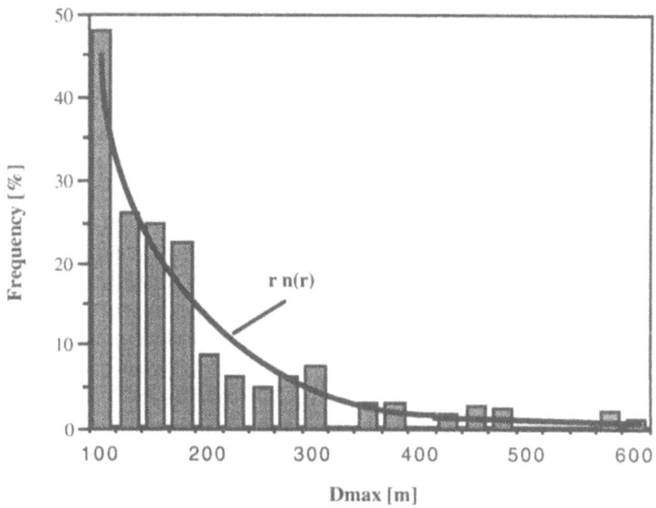

Figure 3.2: Histograms of model cloud diameters integrated over a 30-minute period

The second hypothesis, which is particularly important in Fraedrich's parameterization scheme, is the one which links maximum cloud area to cloud lifetime. Observations of the simulated cumuli have shown that the clouds are characterized by a cyclic growth and decay pattern. The times of growth and subsequent decay for each individual cloud can, on the whole, be considered to be equal, and thus it seems logical to assume that at the time of peak development, the cloud's maximum area is proportional to its lifetime. From the model data, it has been possible to isolate the life cycle of 25 separate cumuli and have traced the distribution of the cloud-lifetime and cloud-area points in Figure 3.1. The empirical regression equation has the form:

$$\Delta t = 3.174 \times 10^{-2} A_c \qquad 3.40$$

where A_c is the maximum area attained by an individual cloud at the peak of its activity and Δt the cloud half-lifetime (equivalent to growth or decay period) in seconds. This relation tends to overestimate the lifetime of an average individual cloud by a factor of 2 approximately, compared to the value obtained from Fraedrich's formula with his own parameters.
Another test of a cumulus parameterization hypothesis concerns the statistical distribution of tropical clouds. Fraedrich (1976) assumes an exponential distribution for the radii of the cumuli which is described by:

$$n(r) = K \exp(-\beta r) \tag{3.41}$$

where n(r) is the cloud number density, r the cloud radius, and K and β are coefficients which can be derived from the total number of clouds and cloud area cover.

Histograms of cloud diameter computed at the level of maximum cloud cover, in this case 725 m, have been contructed and integrated over periods of 30 min, as shown in Figure 3.2. The histogram illustrated here shows a maximum frequency of cloud diameter which occurs in the interval $100 \leq D \leq 125$ m, and a quasi-exponential decrease is observed toward larger cloud diameters. The irregularity of the profile is due to the differing stages of growth or decay in which the clouds find themselves at the time which was chosen for this particular example. Certain clouds grow rapidly by an amalgamation of smaller clouds in their immediate environment; others depend on the evolution of the humidity and mass fluxes and require a certain time of adjustment before growing or decaying. A particular cloud diameter interval is thus made up of clouds which are transient in that interval (either growing or decaying), and clouds which have reached their maximum development stage and are temporarily stabilized in that interval. The exponential curve calculated from Fraedrich's equation for the cumulus radius distribution given by Equation 3.4 has been superimposed upon the histogram. Here K and β are calculated from model data. The profile in Figure 3.2 has been non-dimensionalized by multiplying n(r) by the radius r, in order to be able to compare directly the theoretical and observed data. Statistical F and t tests show no difference between the observed and theoretical profiles at the 95 % significance level.

It should be pointed out, however, that there has been an arbitrary truncation of the theoretical curve at the points which correspond to the observed minimum and maximum cloud diameters in the model. In theory, as the smallest cloud diameter interval is approached, the frequency asymptotically approaches infinity. This arbitrary cutting off of the theoretical curve within reasonable limits is a major drawback of Fraedrich's theory. It seems that Lopez's (1977) assumption of a log-normal distribution of cloud diameters is perhaps more realistic, since the log-normal curve has a positive skew distribution. This means that the distribution tails off on either side of the maximum, the profile to the right of the maximum being very close to Fraedrich's exponential distribution. However, the log-normal distribution suggests that there should be a finite number of frequencies in the interval below the 100 - 125 m interval (i.e., between 75 and 100 m, and between 50 and 75 m). As can be seen from Figure 3.2, this is not the observed case; this may be due to model peculiarities which do not allow for the birth of clouds under 100 m in diameter. The fact that the model does not generate a log-normal profile for cloud size distribution does not in any way refute Lopez's concept of cloud population distributions.

It has been shown in this section that data from a three-dimensional ABL model can be used to verify certain hypotheses which up till now had not been fully tested against real atmospheric data. The results of the testing of cloud morphological features are encouraging and go a stage further toward a more complete understanding of Trade-wind cumulus characteristics. The usefulness of the model is apparent when examining the assumption of a cloud area/lifetime distribution; such an assumption would be almost impossible to test in the real atmosphere at present.

Examples of direct applications of ABL model data will be presented in the following sections. One major area of interest is the development of convection models of various dimensions, and this frequently requires the parameterization of the overall effects of clouds on the mesoscale and large-scale atmospheric environment. The use of a fine grid mesh model can help in verifying certain important parametric schemes, such as those of Betts and Fraedrich. The "validation" of such schemes with the model data set would allow the use of such parameterizations with a considerable degree of confidence in larger-scale models. In these models, the simulation of the dynamic and thermodynamic effects of each individual cloud is not realistic; however, an entire cloud population can have a significant impact on large-scale atmospheric heat and moisture budgets.

Before going on to show how the ABL model can be used to test the Fraedrich (1976) and Betts (1975) models (which have not been fully verified against real atmospheric data), the aims of cumulus parameterization schemes in general, and the main equations used in the present study, will be briefly surveyed.

3.2.4 Use of the model data to verify cumulus parameterization models

Although individual parameterization schemes differ, the ultimate goal of all Trade-wind cumulus parameterizations is to understand and predict the interaction between the large-scale environment and the cloud population. Though cumulus cloud activity is a relatively small-scale phenomenon (compared to global circulation features), a cloud population can produce profound changes in the large-scale environment through moistening or drying, and warming or cooling processes. In order to parameterize the cloud population adequately, three major factors must be known:

- The mass flux, which may be defined by:

$$M = \rho \; |w| \qquad 3.42$$

where ρ is the density and w the horizontal average of the absolute vertical velocity, i.e., $w = N^{-1}\Sigma x(u_3u_3)^{1/2}$, where N is the number of grid points at each level, 1600 in the present model. If the average vertical velocity at each level is zero (in the absence of large-scale subsidence), M is clearly twice the value of the mean upward mass transport. Hence M is a measure of the quantity of air rising to form convection towers. This variable, which is computed directly by the model, should be distinguished from the so-called convective mass flux used by various authors, in particular Betts (1975) and identified by w_*. This mass flux value is described by:

$$w_* = \sigma\rho w_c \qquad 3.43$$

where w_c is the cloud-area averaged ascending velocity and a the fractional cloud cover. Although most of the vertical motion within the cloud is upward, downdrafts are frequently observed on the fringes of the cloud. w_* is that part of the mass flux which is positive within the cloud field.

- Entrainment and detrainment factors. These are quantities which define the turbulent mixing of cloud and non-cloud air at the cloud boundaries. Entrainment *E* and detrainment *D* are defined by:

$$E = -D = M^{-1}dM / dz \qquad 3.44$$

Entrainment and detrainment can be viewed as the change in surface area of the cloud population due to intake of environment air by the cloud, or loss of cloudy air to the environment.

- The characteristics of air detrained from the clouds. Knowledge of the thermodynamic character of detrained cloud air can lead to an estimation of the heat and moisture variations in the immediate cloud environment, and after a much longer time span, in the large-scale environment. As direct in-cloud measurements are particularly difficult, it will be seen that the model is a valuable tool for this kind of study, since in-cloud values of thermodynamic variables are directly available. The large-scale budgets for heat and moisture associated with convective activity take the familiar form:

$$Q_1 = \partial\bar{s} / \partial t + \bar{v}.\nabla\bar{s} + \bar{w}\partial\bar{s} / \partial p \; = Q_R - \partial\left(\overline{w' s'_L{}'}\right) / \partial p \qquad 3.45$$

$$-Q_2 = L\left(\partial\bar{q} / \partial t + \bar{v}.\nabla q + \bar{w}\partial\bar{q} / \partial p\right) \; = L\partial\left(\overline{w' q'}\right) / \partial p \qquad 3.46$$

$$Q_1 - Q_2 = \partial\bar{h} / \partial t + \bar{v}.\nabla\bar{h} + \bar{w}\partial\bar{h} / \partial p \; = Q_R - \partial\left(\overline{w' h'}\right) / \partial p \qquad 3.47$$

where Q_R is the radiative cooling, ω the vertical velocity in pressure coordinates and L the latent heat, and the flux terms on the right-hand side of the above three equations are convective transport terms. Q_1 and Q_2 are the apparent heat source and moisture sinks, respectively with the static energies represented by:

$$s = c_pT + gz \qquad 3.48$$

$$h = c_pT + gz + Lq \qquad 3.49$$

Parallel equations for saturated conditions yield:

$$S_L = c_pT + gz + Lq_L \qquad 3.50$$

$$h_s = c_pT + gz + Lq_s \qquad 3.51$$

where c_p is the specific heat at constant pressure, T the temperature, gz the geopotential, q the specific humidity, q_L the liquid water content, and q_s the saturation value of q.

The flux of S_L, represents the total heat which is responsible for the heating and cooling of the mean atmosphere in the presence of clouds. $\overline{w'h'}$ is the flux of moist static energy which describes the convective activity of the cumulus population; the greatest contribution to the fluxes of static energy comes from condensation processes at the lower levels of the cloud layer.

3.2.4.1 Testing Fraedrich's parameterization scheme

The main points of Fraedrich's (1976) parameterization scheme are briefly summarized here; the reader is referred to the original work for more details.

A simple model of a transient cumulus cloud which can be constructed with direct cloud observations is suggested. Using a Marshall-Palmer type distribution for cloud radii, and the cloud radius to cloud thickness ratio γ, the author shows that the mass budget of an entire cloud population can be computed through the integration of individual cloud mass flux over the entire spectrum of cloud radius and thickness of the given population. Thus the vertical mass flux can be calculated from the equation:

$$Mz = M_B(\beta\gamma\zeta + 1)\ \exp(\beta\gamma\zeta) \qquad 3.52$$

$$M_B = \frac{K}{\beta^2}\frac{c_1}{3\alpha} = \left(\frac{N_0 A_0}{2r}\right)^{1/2}\frac{c_1}{3\alpha} \qquad 3.53$$

Here M_B is the cloud-base mass flux; K and β are the effective cloud number density and exponential number density decrease, respectively, and are calculated from total cloud number N_0 and total cloud area A_0 data, which are generally directly observable features; c_1 is a constant given by:

$$c_1 = \frac{\rho A_{c(max)}}{\Delta t} = \lambda^* m_{c0} \qquad 3.54$$

which is essentially the distribution of maximum individual cloud area versus lifetime, m_{c0} being individual cloud-base mass flux, and λ^* is an entrainment factor, which takes the form:

$$\lambda^* = 3\alpha / r \qquad 3.55$$

with α an empirical constant and r the cloud radius. Final detrainment is dependent on the vertical mass flux gradient, such that:

$$D_f = -\frac{\partial M}{\partial z} = D_{f0}(\beta\gamma z)\exp(-\beta\gamma z) \qquad 3.56$$

with:

$$D_{f0} = M_B\beta\gamma = N_0 c_1 \gamma / 3\alpha \qquad 3.57$$

and lateral detrainment is proportional to the vertical cloud number distribution:

$$D_l = c_l N(z) = D_{l0} \exp(-\beta \gamma z) \qquad 3.58$$

To test the parametric scheme, a time corresponding to a significant number of tradewind cumuli was chosen, at 2.5 h of simulation. Figure 3.3 shows the fluctuations of total cloud as a function of time for the first 3 h of simulation. At 2.5 h a relative peak of cloud activity is observed.

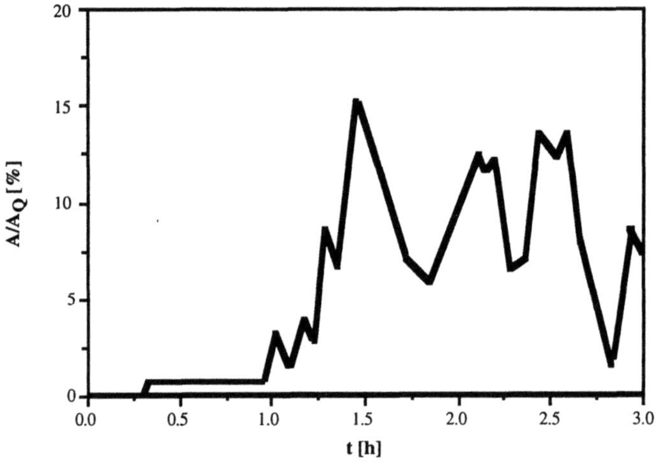

Figure 3.3: Total cloud amounts in time as simulated by the ABL model

It was possible to calculate for the 32 small cumuli which coexist in the model after a simulation of 2.5 h, the value of K, β and α. The value of the constant c_l was computed to be 250 kg/m/s. An analysis of the cloud-base values of mass flux m_{c0} allowed to obtain an average value of λ^*, which, when using the mean of the 32 cloud-base radii, yields a value of 0.185 for the constant 3α. The present magnitudes of c_l and 3α are much smaller than Fraedrich's values, i.e.:

$$c_l = 10^5 \text{ kg/m/s}; \qquad 3a = 0.75 \qquad 3.59$$

This is principally due to the fact that Trade-wind cumuli are being studied in the ABL model, which are among the smallest clouds in the cumulus family; Fraedrich was concerned with much larger cumuli and even cumulonimbi. According to Fraedrich (personal communication, 1977), c_l and 3α values are likely to change from one class of cumulus cloud to another, and even between precipitating and non-precipitating clouds in the same class of cumuli; the values of c_l and 3α which are presented here are indeed representative of populations of small cumuli. From the definition of c_l, it appears that the growth or decay time of larger cumulus clouds is proportionally shorter than that of small cumuli. This is in fact frequently observed in nature. A cumulus of 1 km in diameter will typically have a lifespan of 10-20 min; a cumulonimbus of 5 km in diameter (more than six times the area of the former cloud) generally has a lifespan of 30-50 min.

Figure 3.4 shows the vertical profiles of mass flux M(z), calculated from Fraedrich's model and compares this with the convective mass flux ω_* observed in the model. Also shown are lateral and final detrainment profiles computed from Equations 3.53 and 3.58, all at 2.5 h of model time. The quantities illustrated in Figure 3.5 have been non-dimensionalized by normalizing M_B, D_{l0} and D_{f0}, values in order to have visual in a form comparable to those presented by Fraedrich (1976). The data substituted into Fraedrich's equations are those which have been discussed in the preceding paragraphs (essentially c_1, α, K, and β computed from the model cloud characteristics).

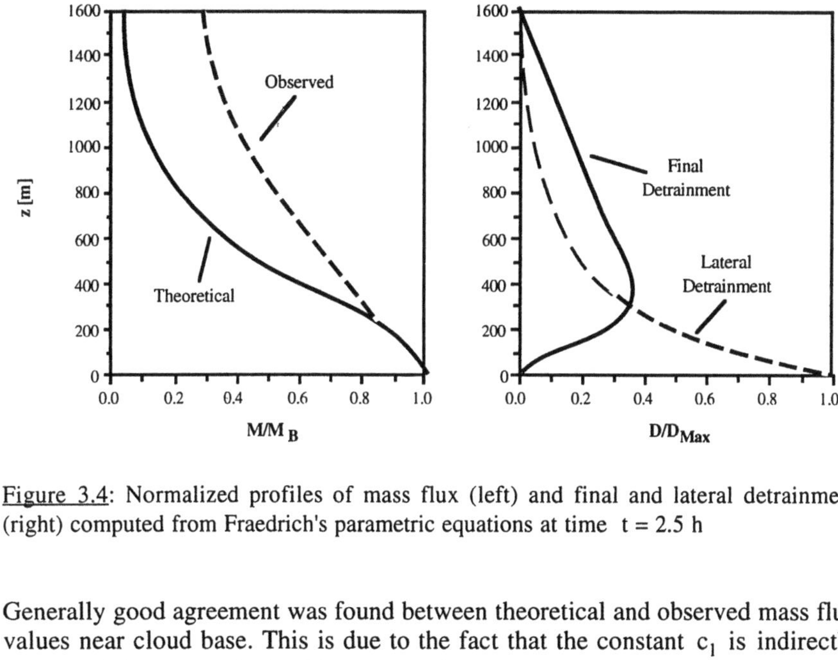

Figure 3.4: Normalized profiles of mass flux (left) and final and lateral detrainment (right) computed from Fraedrich's parametric equations at time t = 2.5 h

Generally good agreement was found between theoretical and observed mass flux values near cloud base. This is due to the fact that the constant c_1 is indirectly computed from model cloud base mass flux. At about 100 m above the 600 m cloud ensemble base height, the two profiles diverge. The reason for this discrepancy is that gravity waves propagate in the stable layers above the mixed layer. In this case, the convective mass flux will have some finite value at the top of the cloud layer instead of approaching some negligible value as is the case of the theoretical profile computed from Equation 3.53.

The detrainment profiles illustrated in Figure 3.4 (right) are computed from Equations 3.56 and 3.58, using model data. The final detrainment profile which represents the turbulent mixing process at the cloud summit has a maximum at approximately 800 m, well beneath the top of the cloud layer and some 200 m above cloud base. The histograms in Figure 3.2 show that the most abundant cumuli in the model are those with diameters in the 100-125 m interval. The ABL model average value of cloud radius to cloud thickness ratio γ of roughly 0.4 yields a cloud top of between 200 and 250 m above cloud base. It is in this region that the most abundant clouds will detrain cloud air to the overlying environment, which is exactly what the theoretical profile in Figure 3.5 (right) is illustrating.

The lateral detrainment profile D_l may not be as realistic as the final detrainment profile. From Equation 3.58, it can be seen that $D_l(Z)$ is proportional to the vertical cloud number distribution, itself an exponentially decreasing function. The observed case in the model (Beniston, 1978) is that cloud numbers do not necessarily decrease with height, exponentially or otherwise, even though maximum cloud numbers are observed at cloud ensemble base. Depending on the intensity of the convection activity, observed relative cloud number maxima occur at middle levels of the cloud layer. Consequently, the exponential decrease of lateral detrainment with height as computed from Fraedrich's model and illustrated in Figure 3.4 (right) do not correspond to the true detrainment profile, which would be expected to follow observed vertical cloud number distribution.

Fraedrich's technique of parameterization from directly observable cloud population features has the advantage of being relatively simple when compared to an essentially theoretical parameterization such as that of Arakawa and Schubert (1974). The Fraedrich model may well be valid for large cumulonimbi and classes of large cumulus clouds, but in the present study the observed vertical distribution of small trade-wind cumuli does not follow the exponentially decreasing pattern suggested by Fraedrich. This explains certain discrepancies found between the theoretical and the actual mass flux profiles, and thereby casts doubt as to the validity of the lateral detrainment profile.

3.2.4.2 Testing of Betts' parametric model

The main point of the paper presented by Betts (1975) which has been tested using the present model is the fact that thermodynamic fluxes can be well represented by a single cloud mass-flux profile rather than by a complex cloud distribution (i.e., Arakawa and Schubert, 1974). This involves the formulation of a model for a single cloud type, which entrains and detrains at all levels during its life cycle. This leads to a parameterization of the fluxes of S_L, Lq and h fluxes (see Equations 3.45 - 3.51) by the product of a convective mass flux, and a perturbation quantity resulting from an entraining cloud parcel rising through the mean environment. This gives:

$$F_x = -\overline{w'X'} = \omega_*(X_c - X_e) \qquad 3.60$$

where X represents thermodynamic variable h, q or S_L; $\omega*$ is the convective mass flux; suffix c represents in-cloud value of X, and suffix e represents the environmental value of X. The above parameterization can be substituted into the budget equation, and using:

$$\frac{\partial h_{sc}}{\partial p} = \lambda(h_{sc} - h_e) \qquad 3.61$$

the definition of the entrainment rate for an ascending cloud parcel, after rearranging certain terms, the following expression is obtained:

$$Q_1 - Q_2 - Q_R = -\omega_* \frac{\partial h_e}{\partial p} + \omega_* \left(h_{sc} - h_e\right)\left(\lambda + \frac{1}{\omega_*}\frac{\partial \omega_*}{\partial p}\right) \tag{3.62}$$

By analogy, Equations 3.45 and 3.46 take similar forms. The original budget equations, which included local time and advection terms, now reduce to a single convective mass flux, an entrainment factor, and in-cloud and environmental values of a particular thermodynamic variable. Though Equation 3.62 is a step toward the computation of heat and moisture budgets from direct atmospheric observations, it should be stressed that even with present-day sophisticated instrumentation, in-cloud meteorological measurements are still very difficult and are often open to question. The ABL model can thus make up for the lack of real atmospheric data, since in-cloud values of thermodynamic variables are accessible, both for individual clouds and for an entire cloud ensemble. The convective mass flux ω_* is the quantity defined in Equation 3.43. The essential terms used in Equations 3.60 - 3.62 are thus easily obtained using model data and allow to compute these equations; at the same time, the model calculates the flux quantities at each time step, and the subgrid, resolved and total (subgrid + resolved) fluxes can be directly accessed and consulted. It was thus possible to compare the fluxes computed directly by the model with those parameterized through Equation 3.60 using model data.

The results appear in Figure 3.5, which compares observed values of L_q and s_L fluxes with calculated values of the same fluxes using Equation 3.60. The two sets of curves are for the entire cloud ensemble, which at 2.5 h simulation is located between 625 and 1175 m.

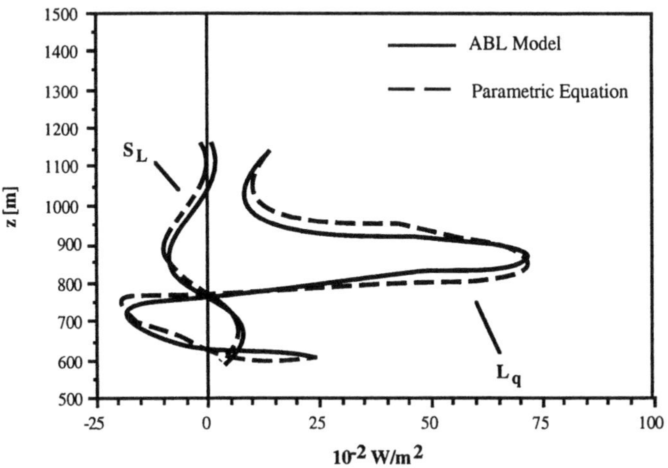

Figure 3.5: ABL simulations of variables L_q and s_L compared to those computed from Equation 3.60

The agreement between observed and parameterized fluxes is remarkable; there is no statistical difference between observed and calculated profiles at the 99 % significance levels. The result is of importance, since it allows, with a simple parameterization, to evaluate with great precision various fluxes which are fundamental in the evolution of the turbulent field in the presence of clouds. The fluxes calculated for Betts' parametric equations can subsequetly be used to evaluate the heat and moisture budgets for the entire cloud layer. Betts' (1975) parametric model which has been verified with the present boundary-layer model allows a description, with a reasonable degree of confidence, of the thermodynamic effect of a population of Trade-wind cumuli on the large-scale humidity and temperature fields. The modification of the fields contributes to changes in the evolution of the cumulus ensemble and the cloud layer. In the long term, these cloud-induced changes can have significant repercussions on large-scale atmospheric processes.

3.4.2.3 Tests of coupled mixed-layer / cloud-layer models

Several authors have underlined the important role of the mixed layer in influencing convective activity in the overlying cloud layer. In particular, Hill (1974, p. 661) states that "*the size and spacing of ... clouds are related to size and strength of the circulations in the subcloud layer just prior to the appearance of clouds. There, circulations provide moisture which is carried above the condensation level.*". Betts (1976) presents a model which couples the mixed layer to his 1975 cloud layer model (which was briefly reviewed in the previous subsection). Betts suggests that the cloud base mass flux and the mass flux gradients can be related to mixed-layer and cloud-layer terms using two closure parameters, α and β.

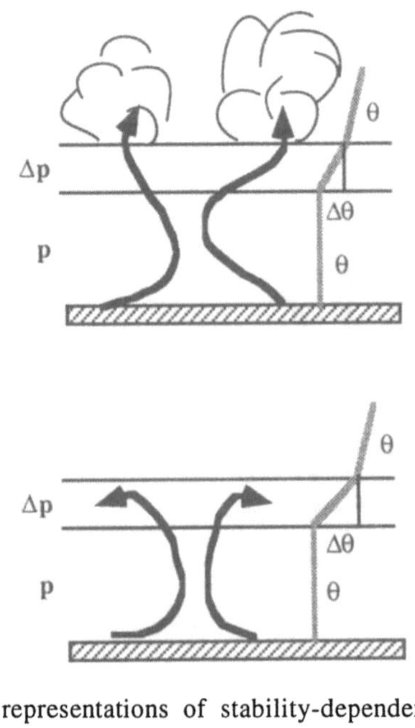

Figure 3.6: Schematic representations of stability-dependent controls on cumulus development

The first closure parameter α is defined as:

$$\alpha = \Delta p / P \qquad 3.63$$

where $P = P_B - P_0$ is the thickness of the mixed layer (subscript B refers to cloud base, subscript 0 to surface values), and Δp represents a model transition layer thickness which separates the mixed layer from the cloud layer. The strength and thickness of the transition layer acts as a control on the upward air currents which may or may not penetrate into the cloud layer to form cloud. The controls exerted by the transition zone are illustrated by the sketches in Figure 3.6. The parameter k (see Equation 3.66) relates the model cloud base static energy flux to the surface flux. The reader is referred to the original paper by Betts (1975) for an in-depth study of the equations which couple the two adjacent layers,the outline of the k-closure theory will only be summarized here. If the strength of the transition layer Δp is represented by:

$$\Delta s = \bar{s}_B - \bar{\bar{s}} \qquad 3.64$$

where s_B is the cloud base value of static energy, and $\bar{s}$ the mean value of s in the entire subcloud layer. It is possible to define a virtual static energy strength as:

$$\Delta s_v = \Delta s + C_p \bar{T} \delta \Delta_q \qquad 3.65$$

where Δ_q takes the same form as Equation 3.64, except for moisture. $\delta = 0.608$ and $\bar{T}$ is the mean temperature. The flux of s_v at the cloud-layer base $(F_{sv})_B$ is related to the surface flux $(F_{sv})_0$ through k closure, giving:

$$(F_{sv})_B = -k(F_{sv})_0 \qquad 3.66$$

Using Equations 3.65 and 3.65, it is possible to parameterize the cloud-base flux of a thermodynamic variable X using the equation:

$$(F_x)_B = -k \frac{\Delta X}{\Delta s_v} (F_{sv})_0 \qquad 3.67$$

Therefore, the ratio of the transition-layer strengths of variable X and virtual static energy s modulates the value of the surface flux of s_v , which when multiplied by the factor k, yields an estimate of the cloud-base flux of X.

Using model data, Betts' closure parameter α (Equation 3.63) has been calculated. In order to determine the value of α, the following equation presented by Betts (1976) is required, i.e.:

$$\Delta p = \Delta s / T_s \qquad 3.68$$

where Γ_s is the vertical gradient of s just above the cloud base. In effect, Δp is an estimate of the pressure difference between the top of the mixed layer and the cloud base. An average value of 0.11 has been found for α, which corresponds to a hypothetical transition-layer thickness of approximately 60 m. This is just within the limits of resolution of the present ABL model, where the vertical grid spacing is 50 m. It has been observed that the change in gradient of the thermodynamic fluxes between mixed-layer and cloud-layer regimes does not occur at one level but gradually over one or two levels near the summit of the mixed layer. This could be interpreted as being due to the presence of a transition layer in the model, whose thickness would be just within the limits of resolution of the present model.

It was possible to evaluate a critical value of α of 0.15, corresponding to a transition layer thickness of about 90 m; above this value, cloud activity is suppressed or is unable to be initiated. In the first 30 minutes of model time, α does indeed exceed 0.15 and no clouds are found. However, the process of inversion erosion through mixed-layer heating reduces the thickness and strength of the transition layer, and after about 1 hour of model time, cloud activity become important. This can be observed in the time evolution of total cloud area illustrated in Figure 3.3.

As was already mentioned in a previous subsection, the most important thermodynamic fluxes form part of the standard ABL model printout. It was thus possible to compare model-computed cloud-base fluxes of s_L, q and h with the same fluxes parameterized using k-closure theory, and in particular Equation 3.67. The mean value of k was found to be in the region of 0.40; this value was determined by substituting model data into Equation 3.66. During the non steady-state period of the model, which lasts roughly 1 hour, the parameter k is observed to vary considerably before stabilizing at this value. Values of k in the range 0.2-0.5 are generally accepted for dry mixed layers (Betts, 1976,). Corresponding values based on virtual static energy fluxes and profiles are lower. Betts (1976) finds a k value of 0.21 and Albrecht (1977) has used a constant value of 0.25 for his Trade-wind boundary-layer model. Laboratory experiments of Deardorff et al. (1969) have yielded a k value of 0.1. The values presented in this work, computed from virtual static energy fluxes, are thus somewhat higher than those found by other authors. Nevertheless, using this constant value of 0.40 for k, cloud base thermodynamic fluxes computed directly by the model and those parameterized by Equation 3.67 have been compared.

For each of the s_L, q and h fluxes, it was necessary to compute the transition-layer strengths of these variables ($\Delta s_{L,}$ ΔL_q and Δh, respectively) from horizontally averaged model data with the help of equations analogous to Equation 3.64.

The results are shown in Table 3.1 for two time periods (2.0 and 2.25 h of simulation). Calculated and observed fluxes are in excellent agreement. The cloud-base fluxes are seen to decrease somewhat during the 15 min period, and this must be linked to corresponding changes in surface s_V flux. It is once again stressed here that a constant value of k was used to compute the fluxes. Had the value of k been computed for each time step, it could have been argued that the good agreement between model-computed fluxes and k-closure parameterized fluxes was forced; the use of a constant k value suggests that this is not so.

The simple parametric scheme used in k-closure theory allows to compute cloud-base fluxes with a knowledge of surface fluxes only; this is especially useful in field experiments where surface fluxes are more readily obtainable than cloud base fluxes. k-closure theory can also be of use in large-scale models where the various layers within the ABL are poorly resolved compared to the present model; in large-scale models, knowledge of mixed-layer and cloud-layer processes is important in order to estimate surface and cloud effects on the large-scale atmosphere, and this type of parameterization can overcome the problems posed by a relatively coarse grid spacing. One problem, however, is the choice of a representative value of k; it was seen that the present value is higher than that used by other authors. It seems that k can vary for different classes of cloud population and type, intensity of mixed-layer turbulence, and the depth of the mixed layer. A single value of k which covers all cases of convection is unlikely to be a valid concept.

	t = 2.00 h	t = 2.25 h
$(F_s)_B$ Calc $(W\ m^{-2})$	– 10.203	-8.090
$(F_s)_B$ Obs $(W\ m^{-2})$	–10.060	–8.115
Difference (%)	**1.7**	**0.31**
$(FL_q)_B$ Calc $(W\ m^{-2})$	117.058	105.418
$(FL_q)_B$ Obs $(W\ m^{-2})$	115.106	105.740
Difference (%)	**1.7**	**0.31**
$(Fh)_s$ Calc $(W\ m^{-2})$	106.856	97.329
$(Fh)_s$ Obs $(W\ m^{-2})$	105.073	7.626
Difference (%)	**1.7**	**0.31**

Table 3.1: Comparison of observed cloud base fluxes of h, s_L, and L_q, and fluxes computed from k-closure parameterization, with k = 0.40

In this section, two important cumulus parameterization schemes have been chosen, which have been tested using data from the ABL model. It has been possible to verify certain aspects of Fraedrich's parameterization scheme using cloud population characteristics observed in the model, and also Betts' representation of a flux by a relation of the form given in Equation 3.60. The closure parameters k and α, which essentially link cloud-base and mixed-layer parameters, were adequately computed from the model data. Some results appear to be less satisfactory; this is in no way indicative of a lack of validity of a particular parameterization scheme, but implies using caution when applying model data to cases which are different from the case under study (such as Fraedrich's parameterization scheme, which he tested against large cumulus and cumulonimbus cloud fields). It appears from this study that tests of an entire range of cumulus parameterization schemes with a simulation of only one particular meteorological situation cannot be envisaged.

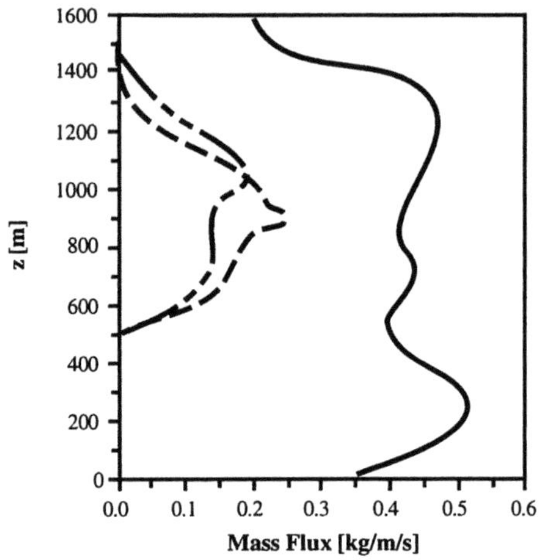

Figure 3.7: Mass fluxes in a case of strong cloud activity

3.2.5 Empirical results which may contribute to further development of parameterization methods

In addition to testing of already existing parameterization methods, a detailed model should be able to infer new methods. The approach consists in deriving empirical relations between various characteristics of the turbulent flow and mean properties of the meteorological fields, when statistical equilibrium is reached. The most important turbulent characteristics are the average vertical fluxes of quantities which are conserved in the condensation process, i.e., horizontal momentum, total specific water content (water vapor + cloud water), and a thermodynamic variable such as θ_L, the liquid water potential temperature defined by Betts (1973) as:

$$\theta_L = \theta - \left(L / c_p\right)(\theta / T) q_L \qquad 3.69$$

The vertical derivatives of these fluxes determine the temporal evolution of the corresponding quantities, which are needed for meteorological forecasting. In a forecasting model, they represent the unknown feedback of small scales into large scales, which has to be derived through parameterization methods from the grid-scale computed field of variables. A detailed microscale model such as this one should ideally provide the link between variables which are considered "grid-scale" in a larger scale model and the corresponding turbulent vertical fluxes.

This requires several numerical simulations under various mean conditions. Instead of using several statistically-steady states, it was necessary to limit the present analysis to natural variations around one statistically steady state. Therefore the results presented here can only be considered as tentative and still have to be generalized for a wider range of study cases. It is interesting, however, to compare those with results expected from a field experiment where the turbulent flow is rarely statistically steady and where horizontally averaged second-order moments are difficult to measure in clear air, and almost impossible to measure inside clouds. Results shown here concern the mass flux and the thermodynamic fluxes. The vertical mass flux averaged at each level for a 20-minute period of relatively strong cloud activity (2.09-2.36 h) is shown in Figure 3.7. It generally shows one relative maximum within the mixed layer, which corresponds to the maximum of vertical velocity variance, and at least a second relative maximum within the cloud layer. It is compared to $2\rho\sigma w_c$ (which corresponds to the case where clouds are approximated by ascending homogeneous towers covering the fraction of the domain. The difference between the two curves is a measure of the subcloud mass flux added to the clear-air contribution. This means that it includes contributions from descending air parcels within clouds and ascending air parcels in clear air. In the subcloud layer, the mass flux is due only to clear-air vertical movements; the contribution of the mean cloud circulation goes up to almost 60% near the middle of the cloud layer.

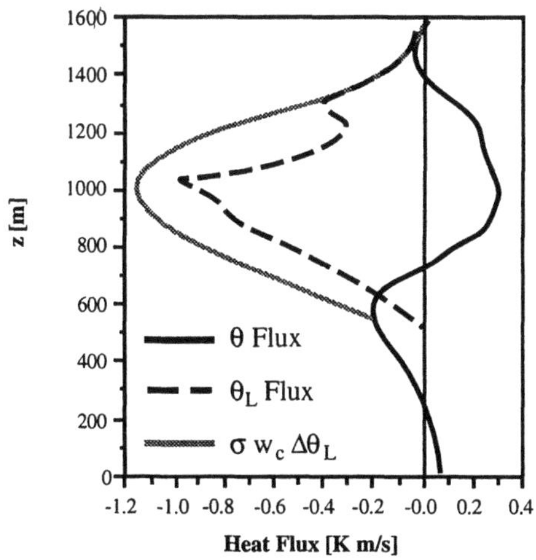

Figure 3.8: Potential temperature fluxes in a case of strong cloud activity

Figure 3.8 shows an example of turbulent fluxes of θ and θ_L averaged for the same 20-minute period. The θ and θ_L fluxes coincide within the mixed layer, the bottom value being determined by the surface conditions. The θ flux shows an absolute minimum at the bottom of the cloud layer where it starts increasing again due to the buoyancy effect of latent heat release; it reaches a maximum near the middle of the cloud layer and gradually approaches a negligible value at the base of the trade inversion. A small negative value can be observed here as the sign of a small overshoot-of one cloud. On the contrary, the θ_L flux shows a continuous decrease from its ground value up to an absolute minimum near the middle of the cloud layer, at the level where cloud droplet evaporation starts prevailing over latent heat release; it gradually reaches zero in the second half of the cloud layer.

As stated by Betts, the vertical derivative of the θ_L flux is equal, with a negative sign, to the tendency of the horizontally averaged temperature, if the cloud liquid water content can be considered as stationary. The shape of the θ_L flux is indeed a good representation of the heat cycle within the cloud layer and the intensity of its minimum is a measure of the upward heat transport by the cloud layer.

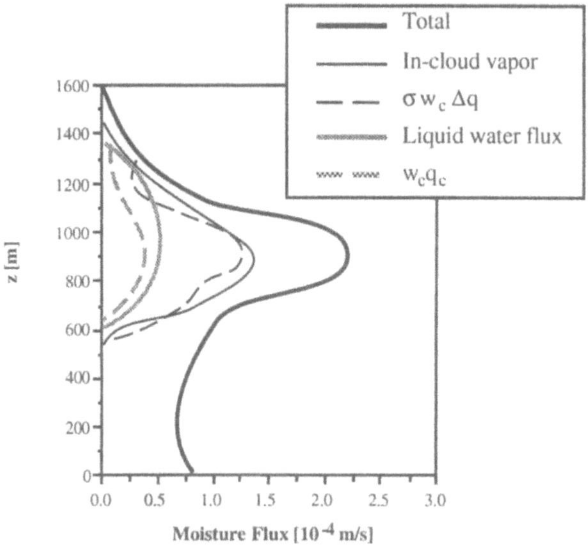

Figure 3.9: Water fluxes in a case of strong cloud activity

The θ_L flux is compared here to the product $\sigma w_c \Delta\theta_L$ which is a simple measure of cloud activity ($\Delta\theta_L$ is the difference between the values of θ_L inside and outside clouds). In this formula, clouds are again approximated by ascending homogeneous towers covering the fraction s of the domain.

The discrepancy between the actual θ_L flux and the approximate formula is a measure of the subcloud-scale transport within cloud added to the θ_L transport occurring in clear air. The schematic expression $\sigma w_c \Delta\theta_L$ however, represents up to 90 % of the actual θ_L flux near the region of their absolute minimum. The difference between this figure and the 60% observed for the mass flux is due to the imperfect correlation of the w and θ_L fields.

Figures 3.9 and 3.10 illustrate examples of the moisture fluxes averaged over 20 minutes. The first one corresponds to the period 2.09-2.36 h, and the second one to the period 1.53-1.81 h of weaker cloud activity.

The total water vertical turbulent flux, the vertical derivative of which would indicate the tendency of the water vapor profile in the case of a stationary cloud layer, shows large temporal variations precisely linked to the cloud activity. Since the surface moisture flux is constant with time, a strong cloud activity corresponds to a moisture pumping from the mixed layer.

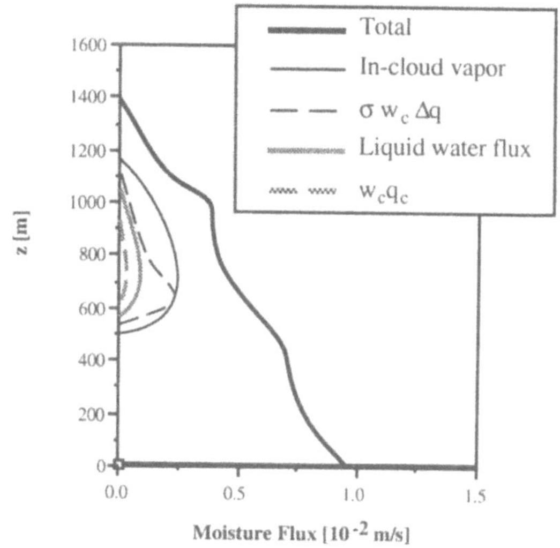

Figure 3.10: Water fluxes in a case of weak cloud activity

It is generally followed by a period of weaker activity allowing moisture to accumulate again in the mixed layer. On the two figures the partial contributions corresponding to liquid water flux and water vapor flux inside clouds are shown along with simple approximations for them. These approximations, $\sigma w_c \Delta q$ (Δq being the difference of q inside and outside clouds), and $w_c q_c$ (q_c being the mean cloud water amount) would correspond, as previously, to homogeneous cloud towers.

The quantities used in the previous approximations $(\sigma,\ w_c)$ and the differences between in-cloud and clear air properties are easier to measure, or to obtain in models through simple parameterization schemes, than the actual fluxes. It would be useful, however, to relate the main characteristics of the fluxes to only one parameter, representative of the cloud cover. In this connection, the correlation coefficient between the horizontally averaged fluxes of q, θ_v and mass flux M has been calculated at each level in the model, and three coefficients representative of the mean cloud cover, namely, maximum cloud cover, cloud-layer thickness and total cloud volume, over a period of 1 hour. The most significant correlations were found to be those which exist between cloud volume and the relative maxima and minima of the fluxes in the cloud layer. It was thus possible to establish direct empirical relationships between cloud volume and the relative maxima or minima of fluxes within the cloud layer. The cloud-base mass flux M_B was found to be the most significantly correlated turbulent transport with the cloud volume, the correlation coefficient being of the order of 0.8. The regression equation between M_B and cloud volume V_T for the particular field of trade-wind cumuli simulated by the model is:

$$M_B = 3.204 \times 10^{-10} V_T + 0.2177 \quad \text{kg m}^{-2}\,\text{s}^{-1} \qquad 3.70$$

The results illustrated in Figures 3.11 and 3.12 imply that the cloud field is an integral part of the turbulent field and not just an independently-acting phenomenon. With empirical relationships such as those described here, and using the k-closure theory of the preceding section, it is possible to construct the profile of a thermodynamic flux by using three types of observable data: temperature and humidity measurements at the surface, fractional cloud cover from satellite or aircraft photography and some estimate of cloud thickness in order to obtain cloud volume. The surface measurements enable to evaluate surface fluxes from bulk aerodynamic methods; k-closure theory then provides cloud-base values of the various fluxes.

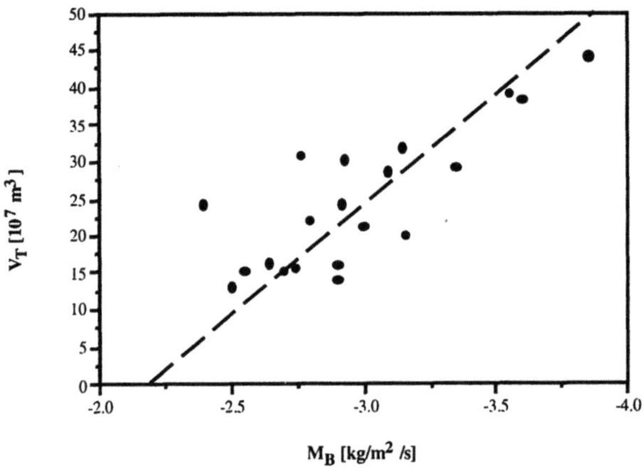

Figure 3.11: Relationship between cloud-base mass flux M_B and total cloud volume V_T

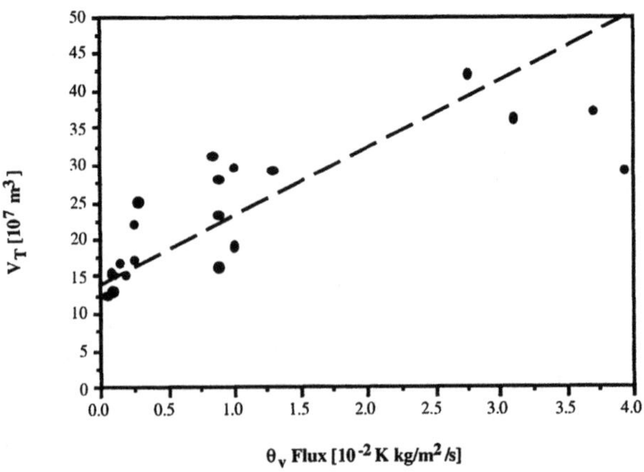

Figure 3.12: Relationship between the relative maxima of $\overline{w'\theta_v'}$ in the cloud layer and total cloud volume V_T

The nephanalysis data yield the values of the relative minimum of $\overline{w'\theta_L'}$ and the relative maxima of $\overline{w'q'}$, $\overline{w'\theta_v'}$ in the cloud layer, as well as the cloud-base mass flux M_B.

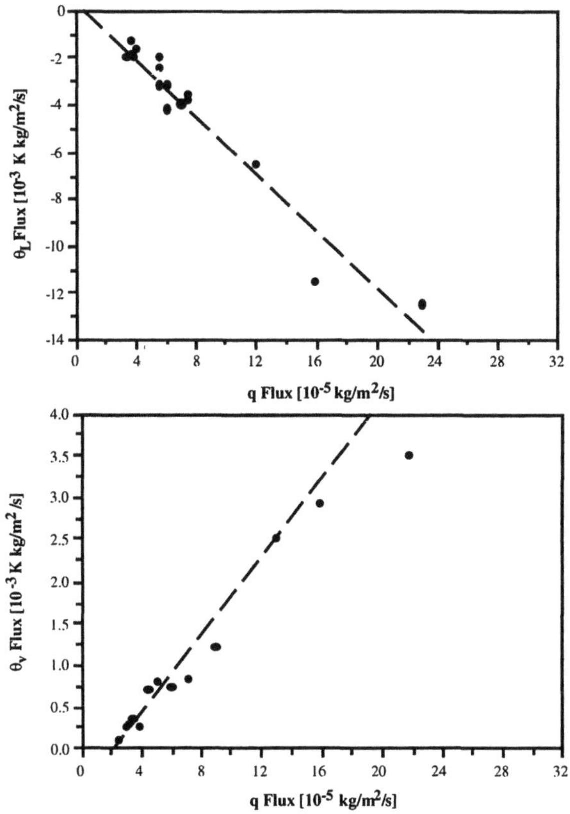

Figure 3.13: Examples of the close coupling of various thermodynamic fluxes

As all fluxes reach a value close to zero at the inversion, the key points in the vertical flux profiles are defined, and a linear profile can be assumed between each set of points. This type of empirical calculation avoids the need of in-cloud and clear-air aircraft or balloon measurements to obtain data to test the parametric equation used by Betts (Equation 3.60).

Figure 3.13 illustrates two examples of the close coupling of the thermodynamic fluxes, in particular, the relations between $\overline{w'\theta_L'}$ and $\overline{w'q'}$. Close coupling of heat and moisture fluxes seems to be a prerequisite to cloud development and evolution, and explains the generally good agreement between cloud volume and turbulent fluxes, represented by the correlation coefficients. The correlation between $\overline{w'\theta_L'}$ and $\overline{w'q'}$ is high, exceeding 0.95. The regression equation has the form:

$$\overline{w'\theta_L'} = -656.42\ \overline{w'q'} + 1.074 \times 10^{-2}\ \text{kg m}^{-2}\ \text{s}^{-1}\ \text{K} \qquad 3.71$$

Other relationships are given in Tables 3.2 and 3.3. This excellent relation between the downward transport of θ_L and the upward q transport helps confirm the concept of a slightly unstable cloud layer, which was briefly summarized earlier in this section. Since the heating and cooling of the layer are linked to the condensation and evaporation of cloud droplets, it seems reasonable that the θ_L and q fluxes also be closely linked.

This section attempted to show the connection which exists between a population of cumuli and the turbulent field. The analysis has shown that the cloud cannot be considered as an individually acting phenomenon; it is as much part of the turbulent field as are the turbulent fluxes. It has been shown that one may easily link observable cloud parameters and turbulent transports of heat and moisture. These transports, while not being readily measurable, are nevertheless governing factors in the development and evolution of a cumulus cloud ensemble.

Flux	Empirical relationship	r
q	$\overline{\rho' w' q'} = 6.53 \times 10^{-13} V_T - 9.796 \times 10^{-5}$ (kg m^{-2}s^{-1})	0.788
q_v	$\overline{\rho' w' \theta_v'} = 9.639 \times 10^{-11} V_T - 1.398 \times 10^{-2}$ (kg m^{-2}s^{-1})	0.759
q_L	$\overline{\rho w' \theta_v'} = 2.512 \times 10^{-11} V_T + 3.265 \times 10^{-3}$ (kg m^{-2}s^{-1})	0.721
M_B	M_B = 3.204 x 10-10 VT + 0.219 (kg m^{-2} s^{-1})	0.804

Table 3.2: Empirical relationships between cloud volume V_T and various fluxes. (Note: V_T has units of 10^7 m^3 in these equations)

Flux relation-ships between	Empirical relationship	r
$q - M_B$	$\overline{\rho w' q'} = 1.29 \times 10^{-3} M_B - 3.161 \times 10^{-4}$ (kg m^{-2}s^{-1})	0.741
$\theta_v - M_B$	$\overline{\rho w' q_v'}$ = 0.32 M_B - 0.084 (kg K m^{-2} s^{-1})	0.790
$\theta_L - q$	$\overline{\rho w' \theta_L'} = -656.42\ \overline{\rho w' q'} + 1.074 \times 10^{-2}$ (kg K m^{-2} s^{-1})	-0.958
$\theta_v - q$	$\overline{\rho w' \theta_v'} = 270.27\ \overline{\rho w' q'} - 5.546$ (kg K m^{-2}s^{-1})	0.834

Table 3.3: Empirical relationships between various fluxes. (Note : M_B has units of kg m^{-2} s^{-1} and $\overline{\rho' w' q'}$ has units of 10^{-5} kg m^{-2} s^{-1})

The empirical results given or implied in the text can represent a useful link between numerical models and atmospheric experiments. Their computation from data of cloud area or volume yields reasonable estimates of the various fundamental fluxes, and provides an alternative to the data obtained from present-day measurement techniques. This reasonable approximation to the true values of fluxes is also a further factor which may eventually lead to simplifications in parameterizations, particularly through the elimination of entrainment factors.

These are at present essential to tradewind cumulus model theories (Arakawa and Schubert, 1974; Betts, 1975; Fraedrich, 1976), but are extremely difficult to quantify.

3.3 Concluding remarks

This study illustrates one type of application of a detailed microscale numerical model: it produces a data set used as a support for hypotheses and schematic relationships related to the problem of parameterizing small-scale convection in large-scale models.

Despite its many limitations, in particular the cyclic boundary conditions which do not allow for low-level convergence, a detailed three-dimensional model appears to be a useful tool to verify relations concerning Trade-wind cumulus characteristics for individual clouds, and also for entire cloud populations.

Most of these rclationships are summalized in the work of Fraedrich (1976) which has been presented in Section 3.2.3. For example, it has been shown that for an individual cloud, there exists a cloud thickness to cloud radius ratio; that there exists a relation between maximum cloud area and cloud lifetime; and that a statistical distribution of cloud radii seems reasonable. The types of hypotheses which have been tested in Section 3.2.3 are important to various cumulus parameterization schemes. Such tests are essential if one wishes to represent the effects of a cloud population on the large-scale environment in a large-scale numerical model.

The substitution of model data into the cumulus parameterization schemes is also useful if one wishes to verify the extent to which these schemes are realistic. Betts' representation of a thermodynamic flux by the product of a convective mass flux and the cloud-environment difference of a particular thermodynamic variable (Equation 3.60) seems to be confirmed by the results obtained in this study; the various parameters used in Betts' (1976) coupled mixed-layer/cumulus layer model which were calculated with the model data also seem to confirm the validity of Betts' scheme in the model. Poorer agreement was found with Fraedrich's (1976) parameterization scheme; this is perhaps because this scheme was formulated for tropical cumulonimbus clouds rather than for Trade-wind cumulus clouds. It is also possible that model inaccuracies are responsible for the discrepancies found between model results of vertical mass flux distribution and that computed from Fraedrich's parameterization scheme. The analysis of the turbulent fluxes and their relationship with cloud parameters is important not only because it underlines the interaction between clouds and turbulence, but also because such a study could be of use in the real atmosphere, where reasonable estimates of fluxes could be obtained from direct cloud observations. Indeed, the cloud population parameters were chosen for the relative ease with which they can be deduced from observations. The estimates of various fluxes from empirical relations such as those presented in Section 3.2 can serve as a useful example of concepts formulated in the context of a numerical model, and the application of these concepts to a situation in the real atmosphere. The possibility of obtaining accurate estimates of turbulent fluxes from direct cloud observations could eventually lead to the simplification of certain cumulus parameterization schemes, in particular through the elimination of entrainment factors which are difficult to quantify.

It remains to be seen, however, whether the empirical relationships are valid for an entire range of cumulus clouds, or whether they are applicable only to the small Trade-wind cumuli simulated by the model. The answer to this question must come either from a set of reliable data obtained in the atmosphere or from results of a large or mesoscale numerical model. It should be stressed that the results of this study are based upon data simulated by one specific three-dimensional model and that all conclusions are valid for this model only, with its particular limitations and arbitrary parameters. In addition, for practical reasons, data have been taken from various states around only one statistically steady state. More general conclusions should come from a wider range of statistically steady states, allowing sufficient variations of the key parameters, such as the thickness and stability of the cloud layer or the depth of the mixed layer. This study, however, has attempted to show the usefulness of a hierarchy of microscale numerical models, from the most schematic to the most detailed ones, as a complement to observations for the improvement of parameterization methods in larger scale models.

4. Investigations with the DREAMS Mesoscale Meteorological Model

4.1 Introduction

DREAMS-1 (Differential-equation REgional Atmospheric Modeling System) is a three-dimensional atmospheric model which makes use of the finite difference technique to solve a series of time-dependent equations for processes acting typically on a regional scale; in literature appearing in the 1980s, this model was referred to as MESOCONV (MESOscale CONVection model). The model was originally developed in the early 1980s to investigate organized maritime cloud convection, such as the Rayleigh-Bénard type convection frequently observed as cold air moves over a warm ocean surface (Beniston, 1983; 1984; 1985). Other applications include interactions between clouds and radiation (Beniston and Schmetz, 1985; Schmetz and Beniston, 1986), as well as the dynamic and thermodynamic influence of a water surface on the regional atmosphere (Beniston, 1986).

The size of the model domain is essentially limited by the computer resources on which the model code is implemented. The characteristic spatial scales of the phenomena to be investigated should also be borne in mind, so that an optimal model domain size can be determined prior to a numerical investigation. For a supercomputer such as the CRAY-2 or the NEC - SX/4, the enormous memory capacity allows a generous domain size; DREAMS-1 has been run with up to 2 million grid-points on such machines. An optimal domain size should generally represent a balance between domain size and computing time. Typical grid interval is 1,000 m and up in the horizontal, and vertical levels are usually unevenly distributed, allowing higher resolution of model levels close to the ground and in the ABL. In general, the first model level is located at 50 m above the reference level, with subsequent even spacing of between 100 and 250 m. The governing equation set is that described in Section 2.2 for a hydrostatic model, namely Equations 2.1, 2.2, 2.4 - 2.9, and 2.11. The text in this chapter draws on material from the following publicatons: Beniston (1982, 1983, 1984, 1985, 1987), Beniston and Schmetz (1985), Schmetz and Beniston (1986), and Beniston et al. (1989, 1990).

4.2 Parameterized physics

The parameterized terms of the DREAMS-1 model are the following:

- *F terms*: Turbulent fluxes
- *C terms*: Cloud microphysics and precipitation processes
- *R terms*: Radiative flux exchanges

4.2.1 Atmospheric boundary layer (ABL) and surface-layer turbulence

Turbulent fluxes, typically a subgrid-scale phenomenon in a mesoscale model, are treated in two manners, according to whether they are acting close to the surface (i.e., surface-layer fluxes) or within the ABL. In the latter case, the subgrid-scale turbulence terms are parameterized through first-order closure, which relates the vertical flux of, for example, u, to its local gradient through a diffusion coefficient. This reads:

$$-\overline{u'w'} = K_M \frac{\partial \overline{u}}{\partial z} \tag{4.1}$$

where the vertical distribution of K is computed through use of the O'Brien (1970) formulation. This method yields maximum values of K at about one-third of the height of the planetary boundary layer, where maximum energy-containing eddies are generally found; a characteristic profile for K is given in Figure 4.1.

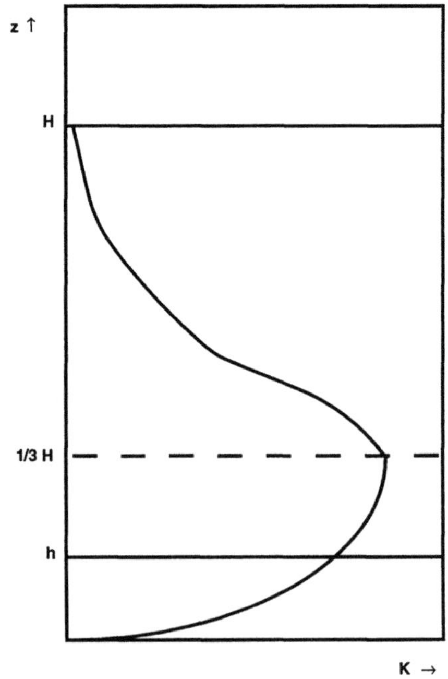

Figure 4.1: Typical K diffusion coefficient profile using the O'Brien formulation. H = Top of boundary layer; h = top of surface layer

Turbulent transfer of heat, moisture, and momentum within the atmospheric layers close to the ground (approximately the first 50 m above the surface) make use of the Businger-Dyer relationships (Businger et al., 1971), where the momentum, heat, and moisture fluxes are given by:

$$-\overline{u'w'} = u_*^2 \qquad 4.2$$

$$-\overline{q'w'} = u_* q_* \qquad 4.3$$

$$-\overline{q'w'} = u_* q_* \qquad 4.4$$

The so-called "friction" terms u∗, q∗ and q∗ are derived from:

$$\frac{kz}{u_*}\frac{\partial u}{\partial z} = \text{Ø}_M\,(z/L) \qquad 4.5$$

$$\frac{kz}{q_*}\frac{\partial q}{\partial z} = \text{Ø}_H\,(z/L) \qquad 4.6$$

$$\frac{kz}{q_*}\frac{\partial q}{\partial z} = \text{Ø}_q\,(z/L) \qquad 4.7$$

Here Ø_M and Ø_H are universal stability functions which depend on the local Monin-Obukhov length L:

$$\text{Ø}_M(z/L) = 1 + 4.7\,z/L \qquad z/L \geq 0 \text{ (stable)} \qquad 4.8$$
$$\text{Ø}_M(z/L) = (1 - 15\,z/L)^{-1/4} \qquad z/L < 0 \text{ (unstable)}$$

$$\text{Ø}_H(z/L) = \text{Ø}_q(z/L) = 1 + 6.4\,z/L \qquad z/L \geq 0 \text{ (stable)} \qquad 4.9$$
$$\text{Ø}_H(z/L) = \text{Ø}_q(z/L) = .74\,(1 - 9\,z/L))^{-1/2} \qquad z/L < 0 \text{ (unstable)}$$

The Businger relationships allow not only the computation of surface-layer fluxes but also the value of the K diffusion coefficient at the top of the surface layer, which is required to compute the distribution for K according to the O'Brien (1970) formulation in the ABL. This profile is a third-order polynomial interpolation for K between values at the top of the surface layer and at the summit of the ABL. In order to take into account the effects of enhanced buoyancy-driven turbulence in unstable atmospheric conditions, and to reflect the fact that thermal turbulent transports are more efficient than dynamic transports in unstable situations, the diffusion coefficient for heat and moisture, K_H, takes the following form:

$$K_H = K_M \qquad z/L \geq 0 \text{ (stable)} \qquad 4.10$$
$$K_H = 1.35\,K_M \qquad z/L < 0 \text{ (unstable)}$$

It should be noted here that the contribution of turbulence to the time evolution of the meteorological variables is not due to the fluxes themselves but to their divergence. Thus, for the term F_u in Equation 2.1, for example, the vertical flux divergence reads:

$$F_u = \frac{\partial}{\partial z} K_M \frac{\partial \overline{u}}{\partial z} \qquad 4.11$$

Turbulence also acts in the horizontal plane, but it is extremely difficult on the regional scale to evaluate what could be the nature of a horizontal diffusion coefficient. In the present model code, horizontal turbulence is dependent on local divergence of air and acts as a numerical control on the intensity of convergence or divergence. Taking example on Pielke (1984), the formulation for the horizontal flux of u has the following form:

$$-\frac{\partial \overline{u'u'}}{\partial x} = K_{horiz}\left[\frac{\partial \overline{u}}{\partial x} + \frac{\partial \overline{v}}{\partial y}\right] / \partial x \qquad 4.12$$

4.2.2 Cloud microphysics and precipitation

Cloud activity for cumulus-type formations is accompanied in nature by profound modifications in the thermal, moisture, and dynamic budgets of the atmosphere in the environment of these clouds, mainly as a result of the release and absorption of latent heat due to phase changes of water. The distribution of heat and moisture is rendered even more complex by the secondary circulations induced by the cumulus cloud activity, which recycle temperature and specific humidity in the vertical. Even if each individual cloud is not explicitely taken into account by the model grid, the global effects of a cloud field on the mesoscale atmospheric environment can have important consequences for regional-scale atmospheric processes.

The convective scheme used in DREAMS-1 is based on work by Augstein and Wendel (1981). The parametric scheme makes use of a one-dimensional cloud model to estimate the thermodynamic budgets related to convective activity. Figure 4.2 illustrates the mechanisms which are taken into account by the scheme. Low-level convergence of air and buoyancy fluxes at the surface raise moist air through to the lifting condensation level (LCL). A convective vertical velocity characterizes the strong ascent inside the cumulus cloud, and turbulent exchange of air, also known as entrainment and detrainment, at the cloud lateral boundaries determine the rate of heat and moisture exchange between the cloud and its unsaturated environment. At the cloud top, detrainment of cloud material leads to the creation of colder, denser air through evaporational cooling, and a subsequent sinking of air.

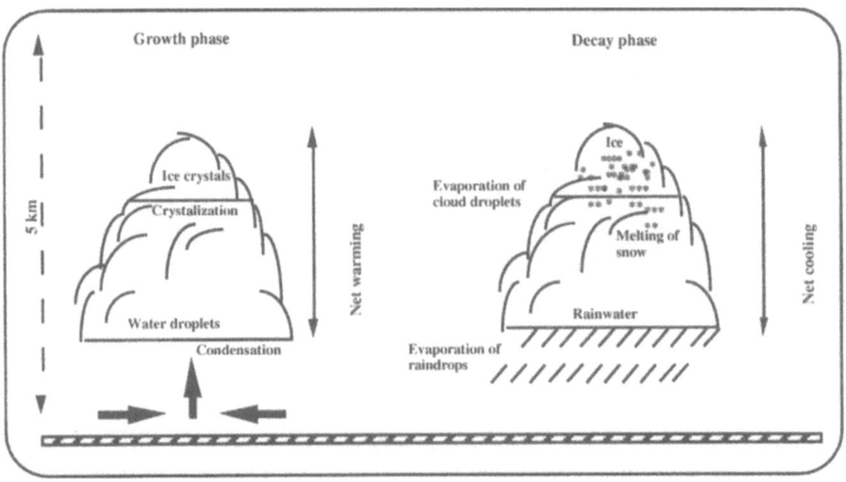

Figure 4.2: Main processes taken into account in the cloud convection scheme

In the model, clouds will form whenever a grid-point threshold value of vertical velocity (around 1 cm/s) is exceeded. The vertical convective velocity is a function of the low-level convergence of air and the buoyancy flux, and can be written as:

$$w_* = -\int_0^{CB} (\nabla . \underline{v}) \, dz + \overline{w'q'_{CB}} \qquad 4.13$$

where CB is the cloud-base level (essentially the LCL). The first term on the right-hand-side of Equation 4.13 represents mass convergence in the sub-cloud layer, and the second term is the cloud-base buoyancy flux contribution to w_*. It is supposed that this convective vertical velocity represents the average uplift of air for all clouds formed within an elementary grid volume of the mesoscale model.

Following the hypotheses of Arakawa and Schubert (1974), Betts (1975) and others, it is supposed that the convective flux of any thermodynamic variable A is the product of w_* and the in-cloud/environment difference of quantity A. This reads:

$$-\overline{w'A'} = w_* (A_c - A_e) \qquad 4.14$$

where c is the in-cloud value of A and e its environmental value. The vertical derivative of Equation 4.14 yields the term which directly contributes to the time rate of change of variable A (i.e., θ, q, q_L, or q_R in Equations 2.4 - 2.6). This is analogous to the turbulent flux divergence terms, and takes the form:

$$-\frac{\partial}{\partial z}(\overline{w'A'}) = w_* \frac{\partial A_c}{\partial z} - w_* \frac{\partial A_e}{\partial z} \qquad 4.15$$

where it is assumed that w_* remains constant within the clouds. The closure problem here is to find the in-cloud vertical profile of A, the environmental profile being known at each grid-point. Augstein and Wendel (1981) have suggested that this profile is proportional to (A_c - A_e). They define a detrainment parameter δ_* which is a function of height above the cloud base, and is a measure of the intensity of turbulent thermodynamic exchange at the cloud boundaries. Consequently, the in-cloud profile of A may be written as:

$$w_* \frac{\partial A_c}{\partial z} = \delta_* (A_c - A_e) \qquad 4.16$$

Substituting Equation 4.16 into 4.15 yields the complete form of the convective flux divergence term:

$$C_A = -\frac{\partial}{\partial z}(\overline{w'A'}) = w_* \delta_* (A_c - A_e) - w_* \frac{\partial A_e}{\partial z} \qquad 4.17$$

The influence of a field of cumulus clouds is therefore to modify thermodynamic budgets through turbulent entrainment and detrainment at cloud boundaries, represented by the first term on the right-hand-side of Equation 4.17, and the downward transport of these quantities by subsiding air in the cloud-free environment (second term on the right-hand-side).

DREAMS-1 also contains a parametric scheme for precipitation, based on the well-known semi-empirical formulation of Kessler (1969). Four precipitation mechanisms are considered in the model:

- *Autoconversion*: the conversion of fine cloud droplets to the much larger raindrops
- *Coallescence*: the growth of raindrops by collision
- *Evaporation*: cooling of the sub-scloud layer as raindrops fall out of their saturated environment
- *Fallout velocity*: this takes into account the fact that, from aerodynamic principles, larger raindrops will fall out more rapidly than smaller raindrops.

The equations which describe the precipitation processes are listed accordingly:

Autoconversion: $A = c_1\,(q_L - c_2)$ 4.18

Coallescence: $C = c_3\, q_L\, q_R^{.875}$ 4.19

Evaporational Cooling: $E = [\ VF\ (1 - q/q_S)\ q_R^{.525})]\ /\ [c_4\ /\ e_S + c_5]$ 4.20

where VF is a ventilation factor.

Fallout Velocity: $F = -w_R\ \partial q_R\ /\ \partial z$ 4.21

where: $w_R = 5.1 q_R^{.125}$ 4.22

The terms c_1 - c_5 are empirical constants. It can be seen in the above equation set that autoconversion and coallescence are a source of raindrops, while evaporation and fallout are sinks for precipitation. All these processes are included as part of the C parameterizations of Equations 2.5 - 2.7.

4.2.3 Infrared radiative fluxes

Solar and infrared radiation has been implemented in early versions of DREAMS in order to investigate interactive properties of clouds and radiation; two principal papers describing these schemes and examples of case studies have been published by Beniston and Schmetz (1985) and Schmetz and Beniston (1986). The radiation fluxes are implicitely included in Equation 2.4 under the R term. Omitting scattering processes, the radiative transfer equation for the longwave flux density M_2 at level z_2 can be written as:

$$M_2 = M_1\{1 - \varepsilon(z_1, z_2)\} + \int_{z_2}^{z_1} B(\frac{\partial \varepsilon}{\partial z})dz \qquad 4.23$$

where z is the height; M_1 is the longwave flux at the boundary z_1; $\varepsilon(z_1, z_2)$ is the flux emissivity of the layer between z_1 and z_2; and B is the Planck-function.

The emissivity depends on the spectral absorption characteristics of the atmospheric absorbers. In the troposphere, the main absorbers for longwave radiation are water vapor and its continuum absorption, carbon dioxide and liquid water. The water vapor continuum absorption within the atmospheric window depends on water vapor pressure and on temperature, and is important in moist atmospheres. Aerosols influence the longwave radiation field only in polluted areas. The emissivity parameterizations by Rodgers (1967) for water vapor and CO_2 are used. The water vapor continuum is accounted for using a parameterization given by Stephens and Webster (1979). These parameterizations relate the broad-band emissivities $\varepsilon_{H_2O}, \varepsilon_{CO_2}$ and ε_{cont} to the equivalent absorber amount. That is to say, no time-consuming integration over spectral intervals needs to be done, which makes the scheme computationally efficient. Allowance is made for the H_2O and CO_2 overlaps by subtracting $\Delta\varepsilon = \varepsilon_{H_2O} \cdot \varepsilon_{CO_2}$ from the total emissivity e which is the sum of the emissivities of the different absorbers:

$$\varepsilon = \varepsilon_{H_2O} + \varepsilon_{CO_2} + \varepsilon_{cont} - \Delta\varepsilon + \varepsilon_{cloud} \qquad 4.24$$

The individual emissivities e_A (A: H_2O, CO_2 or continuum) are computed from a polynomial regression:

$$\varepsilon_A = \sum_{n=0}^{4} a_n \left(\ln u_A\right)^n \qquad 4.25$$

where u_A is the absorber amount. The regression coefficients a_n are different for the upward and downward fluxes due to the different spectral characteristic of these fluxes. The regression coefficients are taken from Rodgers (1967) and Stephens and Webster (1979). For H_2O and CO_2 absorber amounts smaller than a certain threshold, Rodgers provides a different parameterization.

The longwave flux is then calculated from:

$$M_{k+1} = \varepsilon_0 \sigma T_0^4 \left\{1 - \varepsilon(U_k)\right\} + \sum_{j=1}^{k} \sigma T_j^4 \left\{\varepsilon\left(U_k + u_j - U_j\right) - \varepsilon\left(U_k - U_j\right)\right\} \qquad 4.26$$

T_i (i = 1, 2, 3,...) is the mean temperature of a model layer and u_i the absorber amount. The vertically integrated absorber amounts U_k and U_j are calculated from:

$$U_k = \sum_{i=1}^{k} u_i \;;\; U_j = \sum_{i=1}^{j} u_i \quad \text{with } j \le k \qquad 4.27$$

The index k is counted upwards for the upward flux and downward for the downward flux. Thus, the first term on the right side of Equation 4.26 represents either the contributions of the surface (then T_0 is the surface temperature and e_0 the surface emissivity) to the upward flux M^+ at a level k or the contribution of the downward flux entering the top layer of the model (here 10.5 km) to the downward flux at a level k. The influx at the top of the model is derived from a multi-spectral scheme reaching up to 50 km.

The effective emissivity ε_{cloud} of a cloud layer is computed from:

$$\varepsilon_{cloud} = 1 - \exp\{-q.\sigma_{abs}.LWP\} \tag{4.28}$$

where the liquid water path (LWP) is the vertically integrated liquid water content, σ_{abs} is the mass absorption coefficient and the diffusivity factor q converts the optical depth into an equivalent optical depth for the flux density. The value of q is taken as 1.66 (e.g., Kondratyev, 1969). σ_{abs} depends on the actual droplet size distribution and is chosen as 0.075 m^2 g^{-1} which is adequate for the infrared window region (e.g., Platt, 1976; Schmetz et al., 1981). Since the atmospheric window region is the most active in the radiative atmosphere-cloud exchange, it is appropriate to use this value for the broad infrared spectrum.

Direct application of this scheme to the DREAMS mesoscale model requires knowledge of grid-scale specific humidity temperature and cloud liquid water content. These quantities are computed for each time step using the governing equation set and convective parameterization as described in the previous subsection. Because variations of infrared fluxes are small for small changes in q and q_L, it was found reasonable to compute the various terms of the radiation scheme each 30 min. In general, cloud structure does not change so fast as to make computations of infrared cooling at more frequent intervals necessary. The vertical profiles of the radiative cooling rate computed at one 30-min interval are applied at each time step of the half-hour time span, before the infrared schemes are called once again to obtain new estimates of cooling rates. This results in a considerable saving of computer time while still retaining reasonable cooling rates, representative of mature cumulus clouds (Beniston and Schmetz, 1985).

4.2.4 Shortwave radiative fluxes

Since the radiation scheme is applied in the troposphere up to 10 km altitude, it is justified to consider water vapour as the only gaseous absorber. Aerosols are also neglected since the solar absorption by aerosols contributes significantly to the total heating in polluted areas only, on the time and space scales for which DREAMS has been designed. The interaction of cloud droplets with the radiation field is treated in more detail using a scheme which includes explicitly multiple scattering within the cloud.

4.2.4.1 Clear sky conditions

The downward solar flux density F^- is calculated from an empirical expression relating F^- to the altitude z and the solar zenith angle Q_0:

$$F^-(Q_0, z) = S_0\ 0.01 \cos Q_0 \sum_{i=0}^{2} a_i x^i \tag{4.29}$$

where S_0 is the solar constant in W m^{-2} and:

$$x = \exp\left(\sum_{n=0}^{3} b_n z^n\right) / \cos\Theta_0 \qquad 4.30$$

with z being the altitude in km. The empirical coefficients are:

$a_0 = 97.2$; $a_1 = -53.67$; $a_2 = 17.12?$

$b_0 = -0.729$; $b_1 = -0.435$; $b_2 = 0.0356$ and $b_3 = -0.0013$ 4.31

Equation 4.29 has been derived from a least-squares fit to results of a solar radiation scheme with high spectral resolution, assuming a standard midlatitude summer atmosphere (Schmetz, 1981). Equation 4.29 is valid in the ranges $0 \leq q_0 \leq 75°$ and for $0 \leq z \leq 10$ km and it reproduces results from the high resolution scheme with an accuracy better than 5 %.

The absorptance A of solar radiation by water vapour is calculated after Paltridge (1972):

$$A = 0.099\, u^{0.34} \qquad 4.32$$

where u is the equivalent water vapour path length (in g cm^2) accounting for the slant path of the sun rays. The total absorption in an atmospheric column is computed from Equation 4.32 using Equation 4.29 with z = 10 km as input at the top of the model domain. The solar flux incident at cloud top is computed from Equation 4.29 at the level indicated by the convective scheme.

4.2.4.2 Cloudy conditions

While the effects of multiple scattering are neglected for the calculation of the clear-sky solar absorption, this process has to be considered for optically thick media with single-scattering albedos close to 1, because the relative contribution of multiple scattered photons to the total absorption becomes large. In the present solar scheme, the cloud is considered as consisting of several homogeneous cloud layers each having a reflectance ρ_i and transmittance τ_i, determined by the cloud microphysics (liquid water content and droplet size). Denoting the incident radiation at cloud top by F_1^-, the downward flux density at the boundary between layer 1 and 2 is:

$$F_2^- = F_1^- \tau_1 + \rho_1 \qquad 4.33$$

where F_2^- is the upward solar flux at the boundary between layer 1 and 2, t_1 and r_1 are the transmittance and reflectance of the uppermost cloud layer, respectively. Relations similar to Equation 4.33 are valid for the upward and downward fluxes at the boundaries between all adjacent layers. These equations yield a linear equation system:

$$\overline{\overline{M}}\,\overline{F} = \overline{S} \qquad 4.34$$

where $\bar{F} = F_1^-, F_1^+, F_2^-, F_2^+ \ldots, \; F_n^-, F_n^+$ is the vector containing the solar fluxes which are to be determined, $\bar{S} = (F_1^-, 0, 0, \ldots, 0)$ is the 'source vector' here representing the incoming solar flux at cloud top. $\overline{M}$ is the matrix consisting of the radiative properties such as reflectance ρ_i and transmittance τ_i of the individual cloud layers, ρ'_1 and τ'_1 are the characteristics for direct beam radiation. r_s is the system reflectivity of the atmosphere-surface system below the cloud, which is dominated by the surface albedo.

$$\overline{M} = \begin{bmatrix} 1 & 0 & 0 & 0 & \cdot & \cdot & \cdot & \cdot & \cdot & \cdot & 0 \\ -\rho 1 & 1 & 0 & -\tau 1 & \cdot & \cdot & \cdot & \cdot & \cdot & \cdot & \cdot \\ -1 & 0 & 1 & -\rho 1 & \cdot & \cdot & \cdot & \cdot & \cdot & \cdot & \cdot \\ 0 & 0 & -\rho 2 & 1 & 0 & -\tau 2 & \cdot & \cdot & \cdot & \cdot & \cdot \\ 0 & 0 & -\tau 2 & 0 & 1 & -\rho 2 & \cdot & \cdot & \cdot & \cdot & \cdot \\ 0 & \cdot & \cdot & \cdot & \cdot & \cdot & \cdot & \cdot & \cdot & \cdot & \cdot \\ \cdot & \cdot & \cdot & \cdot & \cdot & \cdot & \cdot & \cdot & \cdot & \cdot & \cdot \\ \cdot & \cdot & \cdot & \cdot & \cdot & \cdot & \cdot & -\rho_s & 1 & 0 & -\tau_s \\ \cdot & \cdot & \cdot & \cdot & \cdot & \cdot & \cdot & -\tau_s & 0 & 1 & -\rho_s \\ 0 & \cdot & \cdot & \cdot & \cdot & \cdot & \cdot & 0 & 0 & -\rho_s & 1 \end{bmatrix} \qquad 4.35$$

In the DREAMS mesoscale model, $\rho_s = 0.05$ is assumed, which is a value characteristic of the reflectance characteristic of a sea surface. For n cloud layers, Equation 4.34 implies (2n + 2) linear equations where the first equation is trivial. The remaining (2n + 1) linear equations are readily solved for the $F_i^{+,-}$ using the Gaussian algorithm. It should be mentioned that this method is an application of the more general adding algorithm used for radiative transfer calculations (e.g., Hansen, 1971).

Equation 4.35 does not explicitly contain the direct beam radiation but the algorithm can readily be extended to include the direct solar beam. This implies that the linear equation system becomes inhomogeneous. However, the zenith-angle dependence of ρ and τ for the direct-beam radiation is considered, as will be described below.

The transmittance and reflectance of the individual cloud layers are computed from a two-stream approximation (Schmetz, 1984), using the liquid water path provided by the meso-scale model (op. cit., Equations 28, 29, 32, and 33). In order to account for the solar zenith angle dependence of the cloud radiative properties, the incoming solar flux at cloud top is assumed to be direct-beam radiation and the reflectance ρ_1' and transmittance τ_1' of the upper layer for the downward flux is computed from the two-stream approximation with a direct incoming beam as a boundary condition. All other ρ_i and τ_i are computed using a diffuse flux as a boundary condition. This is a reasonable assumption since the meso-scale model has a maximum vertical resolution of 500 m and cloud liquid water is usually larger than 0.1 g m^{-3}. A direct beam having passed the associated optical depth within the cloud is completely diffused.

Besides the liquid water path, the two-stream method requires knowledge of the mass extinction coefficient σ_{ex}, the single-scattering albedo $\tilde{\omega}$ and the asymmetry parameter g of the phase function of the cloud droplet size distribution. Since these quantities are not given by the dynamic model, they are simply prescribed. In that context, it is to be noted that the prescription of these solar parameters should be consistent with the infrared mass absorption coefficient used in the long wave scheme.

Band	σ_{ex} (m^2g)	$\tilde{\omega}$	g	W
1	0.141	0.9999	0.86	0.880
2	0.141	0.985	0.82	0.064
3	0.154	0.935	0.87	0.035
4	0.156	0.49	0.97	0.021

Table 4.1: Mass extinction coefficient σ_{ex}, single-scattering albedo $\tilde{\omega}$ and asymmetry parameter g for the four spectral bands used for the solar radiation calculations in clouds and the corresponding weighting coefficients W. Parameters pertain to a droplet size distribution with an effective radius of about 10 mm. The choice of a constant droplet size distribution is somewhat arbitrary considering the observed variability. The cloud optical properties do depend on the droplet size spectra; however, the absorptance is mainly determined by the liquid water path (e.g., Twomey, 1976; Slingo and Schrecker, 1982)

The solar transmittance and reflectance for the cloud layers are calculated for 4 bands with different weightings in order to get agreement with high resolution calculations for a range of liquid water paths between about 20 and 800 g m^{-2}, The corresponding values of $\tilde{\omega}$, σ_{ex}, and g and the weighting coefficients are given in Table 4.1.

Two simplifying assumptions of the solar scheme should be noted. Within clouds, only the liquid water absorption is considered. The neglect of absorption by the in-cloud water vapor does not introduce serious errors in the heating rates of low- and mid-level tropospheric clouds since the radiation in the water vapor bands is already diminished by the water vapor column above the cloud. Cloud droplets also absorb in the water vapor windows and therefore they dominate the cloud absorption. This is confirmed in a recent work by Davies et al. (1984), who studied the solar absorption in clouds with a radiation model at a 20 cm^{-1} spectral resolution. Their conclusion is that the in-cloud water vapor gives large contributions to the total absorption only for tenuous clouds or if the vapor column above the cloud top is small.

The agreement between the high spectral resolution scheme and the new parameterization has been further improved by a slight tuning of the single-scattering albedo in Band 4 (see Table 4.1). Originally the cloud parameters were simply derived from the detailed model by averaging, including a proper weighting with insolation. A second simplification is the neglect of below-cloud solar absorption, which is very small anyway. For instance, the detailed scheme gives a heating rate of 6×10^{-3} K/h for the layer underneath the cloud, which is extremely small compared with heating of 0.4 K/h in the upper 200 m of the cloud (Schmetz and Beniston, 1986).

4.2.5 Orography

Terrain is simulated in the model by specifying an altitudinal level at each surface grid-point in the model. Unlike many other mesoscale models existing today, the vertical coordinate system is not terrain-following but Cartesian. This implies that, according to the amplitude of the orography, model levels may intersect the terrain. It is therefore necessary for DREAMS-1 to identify in its three-dimensional domain which grid-points lie above the surface and which points lie beneath.

There are advantages and disadvantages to this method; one of the advantages is that it is often possible to work in terrain characterized by steep slopes despite the fact that DREAMS-1 is a hydrostatic model, as long as the expected vertical accelerations are small and non-hydrostatic effects are not expected to dominate in complex orography; this is deemed to be the case in many of the studies undertaken with DREAMS, some of which will be discussed further in this work. A major disadvantage of the method is that in highly complex topography, such as in the Alps, a substantial number of grid-points will be located beneath the surface, thereby leading to a less efficient use of available computer resources. In order to remain coherent with the turbulence parameterization described above, the model identifies the height above the ground at each grid-point to determine whether the surface-layer parameterization should be applied (for $z \leq h$, where h is the specified surface-layer thickness) or whether the first-order closure of the ABL turbulence scheme is to be used for $z > h$.

4.2.6 Boundary and initial conditions

In order to solve the abovementionned set of equations, it is necessary to supply the appropriate boundary and initial conditions. The latter conditions represent the base state of the model at the start of a given simulation, i.e., at time $t = 0$. The grid-point values of velocity or pressure, temperature, and moisture are required at this initial instant in order to allow the model to proceed forward in time.

A pressure-driven initialization scheme is possible with this version of DREAMS-1; in order to be compatible with Dirichlet and Neumann conditions, if pressure is used as the initial driving mechanism, all initial velocity fields should be set to zero. Pressure initialization consists in speciying the pressure field at the model lateral boundaries at all levels. The model then computes the distribution of pressure at all internal grid-points, taking into account also the "solid" boundaries of nodes lying at or beneath the topographic surface. The most rapid means of obtaining the pressure field at all grid points is to solve the Laplacian Equation for pressure, i.e.:

$$\nabla^2 p = 0 \qquad 4.36$$

The method of solution of this Laplacian is the technique known as "over-relaxation", in which pressure at any point x is described by:

$$p(x,y) = (1 - w) * p(x,y) + .25 * w * (p(x + Dx, y) + p(x - Dx, y) + p(x, y + Dy) + p(x, y - Dy)) \qquad 4.37$$

w is the over-relaxation factor, generally between 1.5 and 2; the method converges for most problems with large numbers of grid points within 120 - 150 iterations. In order to retain vectorization of the code, the so-called "red-black" relaxation scheme is used, where the central p(x,y) terms are computed each second grid-point and the off-center p terms (second terms on the right-hand-side) at the alternate grid points.

Boundary conditions are necessary because of the fact that, as in any limited-area model, the boundaries are mathematical and do not exist in a real continuum such as the atmosphere. Specification of conditions at the boundaries needs to be achieved in such a way that the results remain as unaffected as possible by these conditions. At the material surface defined by the topographic conditions, the three wind components u, v, and w are set to zero (the well-known "no-slip" condition), and temperature and moisture values are set to observed or characteristic values of the region considered.

At the model top, meteorological variables are function of those existing at the level just beneath the model top. In principle, the summit of the model domain should lie sufficiently high above any regions of perturbation, in order to avoid upward-propagating buoyancy-wave reflection which could contaminate the simulated variables. In the numerical experiments undertaken with the DREAMS model, the depth of the overlying stable atmopheric layers serves to attenuate vertical accelerations, so that vertical velocity reaches negligible values close to the model top. In practice, over complex terrain, the model top needs to be located at about twice the altitude of the highest orography and located in a deep, stable layer.

Lateral model boundaries are defined in such a way as to avoid wave reflection as perturbations approach the outflow boundaries. The model has the possibility of imposing cyclic, or periodic, lateral boundary conditions, or open boundaries with a filter to attenuate wave reflection. In the former case, the cyclic condition is adapted to large-scale homogeneous conditions, such as ABL processes over an ocean surface; internal wave reflection is eliminated by this method, but it cannot be applied to complex terrain in a limited area model for obvious reasons.

In this case, open boundaries are necessary to allow inflow and outflow of air, while retaining mass consistency in the 3-D model domain. In order to control wave reflection at the lateral margins of the model, a filter function proposed by Perkey and Kreizberg (1976) has been applied to the first three inside grid points adjacent to the model borders. The filter has the following form for any variable X:

$$X_{t+\Delta t} = X_t + a\,\Delta t \left[\frac{\partial X}{\partial t}_M + (1 - a) \frac{\partial X}{\partial t}_{LS} \right] \qquad 4.38$$

with the filter function a taking on the value .4 at ± 1 grid points, .7 at ± 2 grid points, and .9 at ± 3 grid points. M is the model-calculated tendency and LS the large-scale tendency. The Perkey-Kreizberg filter introduces diffusive effects at the lateral boundaries which damp out the tendency of internal wave reflection. The filter also damps out the meteorological signal, so that the analysis of model results close to the boundaries should not be attempted.

4.2.7 Methods of solution

The finite-difference techniques used in the model are described in Section 2.3 and are based on the widely-used centered-in-space, centered-in-time scheme (otherwise refered to as the "centered-leapfrog" scheme), which has the numerical advantage of neither amplifying nor damping meteorological wave modes.

4.3 Mesoscale model simulations

The following results of simulations with the DREAMS modeling system are designed to illustrate the manner in which small-scale features, such as subgrid cloud fields and cloud-radiation interactions, are capable of significantly influencing the larger-scale atmospheric fields. A common theme for many of the studies described below is related to the study of organized moist convective processes over a maritime atmospheric boundary layer. Atmospheric mesoscale open cellular convection is characterized by fields of regular-shaped structures (ideally hexagonal configurations), with the cloud walls enclosing a core of clear air in which sinking motion induced by cloud activity - the secondary circulation - is predominant. Open cells are frequently observed over a warm ocean surface during episodes of cold air outbreaks. Cold air advected over the milder ocean results in a very unstable situation close to the surface; strong upward fluxes of temperature and moisture, and significant vertical velocity generated by the surface instability, result in cumulus formations which organize themselves into cells or rings. These open cells, which often cover large areas of ocean during favorable conditions, are reminiscent of patterns associated with laboratory experiments of Rayleigh-Bénard convection. It is observed in such experiments that when a fluid between two rigid horizontal surfaces is heated from below, a circulation pattern is set up in such a way that cellular features occur. It is therefore tempting to describe the patterns associated with mesoscale open cells as a manifestation of Rayleigh-Bénard convection in the atmosphere.

However, despite many similarities between laboratory and atmospheric cells, there are many discrepancies also. In particular, there is no clearly-defined upper boundary in the atmosphere, and even a relatively strong temperature inversion (which could be considered analogous to the upper rigid surface in the laboratory experiments) can be eroded by dynamical processes, thus increasing the vertical domain of the circulation and thereby modifying its original configuration. It has also been suggested that cellular convection in the atmosphere cannot occur without the dynamic/thermodynamic feedback effects induced by the phase changes of water. This fact is supported by model results, as will be discussed later.

As in general open cells occupy vast areas of ocean, it seems reasonable to suppose that there are substantial feedback effects between the clouds and the clear atmosphere. It is consequently of interest to study these phenomena in order to understand the mechanisms inherent to them, to estimate their relative importance in large-scale atmospheric processes, and eventually to parameterize their collective influence for application to forecast models. With such objectives in mind, the 1981 German KonTur field experiment (Hoeber, 1982) was set up with the explicit purpose of obtaining data on extratropical marine convection (in particular boundary-layer rolls and open cells), both for analysis in order to understand certain features of the convection, and also in order to build up a comprehensive data set for input into numerical models. The original versions of the DREAMS model were developed on the basis of the KonTur experiment (Beniston, 1982; 1983; 1984) as a complement to the *in situ* observations over the German Bight and the North Sea.

4.3.1 Simulation of cellular cloud convection

The discussion focuses here on two simulations, each of 12 h duration, using identical initial data (Beniston, 1984); the principal difference between the two simulations lies in the fact that one allows cloud activity to occur, while the other does not. In this manner, the effects of the cloud-induced secondary circulations should be readily observed by comparing results with those from the no-cloud simulation. As initial data input for the model, a particular sounding from the KonTur experiment has been chosen (October 13, 1981). This sounding was made on a day in which a considerable number of open cloud cells were observed over the North Sea, during a period of cold air advection from the northwest over a relatively warm surface. To drive the system, a temperature perturbation of 0.1 K in the form of a single open cell was imposed in the center of the model domain.

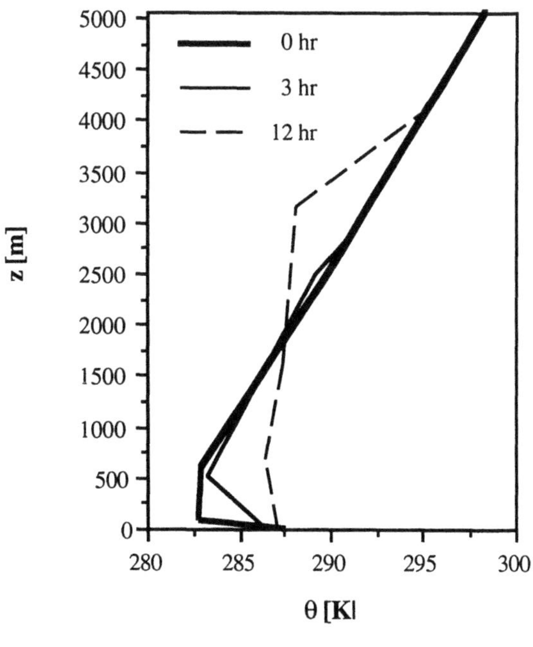

Figure 4.3: Mean vertical profiles of potential temperature at initial time, 3 h, and 12 h of model time

The profiles in Figures 4.3 and 4.4 illustrate the initial conditions of potential temperature and horizontal wind vector. Potential temperature is characterized by a strong air-sea temperature differential (4 K), surmounted by a quasi-neutral mixed layer some 500 m in vertical extent, followed by stable stratification to the top of the troposphere. The wind vector shows little shear with height (and also—not shown here—little turning with height), thus creating favorable conditions for the maintenance of cumulus "towers" over long periods of time. The initial specific humidity profile is an almost monotonically-decreasing function of height, except near the surface where the ocean-surface yields high humidity values. Figures 4.3 and 4.4 also show the profiles after 3 and 12 h of integration, in the case of cloud activity. Near the surface, in Figure 4.3, it is seen that upward transport of heat, primarily owing to surface layer buoyancy fluxes, is responsible for warming the lower layers. After 12 h, the air-sea temperature difference has diminished to less than 1 K, which has significant implications on the nature of cellular activity. The regions of cloud condensation warming are observed between 600 m (cloud base) and 1500 m, while the effects of cloud droplet evaporation in upper cloud layers is readily seen above 2000 m. At 12 h, the combined effects of surface heat exchange and cloud-induced subsidence warming have warmed the entire boundary layer by as much as 2.5 K in the lower 500 m.

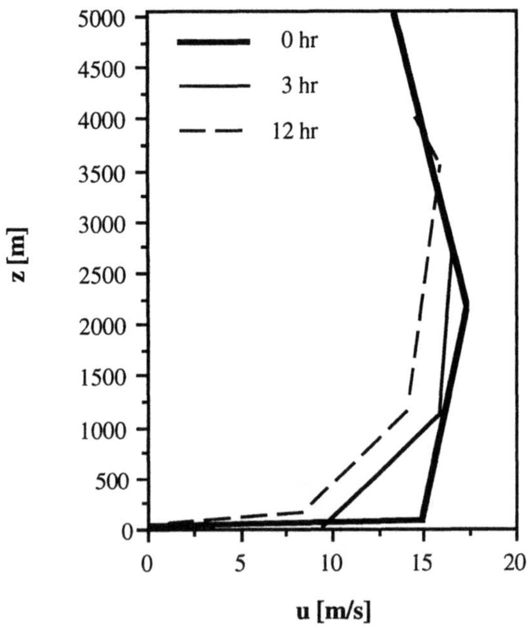

Figure 4.4: As Figure 4.3, except for the horizontal wind vector

Cooling in the upper cloud layers has resulted in a temperature decrease at 3000 m of over 3 K. The resultant profile is quasi-neutral in the first 3000 m (more precisely, it is on average less stable than at the initial time step), and the strong capping inversion above 3000 m acts to suppress any further vertical development of the clouds. The clouds are thereby seen to exert their own controls on their development, even in the absence of large-scale flow constraints.

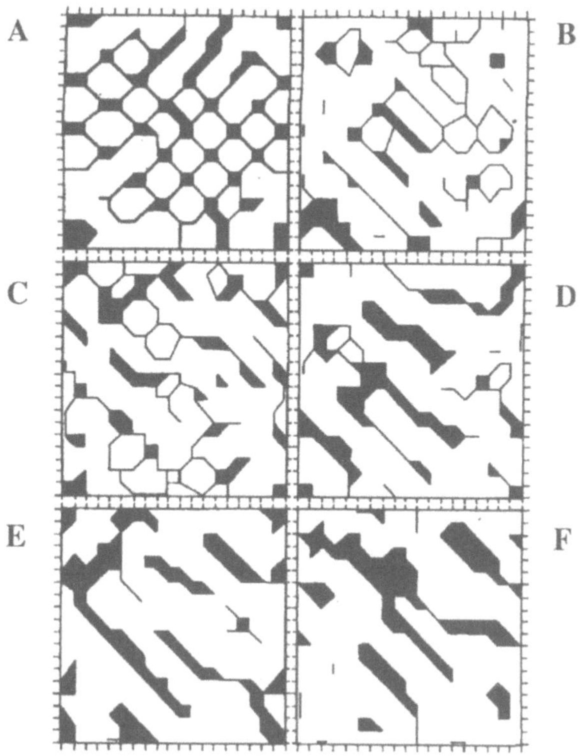

Figure 4.5: Cloud patterns at first model level above cloud base (clouds are black regions), illustrating the transition from open cells to mesoscale streets. Intervals between dashes on axes: 10 km
(A) t = 1 h; (B) t = 2 h; (C) t = 3 h; (D) t = 6 h; (E) t = 9 h; (F) t = 12 h

The wind vector profile after 3 and 12 h of simulation is essentially dominated by friction effects with the surface (Figure 4.4). Since the large-scale pressure gradients have been neglected, there is no force which can counteract the deceleration influence of the ocean-surface; after 3 h winds have been reduced by one-third of their initial value, and decrease by a further 2 m/s between 3 and 12 h of simulation. The effects of surface friction are expected to be felt through to the top of the planetary boundary layer (set at 1600 m in the present study), and this can already be seen after 3 h. However, at 12 h, deceleration is observed even above 3000 m, which implies that boundary layer turbulence is not the only factor influencing the wind profile. The change in the wind profile above the top of the boundary layer can be attributed to the dynamics of the cloud-induced secondary circulation. This appears to couple the vertical cloud circulation above the boundary layer with the eddies within the mixed layer, thus enabling surface friction effects to extend beyond 1600 m. This lower troposphere-boundary layer circulation interaction acts as a momentum sink through to the top of the cloud layer.

The humidity profile changes in time according to the surface moisture flux effects and cloud thermodynamic sources and sinks. At all levels in the mixed layer, a rise in specific humidity is observed, and also in the upper cloud layers where cloud droplet detrainment to the environment acts as a moisture source. A minimum increase in specific humidity is seen in the lower cloud layer, as uptake of moisture by clouds acts as a cloud sink. However, this sink is compensated for by strong upward turbulent transfer of specific humidity from the surface and mixed layers.

Figure 4.5 illustrates the patterns of clouds for different times of simulation. A vast field of open cells is observed in the early stages of the simulation. The cells change in shape and size as time goes on, and after 6 h a few open cells remain in the domain. Later on, banded structures characteristic of mesoscale "cloud streets" appear. These structures appear to be longer lasting phenomena than the open cells. The transition from cells to banded cloud distributions can be explained in part by the decrease in the difference in temperature at the air/water interface. The instability, which is essential to open cell formations, is no longer present after some time, and as such the cells no longer have the energy available to persist. However, there is more than enough cloud material present above the LCL, so that the clouds regroup themselves into mesoscale structures, as seen in the sequence at 9 h and at 12 h. The exact threshold value of air-sea temperature difference below which open cell formations can no longer exist is not obvious from the present simulations, but a figure for a critical value of 2.5 K does not appear unreasonable.

However, the surface instability may not be the only factor which influences the life cycle of cloud cells. Figure 4.6 shows the convective heating rates (combined latent and sensible heating) for cases of open cells and mesoscale streets for 3 h and 12 h for simulation, respectively. The cell case displays a remarkable asymmetry between condensation warming in the lower 1000 m of the cloud layer, and evaporation cooling which has a maximum at an altitude of 2500 m. The cooling is almost three times as large as the low-level warming, and this results in the formation of large amounts of cold, heavy air which sinks rapidly. On encountering the surface, the subsiding air diverges outwards and rises once again when encountering air from neighboring cells. Enhanced by surface layer buoyancy forcing, the rising air passes above the LCL, thus completing the cycle of the cloud circulation, and thereby maintaining the cellular structure. It is only when condensation warming at lower cloud levels becomes more important that the instability near the surface is no longer the dominant mechanism in the cellular convection. A new cloud-induced circulation may be set up, in which the sinking air does not necessarily interact with the surface, and therefore no longer contributes to the low-level convergence-divergence patterns associated with open cells. If, in addition to this new secondary circulation, the surface instability itself vanishes, then favorable conditions which maintain the cellular activity are no longer present.

In the case of banded cloud features, Figure 4.6 indicates that the area covered by condensation warming is approximately equivalent to that covered by evaporation cooling. This distribution leads to a quasi-steady state convective overturning, in which the upward branch of the circulation is driven by the warming near cloud base, and the downward branch is triggered by upper cloud-level cooling. As this process is more or less independent of the surface and mixed layer (the air-sea temperature difference now being of the order of 1 K or less), this type of circulation can maintain itself over long periods of time. Probably only large-scale perturbations would be able to modify the equilibrium state of this cloud-induced circulation pattern.

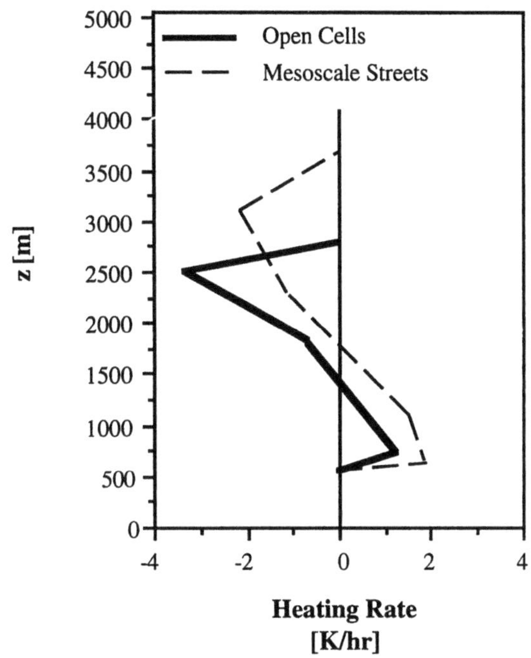

Figure 4.6: Convective heating rates (latent and sensible) during periods of cellular convection (3 h) and mesoscale street formations (12 h)

Figure 4.7 illustrates cross sections of the perturbation fields of potential temperature and zonal wind after 2 h of simulation. The sections are located in an x-z plane through the domain centre; cloudy walls are seen as hachured regions. The u' section reflects the well organized patterns of low-level convergence beneath the clouds and divergence beneath the cloud-free cases (this is also observed in nature during cellular convection, and in particular was seen throughout the cell episodes during the KonTur experiment). The indirect influence of clouds on the dynamic fields (through temperature-induced pressure fluctuations) is seen in upper cloud regions, where cooling by cloud evaporation creates a divergence pattern at cloud top. It can be readily deduced from the continuity equation that such patterns of low-level convergence and higher-level divergence must result in a cloud-induced circulation. This is effectively characterized by strong updrafts in the cloud walls and slower sinking motion in the cloud cores.

Analysis of the potential temperature patterns along the same cross-section as u' shows regions of warmer than average air beneath the cloud walls of the open cells. The temperature excess reaches some 0.7 K over the average temperature of that level, and is a result of both surface heat flux transports and low-level convergence of heat by the flow. This result is supported by Bakan (1982) who reported from KonTur observations that surface air temperature increases up to 1 K beneath the cloud walls of a cell. In the cell cores, adiabatic compression warming by the sinking air is counteracted by the production of cool air in the cloud detrainment region. This latter layer is seen as a band of negative θ fluctuations at and above the cloud tops.

The influence of precipitation in the model is to limit cloud growth (as autoconversion to rainwater is a sink of cloud liquid water), and to cool the subcloud layers. It is difficult to distinguish this evaporational effect in light of the strong positive fluctuations which occur beneath cloud walls, but the raindrop evaporation is likely to limit the magnitude of the positive θ. Typical average values for rainwater specific humidity in the range 0.1 -0.40 gm/kg beneath open cell cumuli correspond well to aircraft measurements during KonTur.

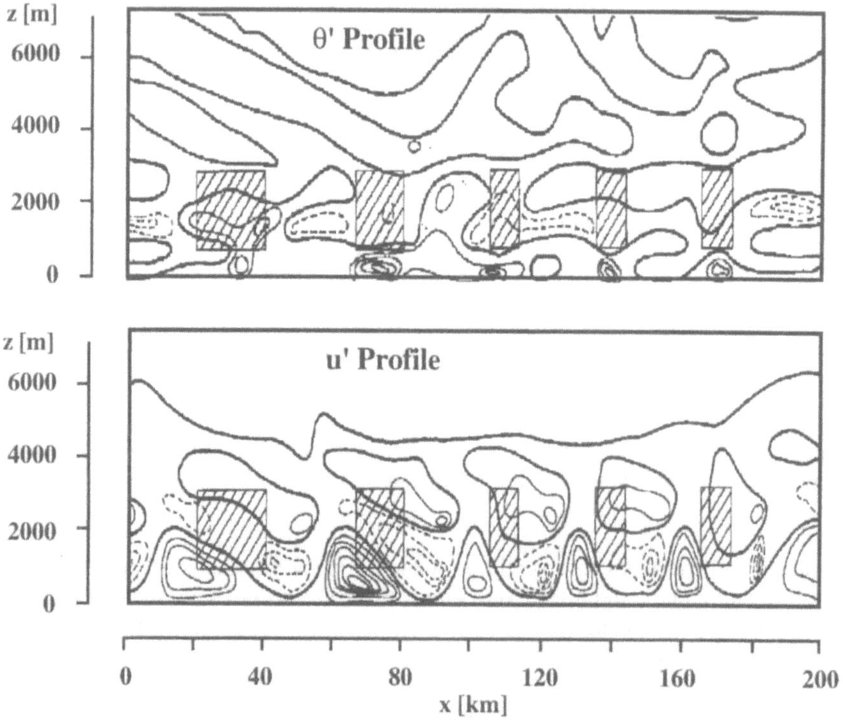

Figure 4.7: Vertical x-z plane section at t = 2 h of simulation for zonal wind and potential temperature fluctuations. X = 0 - 190 km; Y = 180 km; Z = 0 - 7 km. Isoline intervals: 0.2 K for θ; 0.1 m/s for u'

The parallel simulation in which cloud activity was artificially suppressed shows that in the absence of cloud-induced secondary circulations, processes are far less vigorous, fluctuations of dynamic and thermodynamic variables much less intense, and the vertical extent of motions and transports considerably reduced.

Figure 4.8 displays this conclusion adequately. It has simultaneous plots of the initial vertical profile of specific humidity, and the profiles after 12 h of simulation in the cloud and no-cloud cases. It is obvious that the dominant processes acting upon the mesoscale environment in the no-cloud case occur in the ABL; the lowest 1500 m of the model atmosphere have all been moistened through turbulent flux effects. Grid-scale vertical advection of humidity has played an insignificant role above the boundary layer, and the humidity profile at 12 h is coincident with the initial profile at around 1700 m.

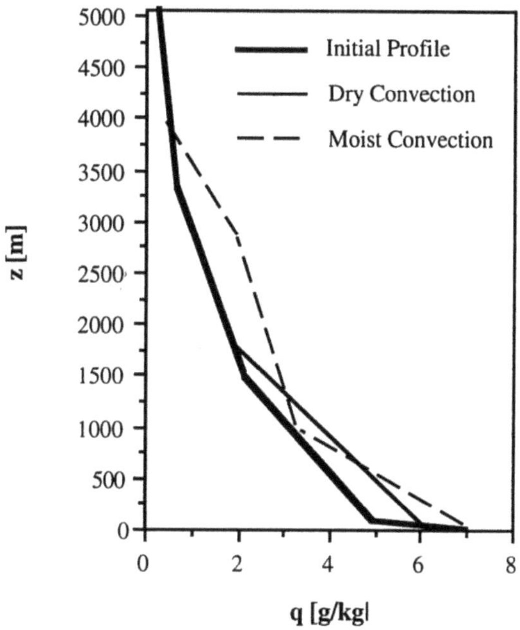

Figure 4.8: Profiles of specific humidity at 12 h of simulation for cases with and without moist convection

This leads to the conclusion that in a "dry" mesoscale atmosphere, a temperature fluctuation of 0.1 K supplies a negligible amount of energy to the dynamic and thermodynamic fields and is rapidly damped out. When this same minor perturbation initiates a cloud field, however, the results are vastly different, as is seen from the third curve in Figure 4.8. The vertical extent of the region where humidity has changed in time now stretches to beyond 4000 m, with specific humidity increasing at the 3000 m level, even greater than near the ocean surface, where latent heat fluxes are also very strong. It is interesting to note that between cloud base at 600 m, and 1200 m approximately, the humidity increase in the cloud simulation is less than in the no-cloud simulation. This is a direct consequence of the uptake of environmental moisture by the condensation process in this layer.

Figure 4.9 gives further illustrations which emphasize the differences between the cloud and no-cloud simulations. These show the vertical profiles of zonal velocity and potential temperature variances. Both sets of curves confirm the preceeding conclusions concerning intensity and extent of the cloud and mesoscale circulations.

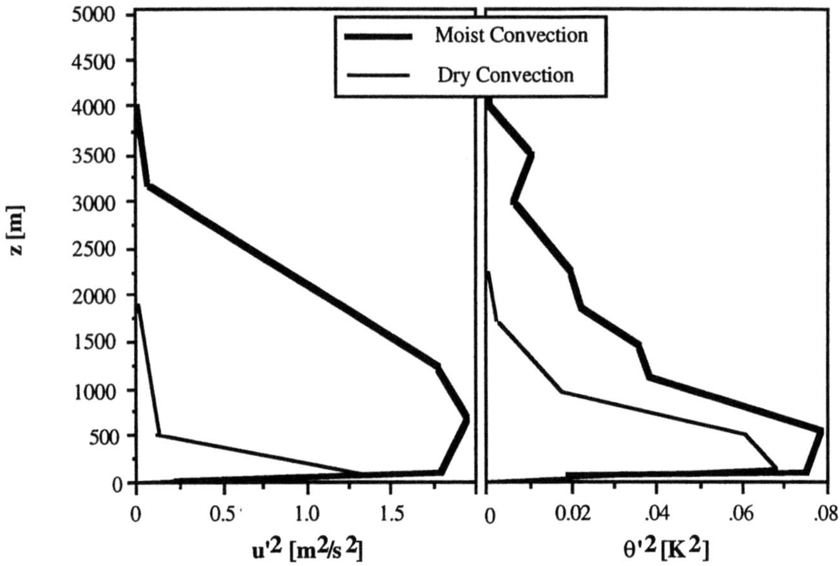

Figure 4.9: As Figure 4.8, except for variances of zonal velocity and potential temperature

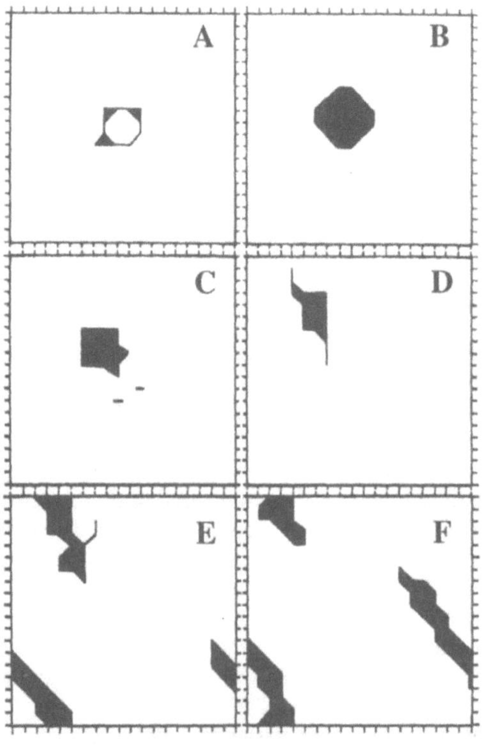

Figure 4.10: As Figure 4.5, except for vertical velocity in the dry convection case

In both moist and dry convective cases, variances of u and θ in the mixed layer are of comparable magnitude, showing that the surface layer is the dominant forcing region; the slightly larger values in the moist convective case are probably indicative of stronger fluxes induced by the secondary circulation, which enhances the friction velocity (a component of all surface layer fluxes) through higher velocities near the surface. Above the mixed layer, the profiles in the no-cloud case rapidly reach negligible proportions, while those in the cloud situation remain significant up to and above 3000 m.

Perhaps the most impressive difference between the two simulations is seen by the comparisons between Figure 4.10, which shows regions of positive vertical velocity in the dry convective case, and the cloud patterns of Figure 4.5. As cloud distributions in Figure 4.10 correspond to regions of upward vertical velocity, a direct comparison can in effect be made. It is immediately apparent that at no time, other than 1 h when initialization effects have not yet been damped out, do cellular features appear in the dry convective case. This suggests that the thermodynamic and dynamic influence of the phase changes of water are essential to the development and maintenance of open cellular activity in the real atmosphere.

4.3.2 Influence of model initial and boundary conditions on cellular convective processes

4.3.2.1 Effects of large-scale temperature advection

As described in the previous section, the DREAMS model is initialized by introducing mean profiles of wind, temperature, and humidity as observed during a particularly interesting KonTur case (13 October 1981) The dynamic and thermodynamic fields are initially horizontally homogeneous; in order to drive the system, a small perturbation of temperature of order 0.1 K has been applied at the first time step (20 seconds), which disturbs the dynamic fields through temperature-related pressure fluctuations. Vertical transports progressively result in cloud development, and feedbacks between clouds and the mesoscale environment lead to secondary circulations and complex scale interactions. Most of the analyses to be presented in this section pertain to a time 12 hours after the start of the integrations, at which time the model solutions have reached a quasi-steady state. In particular, the patterns of secondary circulations induced by the clouds are already well established after 6 hours. Little change in the temperature, humidity, or momentum profiles is observed after 9 hours of simulation, pointing to the existence of an equilibrium between the various dynamic and thermodynamic sources and sinks. In the lowermost levels, where strong surface buoyancy leads to considerable turbulence, slow time changes are to be observed in the latent and sensible heat fluxes. However, these are of too small a magnitude to greatly influence the mesoscale fields on the time scales considered.

In early numerical experimentation (Beniston, 1983; 1984), it was thought that a perturbation in the form of a single open cell located at domain center should force a field of open cells of the same shape and size (30 km diameter). This was to great extent succesful, but at the same time such a form of initialization might in fact be generating too great a bias to the cell shapes and sizes by reproducing only the initial perturbation configuration. As a consequence, various shapes of the initial disturbance were tested and will now be described.

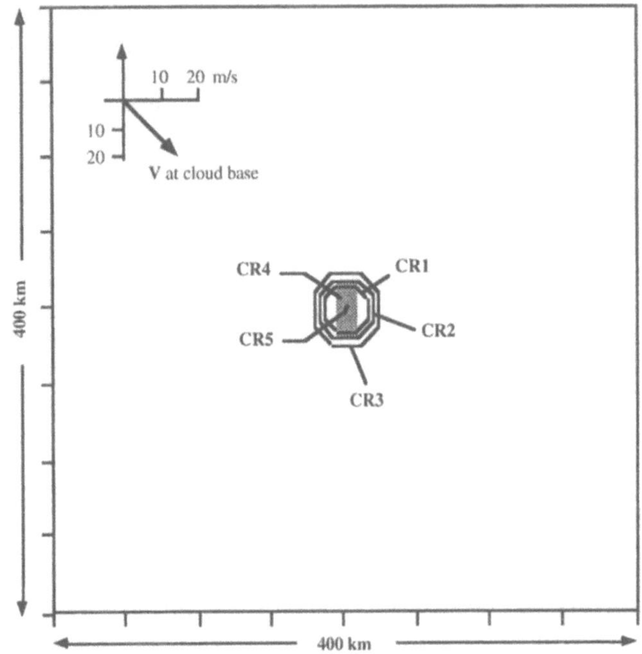

Figure 4.11: Different shapes of the 0.1K temperature perturbation used to initialize the model dynamics in experiments CR1-CR5. Model domain characteristics and mean wind vector are indicated on this horizontal planform

The different configurations of the initial perturbations are illustrated in Figure 4.11. The original 30 km cell is once more included for comparison with the same initialization in the absence of the large-scale temperature advection. Temperature disturbances in the form of rings of 50 km and 70 km diameter are also tested to see whether the model will reproduce only these cell sizes or whether a preferred cell size mode will emerge from the simulations. Furthermore, a perturbation in the form of a N-S oriented band is applied, again to determine whether the model will respond by creating open cells of a characteristic size and shape. Lastly, a one-point temperature disturbance is tested; this is the configuration which gives the least bias to the model results, by not imposing any particular shape or size to any cells which may subsequently develop.

Estimates of temperature gradients obtained from budget analysis of KonTur data on the relevant days were used to simulate the large-scale advection of temperature. Temperature gradients are on order 0.4 K per 100 km in both the N-S and E-W directions, leading to a cooling of the planetary boundary layer (ABL) of some 5 K/day for a mean wind at low levels of about 15 m/s. Such cooling is generally offset by surface buoyancy fluxes and cloud-induced secondary circulation effects. The advection terms are included in the model to the top of the planetary boundary layer (1500 m) in order to maintain the air-sea temperature difference and to balance low cloud-level condensation effects.

Figure 4.12 shows the horizontally-averaged initial profile of temperature as was observed during KonTur and which has been used as input data into the model. Profiles are also given for cases after twelve hours of simulation for cases with and without the inclusion of the large-scale temperature advection term.

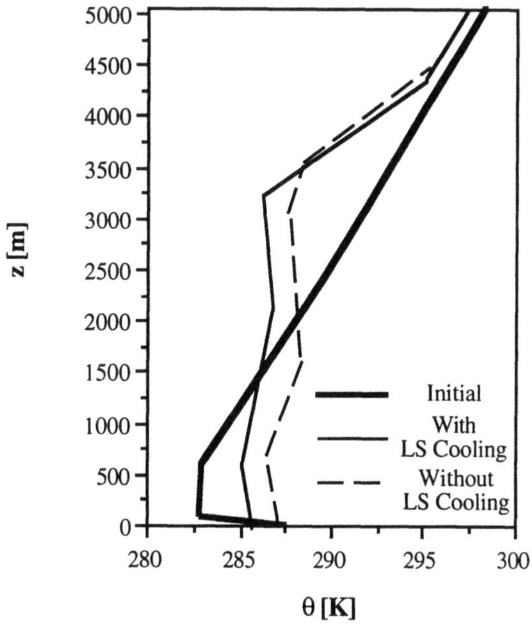

Figure 4.12: Vertical profiles of potential temperature for the initial situation, and after 12 hours for the cases with and without large-scale temperature advection

It is seen that in the case without advection, strong warming has taken place within the ABL, due to surface effects and a number of interactive physical processes—subgrid turbulent transports, condensation warming, cloud-induced subsidence warming, and so on. In upper cloud layers, above 2200 m, cooling by cloud detrainment to the environment is the dominant term, resulting in a temperature drop of over 5 K in 12 hours. This leads to a strong inversion, capping the cloud field, which inhibits further vertical cloud development, as already seen in Beniston (1984; cf. Section 4.3.1). A simulation of the same case but with the advective term included demonstrates the counteraction of the low-level warming terms by large-scale inflow of cold air into the model domain. The KonTur estimates of temperature advection are not of sufficient magnitude to counteract all warming effects, and the lower levels have warmed by 1.5 K, though as will be seen subsequently, this has not destroyed the required instability for maintaining open cell formations. The fact that a warming does occur despite strong synoptic cooling is due the fact that this feature has an influence on the cloud parameterization, at least in the formative stages of the cloud field. As the cloud entrainment and detrainment depends on cloud-environment temperature differences, a reduction of cloud base temperature due to the large-scale flow will lead to larger cloud-environment temperature differences. Though it is usually very difficult to predict what will happen in a strongly non-linear system, it is believed that the synoptic cooling in fact stimulates a stronger secondary circulation than in the case where this cooling is absent, thereby resulting in the observed warming through adiabatic subsidence and enhanced surface heat flux effects.

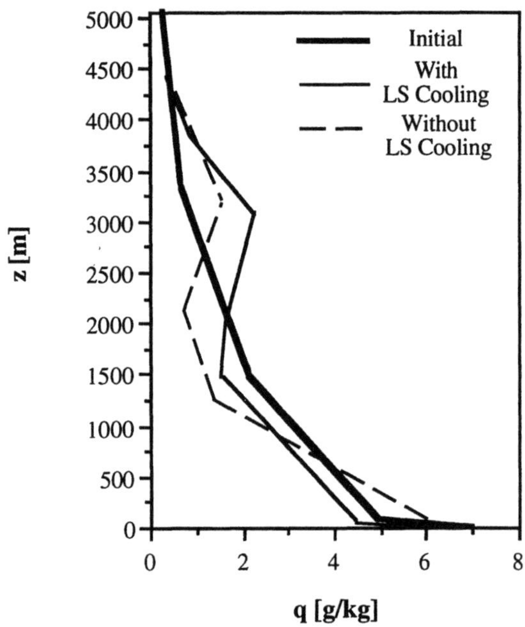

Figure 4.13: As Figure 4.12, except for mean specific humidity profile

This more intense cloud activity is also observed in Figure 4.13, which shows the profiles of specific humidity, again for the initial case and for the simulations in which large-scale advection of temperature is included or neglected. The more intense cloud activity in the case with advection is reflected by the fact that, despite strong moisture fluxes from the ocean surface and raindrop evaporation into the subcloud layer, this layer is in fact dried out through uptake of moisture by the clouds. Interestingly, this is a feature which was actually observed during the open cell phase of the KonTur experiment. This drying is not apparent in the case where the temperature advection is absent, except within the lower cloud layer where condensation effects are sufficiently strong to result in drying at these levels. In upper-cloud regions, the cloud detrainment processes work in such a way that cloud material is mixed into the environment; strong moistening of the environment is seen at these levels in both cases.

The profile of the wind vector is modified in different ways by the cloud activity according to whether the temperature advection term is present or not, as seen in Figure 4.14. In all cases the wind has decelerated even beyond the upper limit of the ABL, indicating that surface friction and subgrid-scale momentum fluxes are not the only processes operating here. The secondary circulation acts to smooth out and slow down the wind up to 3000 m by transfering upwards the momentum defect which occurs within the ABL. The deceleration of the wind is more noticable in the simulation with the temperature advection included, confirming the more vigorous nature of the cloud activity.

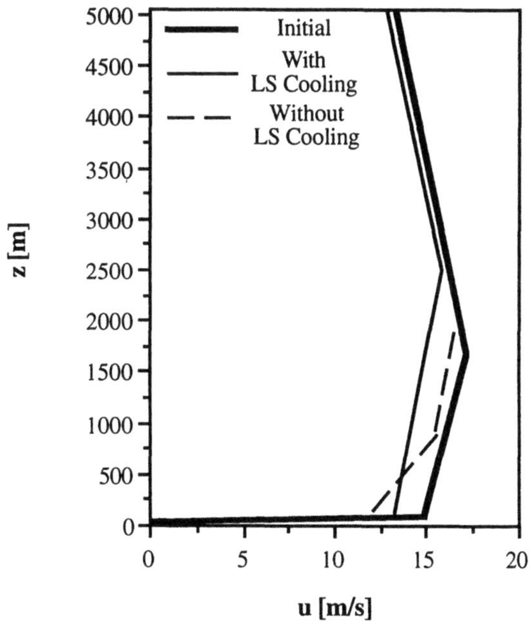

Figure 4.14: As Figure 4.12, except for mean vertical wind velocity profile

Figure 4.15 gives time changes of sensible and latent heat fluxes in the surface layer for the first 12 hours of simulation, for both experiments. The negative values indicate fluxes directed upward from the surface. Sensible heat fluxes have an absolute initial value of 56 Wm^{-2} and decrease slowly in time due to the input of heat into the lower ABL, thereby reducing the surface gradients of temperature. The reduction of sensible heat is also due to the slowing down of the wind through surface friction, thus decreasing the friction velocity used to parameterize surface layer fluxes. Where large-scale advective processes are included, the flux after 12 hours is 25 Wm^{-2} directed upwards, compared to the 11 Wm^{-2} in the case without advection. Latent heat fluxes show somewhat different time profiles; in the case without advection, the moisture flux in the surface layer drops from an original absolute value of 162 Wm^{-2} to 92 Wm^{-2} after 12 hours. With temperature advection included, however, the flux initially decreases then increases again after 3 hours, when cloud activity becomes important, and the low-level drying increases surface layer moisture gradients which conteract to some extent the lower friction velocities. From 7 to 12 hours of simulation, a slow decrease of less than 10 Wm^{-2} is observed, with the 12 hour absolute value of 140 Wm^{-2} substantially greater than that resulting from the case without advection.

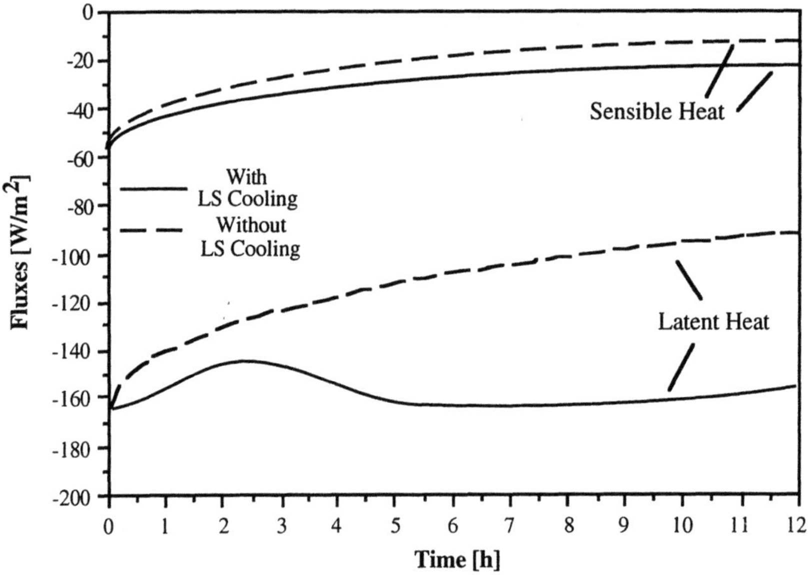

Figure 4.15: Time changes of sensible and latent heat fluxes in the surface layer, for the cases with and without large-scale temperature advection. Negative values indicate upward-directed fluxes

The inclusion of a large-scale cooling term due to temperature advection in the model has the principal effect of enhancing the intensity of the cloud activity and cloud-mesoscale interactions in the layers where the advective term is included; this is reflected through stronger mesoscale dynamic and thermodynamic responses at lower levels. In upper layers, and in the free atmosphere above the clouds, differences in mean profiles are to be noted after 12 hours, mainly due to the effects of grid-resolved transports which redistribute heat, moisture, and momentum in the vertical.

4.3.2.2 Cloud patterns as a function of the initial perturbation configuration

Many micro and mesoscale models initialize atmospheric motion through small random perturbations in order not to bias the system and force a given wave mode. In the present case, however, it was of interest to see whether open cellular convection in the model required forcing through a specific type of initialization. Observations of open cells during the KonTur experiment yielded approximately regular shaped cells of about 30 km in diameter. Therefore the first experiment conducted on the Max-Planck-Society CRAY-1 supercomputer in Garching (Bavaria) attempted to force a cloud field by initializing a temperature perturbation in the shape of a single cell of 30 km diameter. The initial perturbations are in each case only applied at the center of the array. This simulation will subsequently be denoted by CRl, and the initial pattern is shown in Figure 4.11. Simulations with disturbances of 50 km, 70 km diameters, linear initial perturbation, and single point initialization are denoted by CR2, CR3, CR4, and CR5 respectively.

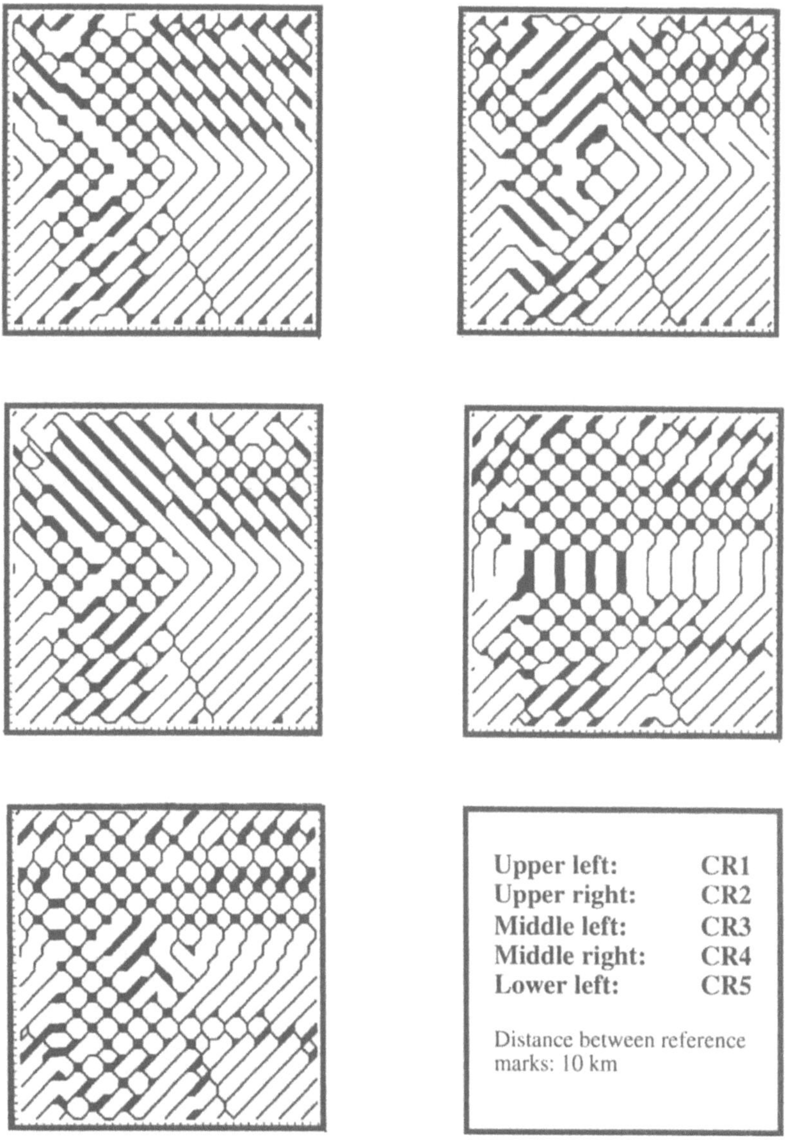

Figure 4.16: Cloud patterns after 12 hours of simulation

Figure 4.16A illustrates a plan view of the cloud patterns after 12 hours for CRl. The black shading links cloudy grid points at 1050 m (i.e., the first model level above cloud base). Here the term "cloud" refers to points where the liquid water content is above zero. The initial cell is still present at the domain center, as air is continuously recycled by the clouds; it should be stressed that the model domain is allowed to translate with the mean horizontal wind vector at cloud base, thus enabling cloud development and cloud circulations to be analyzed as they move along with the mean wind. It is for this reason that the initial perturbation, which has developed into a cloudy open cell, can still be observed after 12 hours. The initial perturbation acts to generate wave-disturbances which propagate outward towards the boundaries, much in the same way a stone thrown into a calm pool will generate concentric waves. In this case the "wave pattern" is characterized by regions of alternating upward and downward vertical velocities, with clouds forming along zones of upward motion. In the early stages, the outward-propagating waves are seen as parallel bands moving south-east and north-east from the central disturbance (this prefered direction is governed by the stronger NW flow above the cloud field, in much the same way as the 500 mb level is considered the governing flow for synoptic scale disturbances). As these bands fill with cloud, they are seen as mesoscale streets, with horizontal wavelengths of 30 km (3Δx or 3Δy, with Δx, Δy being the horizontal grid lengths). In a general upstream direction from the central open cell, a series of identical cells have been formed, each cell interacting with each other and with their mesoscale environment, as will be seen later. A series of more elongated open cells are present in the upper right quadrant, indicative of some flow interaction between the regular-shaped cells and the parallel mesoscale streets. Though this result is encouraging due to the fact that a number of open cells have been generated by the model, a possible criticism is that the initial perturbation has forced the entire system more numerically than physically by imposing a 3Δx wave mode in both the cell and street cases.
Further tests with different initializations are therefore essential to estimate to what extent the model numerics are generating open cells, rather than purely physical principles. It should be emphasized, however, that one significant difference between the present simulations and previous studies by Beniston (1983, 1984) is that without large-scale advection of temperature, the open cell formations were seen to disappear within the first six hours of simulations.

In the present analysis, it is clear that after 12 hours of integration in the presence of continuous surface instability, open cells are still very much present and vigorously interactive. This is clearly a physical process and not solely a numerical effect on the convective activity.

The CR2 simulation results in a cloud pattern as shown in Figure 4.16B. The initial 50 km diameter cell is still present after 12 hours, the core of subsiding air very prominent in the domain center. This configuration, having rather lengthy diagonal branches, has generated cloud bands in concentric patterns surrounding the domain center. What is interesting, however, is that the model has not reproduced the typical configuration of the initial disturbance, but as in CR1, open cells with a diameter of 30 km. In the upper section of the domain, a mixture of 30 km open cells, elongated cells as in CR1, cloud bands, and some hexagonal structures can be seen. This is a pattern which can often be observed on satellite pictures during periods of organized convection, where cells and band structures coexist, and where the cells themselves do not always have the same configuration. Hexagonal features are typically encountered in laboratory experiments (Bénard's original studies, or more recent work by Krishnamurti, 1975, have shown that in dense fluids, hexagons are the most likely configuration to occur in convection forced by a surface instability). Purely hexagonal shapes are rarely observed in the atmosphere due to turbulent activity in the cloud field and general inhomogeneity in mesoscale or synoptic-scale conditions; a numerical model, however, is more capable of generating such features because of its grid structure which only allows certain regular structures to occur. A finer grid than the one used in the present model would probably create irregularities in the cell shapes, as a result of more efficient advection or turbulent diffusion from one grid-point to another, though the dominant shape would still be retained and be recognizable.

Increasing the size of the initial open cell to 70 km, as in run CR3 (Figure 4.16C), results once again in the formation of what may now be termed the "preferred open cell mode" in the model: cloud rings of 30 km in diameter. The coexistence of streets and cells of different shapes is to be observed as previously, but the distinctive feature are definitely the open cells as those forced by the initialization in CR1. What is also noteworthy is that, unlike the 30 km or 50 km cells in CR1 and CR2, the 70 km cell no longer exhibits a stable mode and has broken down into smaller cell modes, within the boundaries of the initial cell, while simultaneously generating the small open cells elsewhere in the domain. Suggestions by Agee (1976) as to an inverse square relationship between convective depth and aspect ratio (characteristic depth to length scale ratio of a cell) are one possible method to explain this destabilization of the 70 km cell. In all the experiments conducted so far by the model, the clouds have influenced their mean environment in such a way that the strong capping inversion observable in Figure 4.12 occurs in the layer between 3500 and 4000 m. Since a limit on the convective depth is imposed by the clouds themselves, it is clear that if a convective depth-aspect ratio relation exists, then model cell diameters will be unable to exceed a certain size, as demonstrated in CR3. A cell diameter of 50 km appears to be the largest stable mode of cellular convection in the mesoscale model, with the 30 km mode appearing to be the one most easily achieved.

The three simulations just described had one feature in common: all cells were forced by a perturbation which itself was in the shape of an open cell. CR4 is different in that a linear temperature perturbation in the form of a double band 60 km long and oriented in a north-south direction was used to drive the system. Such an initialization is expected to give a smaller bias to the generation of a preferred cell size, and also will not necessarily force diagonally oriented cloud bands as were observed in CR1 - CR3.

What can be seen after 12 hours in Figure 4.16D are three distinct regimes in the model: a central band where parallel cloud lines oriented NS are separated by clear regions 30 km wide; below this central region, a zone with a mixed open cell mesoscale band structure appears, though the diagonally-oriented bands are far less numerous than in the CR1 - CR3 simulations; and above the central region, a domain where a large number of 30 km cells ("preferred size cells") and hexagons coexist. Whereas in the previous experiments, it could be argued that all cells were somehow forced by a bias in the initialization, the cloud patterns resulting from the CR4 perturbation are a true indication that surface forcing and cloud-mesoscale interactions (i.e., predominantly physical processes) are acting to create cells. The physical forcing is also emphasized by the single-point initialization of simulation CR5, where the cloud patterns are illustrated in Figure 4.16E. Here again a mixed mode (bands, hexagons, and 30 km cells) occurs, with cells the most predominant feature in this figure. By using the simplest initialization with the least bias possible - excepting random forcing - the model has in fact created the largest number of open cells of any of the five experiments CR1 - CR5. Over 50 open cells of the "preferred size mode" have been generated in this picture at 12 hours of simulation, along with some 30 hexagons and 15 cells of different shapes and sizes. These latter types of cells are in fact the result of the linking of two cells to form only one larger cell, or alternatively the early stages of the destabilization of a larger cell into two smaller cells.

The question which arises in these observations of cloud patterns is why the most obvious forcing - the 30 km diameter cell in CRl - yields fewer cells in the domain than the simplest type of forcing, i.e., the single point perturbation in CR5. The principal reason is that the CR1 initialization favors not only cells but also the formation of long-lasting mesoscale cloud streets, which quickly develop their own characteristic circulations and which feed on the energy provided by the surface air-sea temperature difference. This maintains the simple convergence-divergence patterns at low levels and outflow in the opposite direction in upper cloud regions. The open cells also require the surface forcing to develop their activity, but the cloud interactions are not significantly stronger than in the banded structures, so that open cells are unable to develop at the expense of the cloud bands. This results in a quasi equilibrium between the cell and street regimes, where interactions between both regimes exist, but no one feature will become dominant. In the case of the CR5 simulation, however, no particular mode is favored in the initial stages of the integrations, and therefore the appropriate mean and lower boundary conditions which are required for open cell formations will result in open cells being the major dominant feature in the model domain. The fact that banded structures are also observed is perhaps indicative of the reduction in surface heat fluxes despite the inclusion of cold air advection by the synoptic flow. However, it also accentuates the physical realism of the model simulations, whereby band and cell modes exist simultaneously as they do in nature.

The height to diameter ratio of the convective cells is considered to be a key dimensionless parameter describing open or closed cellular activity (see for example Agee et al., 1973). For the model-simulated cells, this parameter is defined as the ratio of average cloud thickness to typical cell diameter. For the 30 km cells, this yields a figure of about 1: 10, while for the 50 km cell it is nearer 1: 15, which appear to be the upper limit of aspect ratio which the model can handle. Future work with a reduction in grid size may enable more precise figures to be estimated, though they would still be expected to lie within the range 1:10 -1:20. Such model aspect ratios are, incidentally, well within the limits given for atmospheric mesoscale cells.

4.3.2.3 Kinematic and thermodynamic fields in the presence of cells

In order to analyze the model results not only from the point of view of whether "correct" cloud patterns are created by the model, it is also of interest to examine features of the dynamic and thermodynamic fields. These can shed light on how the numerous physical processes and parameters are working together during a situation of near stationarity in the cloud field. Unreasonable fluctuations in wind or temperature fields would be indicative of noticeable failures in the model, even though the cloud patterns may in fact ressemble those in nature.

The convergence-divergence patterns described in Section 4.3.2 ensures the organisation of regions of upward motion and subsiding air into cellular structures. It has been observed from careful analyses of the flow pattern in the model that no two eddies or core divergences are exactly the same: the magnitude and direction of the flow varies slightly from one open cell to the next. This implies that small temperature inhomogeneities, differential availability of moisture, and dynamic fluctuations, which are all linked in a highly non-linear system, are able to generate a cloud field whose gross characteristics are similar but where the finer detail differs from one convective feature to another. Fluctuations in the wind flow in the vicinity of hexagonal features are more difficult to generalize, because some hexagons are embedded within the 30 km cells while others lie at the edges of these cells and thus interact with their mesoscale environment in a different manner. It is observed, however, that the clear core contains air which rotates cyclonically in its eastern (right hand) part, with anticyclonic rotation in its northern (upper) part, contributing to the sub-cloud convergence as the core circulation meets flow from neighboring hexagons.

In the lower cloud layer, the mesoscale circulation corresponds to the expected cloud-scale circulation; in the upper cloud layer, the sinking motion on the mesoscale is generated by the cloud top divergence of air, and a reverse circulation is observed. Such a situation may be confusing if one does not bear in mind the differences between the cloud-scale (subgrid scale) and the mesoscale processes. The typical picture of open cell circulation is in fact most noticeable in the lower cloud and subcloud layers, where the circulation vortices demonstrate characteristic updrafts in the clouds and downflow and low-level divergence in the core and subcloud layer.

In terms of temperature, low-level convergence of temperature flux leads to warming beneath the cloud walls, the temperature fluctuations often exceeding 0.5 K and occasionally above 0.7 K (Bakan, 1982, estimated surface layer fluctuations to be about 1 K beneath open cell cloud walls). Condensation processes show the cloudy part of the cells (located in the left half of the picture) to have warmer than average fluctuations, while the cell cores within the cloud layer are cooler than average despite adiabatic compression warming; such model results are supported by the aircraft measurements presented by Brummer et al. (1985) for KonTur. The cloud streets do not exhibit the same temperature fluctuations, mainly because they are oriented in a diagonal direction and therefore the effects of flow from the third dimension are more strongly felt. As previously mentioned in the discussion of the wind field, the temperature fluctuation patterns differ from one cell to the next, showing that no two cells are alike even though they exhibit the same cloud ring structure. This is a feature which was commonly observed from analysis of the KonTur aircraft data set. At cloud tops, strong cooling due to the detrainment of cloud material to the environment lead to negative fluctuations of temperature at these levels. The perturbations can be lower than -0.5 K, which is of the same absolute magnitude as the lower ABL fluctuations; this reflects some form of "continuity" in the perturbed temperature field which appears reasonable in the absence of atmospheric data.

In horizontal cross sections, both the cells and the cloud bands have their distinctive signatures, from the top of surface layer to the free atmosphere above cloud top. The mesoscale streets are reflected by elongated bands of temperature fluctuations, i.e., relatively simple structures. Open cells, on the other hand, are seen as “pools” of warm air beneath clouds and in the lower cloud layer, no more than 10-20 km in diameter, whereas the cores are seen as larger circles (25-30 km) of positive fluctuations, though these are smaller than beneath the clouds or in the clouds. Above cloud top, negatively correlated "polls" of air carry the effects of cellular convection up to 3500 m or more. The band structures are also apparent at these heights, but the magnitude of the fluctuations are somewhat lower than for open cells.

Variables	**KonTur Experiments**		**Mesoscale Model**	
Surface layer (50 m):				
$\overline{u'w'}$ [m^2/s^2]	-0.15 ±0.035		-0.12	
$\overline{v'w'}$ [m^2/s^2]	-0.06 ±0.065		-0.053	
$\overline{\theta'w'}$ [K m/s^2]	0.39 ±0.02		0.044	
$\overline{q'w'}$ [g/kg m/s^2]	0.06 ±0.02		0.066	
Mixed layer (550 m) :				
$\overline{u'w'}$ [m^2/s^2]	-0.015 ±0.05		-0.021	
$\overline{v'w'}$ [m^2/s^2]	-0.025 ±0.125		-0.020	
$\overline{\theta'w'}$ [K m/s^2]	0.015 ±0.02		-0.004	
$\overline{q'w'}$ [g/kg m/s^2]	0.060 ±0.02		0.045	
	Maxi	*Mini*	*Maxi*	*Mini*
Cloud layer (1350 m) "Grid-scale" flux:				
$\overline{u'w'}$ [m^2/s^2]	10.0	-8.75	9.0	-6.5
$\overline{v'w'}$ [m^2/s^2]	6.0	-7.50	3.5	-5.5
$\overline{\theta'w'}$ [K m/s^2]	1.5	-7.5	1.3	-0.01
$\overline{q'w'}$ [g/kg m/s^2]	3.0	-1.0	1.1	-0.03
Typical fluctuations; Surface layer :				
u' (m/s)	2.50	-2.50	1.98	-1.88
v' (m/s)	2.50	-2.50	2.72	-2.41
w' (m/s)	?	?	0.02	-0.2
θ' (K)	0.90	-0.90	0.83	-0.79
q (g/kg)	?	?	0.98	-0.63
Typical fluctuations; Cloud layer :				
u' (m/s)	4.0	-3.0	2.52	-1.98
v' (m/s)	3.5	-4.0	2.61	-2.69
w' (m/s)	?	?	0.16	-0.16
θ' (K)	0.60	-1.00	0.73	-0.88
q' (g/kg)	0.50	-0.50	1.65	-1.14

Table 4.2: Comparison between model-computed and measured fluxes during conditions of cellular convection

These descriptive remarks on the dynamics and thermodynamics of the mesoscale fields in the presence of organized convection give some idea of the complexity of the processes and interactions involved with these phenomena. No experimental data would be capable of producing such an insight into the three dimensional fields. Uncertainties in the realism of the model simulations are to some extent reduced by the comparisons which are possible with available data from the KonTur experiment.

Though in many respects, the model simulations differ from the actual KonTur meteorological situations, typical magnitudes for wind and temperature fluctuations near the surface and in the subcloud layer predicted by the model are within the uncertainty limits of the KonTur data. Table 4.2 gives an idea of the variability of dynamic and thermodynamic terms in the model in comparison to measurements or estimates provided by the KonTur experiment, for representative cases of cellular convection (see also Bakan, 1982; Brummer et al., 1985).

4.3.2.4 Influence of surface instability on cellular convection

As a further test of the physical realism of the present model, it was thought to be of interest to alter the lower boundary temperature conditions in order to analyse what changes would occur in the cloud activity. In experiment CR6, the surface instability was cancelled by neutralizing the surface layer temperature gradient. It was thought that if the same initialization as in CR1 were applied to this simulation, and that identical long-lasting open cells were also created despite a change in the surface instability, then some doubt could be cast as to the importance of the physics in the model with respect to the numerics. The preceeding discussions have focussed upon preferred cell size modes and mesoscale fields in the presence of cells; by changing the lower boundary condition, it is interest to analyze whether significant differences will result from the modified physical processes.

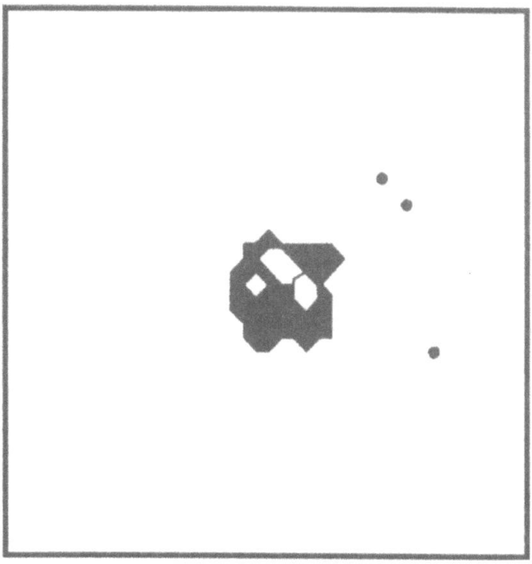

Figure 4.17: Cloud patterns after 12 hours of simulation in Experiment CR6

Figure 4.17 gives a plan view of the cloud field obtained after 12 hours of integration, initializing with a single cell perturbation as in CR1, and with no air-sea temperature difference. The contrast between cloud fields simulated by experiments CR1 (Figure 4.16A) and CR6 (Figure 4.18) is quite obvious. Banded and cellular structures have not been generated in CR6, but the entire central area of the model domain has filled inwards and outwards with a shallow cloud field, about 1500 m in vertical extent. The cloud cluster is characterized by slow ascent from the surface to cloud top (values of mesoscale w do not exceed 6 cm s^{-1}, compared to 20 cm s^{-1} or more in the case of open cell convection), and subsidence at the boundaries of the feature.

In many respects, this cloud cluster exhibits the characteristics typical of a closed cell: a larger aspect ratio than an open cell of similar diameter; dynamic characteristics typical of closed cell; and as a consequence of the weaker secondary circulation, the clouds do not interact as strongly with their mesoscale environment as open cells do. This is illustrated in Figure 4.18 by the plan view of the temperature perturbation at 1050 m, whose structure is far simpler than the perturbations observed during the open cell and cloud band cases. In this figure, the temperature fluctuations reflect the weak nature of cloud-mesoscale interactions; maximum positive fluctuations occur within the cloud field and are a result mainly of condensation effects. The concentric ring bounding the region of positive temperature fluctuations extends some 50 km outward from the cloud cluster, with fluctuations outside of the cloud field lower than one quarter of a degree. The rest of the model domain is characterized by minor values of negative temperature fluctuations; in other words, the cloud cluster is so inactive that its effects to not even reach the boundaries of the model.

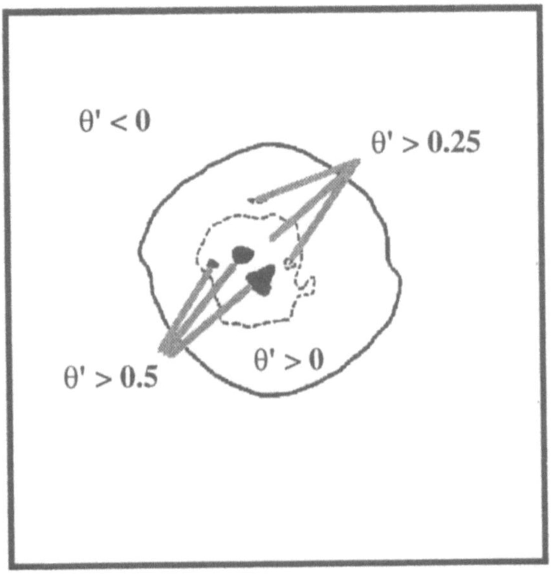

Figure 4.18: Horizontal cross-section of temperature fluctuations at cloud base for Experiment CR6

The obvious differences in the structure shown in Figure 4.18 and those of the open cells illustrated in Figures 4.16 are directly linked to the influence of the air-sea temperature difference on the convection, which is the only parameter which was changed between simulations CR1 and CR6. As descending branches of the cloud circulation encounter the rigid ocean surface, they begin to diverge in order to satisfy principles of continuity. If a substantial heating occurs in the surface layer, then this will add sufficient buoyant energy to the diverging air to force it to rise once again and thus result in a field of clouds. In the absence of surface forcing, only the cloud field itself, through hydrostatic pressure modifications, can force a circulation in the immediate vicinity of the cloud ensemble (i.e., the clouds act to "pull" air upwards; with surface heating, this effect is enhanced to give much more vigorous circulations). Diverging air, which flows outside of the region of direct cloud influence does not receive the impulse from the air-sea heat differential which would force this air to rise and form new clouds elsewhere. The surface instability is therefore seen as a feature which enables the propagation of multiple circulations outwards from the initial disturbance. The absence of surface instability cuts off much energetic input into the model atmosphere, and thus the shallow, weakly interactive cloud cluster is the only significant feature to be observed after 12 hours. The fact that the initial perturbation ring has been filled in with clouds, and that the clouds have spread outwards from the central region is due to grid-scale advection processes and subgrid-scale horizontal transports operating at this level.

Experiment CR6 was undertaken as a test of the response of the mesoscale model to a change in a significant physical parameter. The present simulation indicates that the model is indeed responding in an intuitively reasonable manner, by not simply reproducing the cellular structures as in CR1 but by yielding an entirely different cloud regime. Such experimentation is in fact one of the principal objectives of numerical models, and make these a powerful tool for investigating a wide variety of atmospheric phenomena in the absence of adequate in situ observations or well-documented meteorological situations.

4.3.3 Mesoscale response to interactions between subgrid-scale clouds and infrared radiation

4.3.3.1 Importance of cloud-radiation feedbacks

Much attention has been paid to the representation of convective heat and moisture sources in limited-area grid-point models in order to take into account the effects of a cloud population on the large-scale thermodynamic fields. Quite often, however, radiative processes are rather crudely treated, due in part to the impossibility of solving the radiative transfer equation at each grid-point of a numerical model. In many models, longwave radiative cooling is included as a constant cooling rate at all grid-points. This simple representation of radiation obviously cannot distinguish between cloudy regions and clear air regions, thereby misrepresenting radiative forcing in the presence of clouds. Clouds themselves are generally destabilized by strong infrared cloud-top cooling which is not locally balanced by shortwave heating. Consequently, the interaction studies of cloud and radiation with the atmospheric thermodynamic fields are important at various scales.

The cloud-capped atmospheric boundary layer has been studied experimentally and numerically for varying degrees of complexity. Chen and Cotton (1983) have used a one-dimensional second-order boundary layer model which includes a detailed radiation scheme, and they were able to reproduce details of Deardorff's (1980) three-dimensional modeling results. Schaller and Kraus (1981) and Fravalo *et al.* (1981) have demonstrated the sensitivity of the cloud-capped boundary layer to radiative properties of clouds.

Relatively little work has been undertaken so far concerning the radiative influence on mesoscale processes. Veyre et al. (1982) have included an adequate longwave radiation scheme into the Deardorff-Sommeria (1976) three-dimensional planetary boundary-layer model and studied the effect of local radiative forcing on the development of small cumulus clouds.

Probably the most dominant feature of cloud-radiation interaction is the strong cloud-top cooling. This cooling can lead to dissipation of a cloud by enhanced turbulent mixing of liquid water at cloud top. which erodes the cloud from its upper boundary downward. Alternatively, when the cooling due to radiative processes significantly perturbs the dynamic fields, through temperature-induced pressure fluctuations, cloud activity can be increased due to greater local thermal instability. It is consequently clear that neglect of a radiative scheme which interacts with a cloud field will lead to an unrealistic representation of cloud activity in mesoscale or small-scale models. Furthermore, poor representation of these processes may lead to an overall over- or underestimation of cloud forcing on the large-scale fields, thereby potentially resulting in erroneous dynamic and thermodynamic structures throughout a model domain.

4.3.3.2 Mesoscale simulations of interactions between clouds and infrared radiation

Veyre et al. (1982) have outlined the outstanding questions concerning radiation interactions with atmospheric processes. Their study, in which a radiation scheme was introduced into a detailed microscale planetary boundary-layer model, addressed the manner in which cloud-radiation interactions modify the convective activity and cloud cover and turbulence. These include:

- Ways in which cloud-radiation interactions modify the convective activity and cloud cover and turbulence;
- Relative importance of clear air and cloud radiative effects in cumulus cloud development;
- Relative magnitudes of the heat sources in an atmosphere which includes moist convective activity; the heat sources include subgrid-scale fluxes, convective sources and sinks, and radiative flux divergence.

Although the investigation presented in this section (cf. Beniston and Schmetz, 1985) is undertaken for much larger time and space scales than in the work of Veyre et al., the three principal questions above remain valid. In order to study the impact of a comprehensive parametric model of radiation, it is necessary to study the relative changes in various terms. These include thermodynamic quantities, essentially temperature, humidity, rainwater, representative dynamic terms in order to determine whether a broad interaction exists between clouds, radiation, and mesoscale flow; and finally cloud morphology such as cloud cover and average vertical extent of the clouds. Because the mesoscale model is detailed enough to include processes such as mesoscale advection, subgrid-scale turbulence, surface forcing, cloud condensation and evaporation, and precipitation mechanisms, there exist complex non-linear interactions between all these processes. As a consequence, the study of the model response to radiative forcing embedded within the other forcing terms is by no means a straight-forward matter. For this reason, a control experiment has been set up in order to compare a case in which only a uniform cooling rate of 1 K/day is specified at each grid point, and simulations where the radiation scheme has been included. The control run is based on previous numerical studies of organized convection in extratropical regions, as discussed by Beniston (1983, 1984). Initial data for the mesoscale model are based on the same case-study period of the KonTur experiment (Hoeber, 1982), already described in Sections 4.3.1 and 4.3.2.

In addition to the control run (hereafter denoted by CR), two simulations which include the radiation scheme were made. The two runs differ by the distribution of radiative cooling; in the first case, maximum cloud-top cooling is allowed (denoted by R1) whereas in the second case, this cooling is smoothed vertically (denoted by R2). It was thought initially that the vertical grid separation in the model (maximum $\Delta z = 500$ m) was too coarse to adequately represent the interactions between radiation and the cloud and mesoscale fields. Consequently, it appeared justified to apply a simple filter to the cloud top cooling, thus spreading this cooling to neighboring vertical grid points. This in some way implies a cloud-top height distribution which is not predicted by the convective scheme. This latter parameterization computes only a single value of cloud ensemble-top height within each cloudy column. It should be noted here that in the discussion of results, a vertical profile refers to the horizontal average at each level of a particular variable.

4.3.3.3 Influence on thermodynamics

Figure 4.19 shows the vertical profile of potential temperature in the first 5000 m of the model atmosphere, for the initial time (i.e., the observed Kon Tur profile) and after 12 hr of simulation for the control run and both runs which include the radiation model. The initial profile for temperature is characterised by a strong surface instability, with an air-sea temperature difference of almost 4 K, capped by a quasi-neutral mixed layer. Above the mixed layer, whose vertical extent is roughly 500 m, stable stratification stretches through to the tropopause.

After 12 hr, it can be seen that convective processes have eroded the stability of the layer between 1200 m and 3500-4000 m. This erosion is primarily driven by the strong cooling by cloudy air as it is detrained into the environment. When examining R1 and R2 runs, it is seen that the deepening of the convective layer is a result of the combined evaporation cooling and the strong cloud-top infrared cooling. There is no obvious dissimilarity between R1 and R2 in this diagram, essentially because this is an instantaneous picture at 12 hr, at which time the cloud tops in both cases are approximately equal. As this has been the case after 10 hr of simulation (as will be seen further on), the various heat sources have been able to operate in much the same way, leading to the very similar profiles for R1 and R2 . The powerful cloud-top cooling results in a strong inversion, in all three cases, which efficiently suppresses further growth of the convective layer.

From the surface through cloud base and up to 1200 m, little change of temperature from the initial state is observed. In the lowermost layers, the surface turbulent heat fluxes are compensated by radiative cooling. It may appear surprising that a 4 K temperature difference between the surface and the top of the Prandtl layer has not generated strong upward heat fluxes which would overcome radiative cooling effects. This is mainly a result of the slowing down of the wind near the surface by friction effects, which has reduced the friction velocity. Because this u_* term is used to compute the buoyancy flux at the surface, a small value of u_* even in the presence of a large θ_* value will lead to relatively small buoyancy fluxes. This effect has implications for the moisture fluxes as well: the reduction of u_* will lead to a diminishing of the moisture supply required to maintain the cloud system's activity. As a result, clouds will tend to dissipate in time. However, cloud dissipation is not a problem in the present 12-hr simulations, for reasons to be discussed further on in the text.

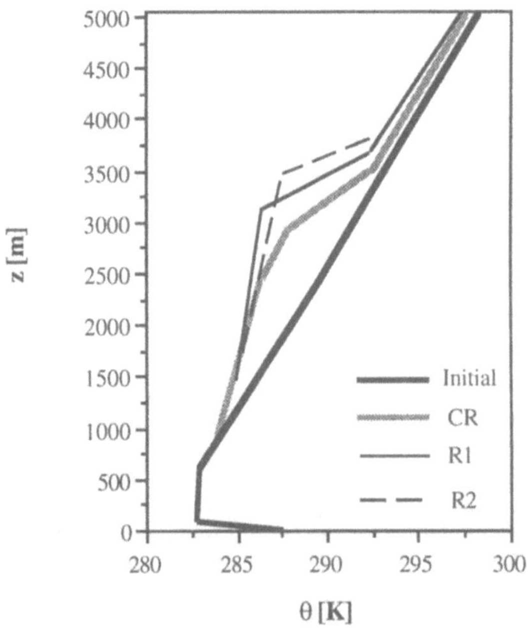

Figure 4.19: Vertical profiles of potential temperature at the initial time and after 12 hr of simulation for the control run and the two simulations which include the radiation scheme. These profiles are horizontally averaged at each level

Near and above cloud base, condensation warming and in-cloud transports largely dominate radiative cooling/warming, but the rather strong grid-scale vertical velocities (approximately 10 cm/s) transport lower potential temperature upward, thus balancing convective heating (combined condensation warming and subsidence adiabatic compression).

Although a mean profile shows little change of temperature with time at lower levels, considerable perturbations can occur from one grid point to another. This is illustrated in Figure 4.20, which gives vertical profiles of potential temperature variance, $\overline{\theta'^2}$, at 12 hr of simulation. The variances shown here are in fact a gross measure of the grid-resolved fluctuations over a grid mesh at each level at a given time. This allows some estimate of where regions of peak activity are located. The lowermost layers in all three cases are affected by surface buoyancy fluxes. A relative maximum occurs around 1500 m, which is where convective flux divergences play a dominant role; this maximum reflects the transport of heat from the condensation level higher up into the cloud layer.

The decrease in $\overline{\theta'^2}$ above 1500 m occurs because at intermediate levels of the cloud layer, a relative constancy with height of convective fluxes is observed between regions of low-level condensation heating and upper-level evaporational cooling.

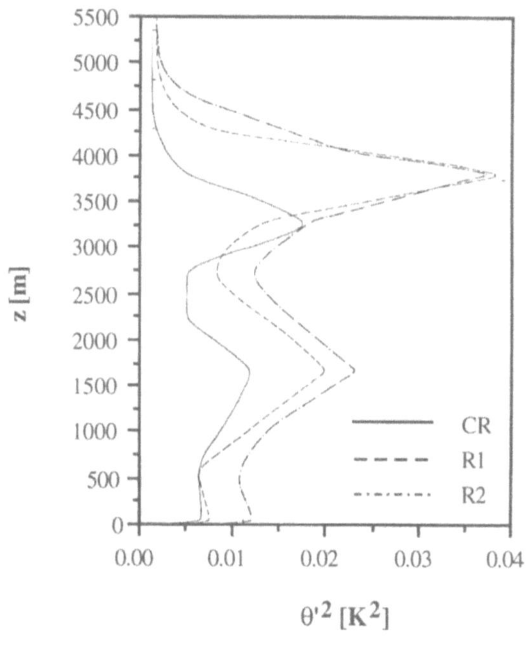

Figure 4.20: Vertical profile of potential temperature variance at 12 hr, for all three simulations

In the upper regions, the most significant signal in the $\overline{\theta'^2}$ profiles is due to the strong evaporational cooling in the case of CR, which is enhanced by the cloud-top radiation cooling in R1 and R2. The separation of levels of maximum variance between the CR and the R1, R2 cases demonstrates the effectiveness of the radiative destabilization in the development of the cloud layer. At all levels, the R1 and R2 profiles are greater than the CR profile, indicating the stronger nature of the circulations forced by the interactive convective-radiative processes.

The differences are most noteworthly in the cloud layers; in the mixed layer, $\overline{\theta'^2}$ profiles for CR and R1 are almost identical, which can perhaps be explained by the fact that at this stage of the simulation, cloud-layer and subcloud-layer processes are only weakly coupled. In the R2 case, however, even mixed-layer temperature variance is stronger, which means that locally, cloud influence on the dynamic fields is enhancing buoyancy fluxes. The reason for the larger $\overline{\theta'^2}$ in R2 is that the vertical smoothing of the infrared cooling at cloud top implies a deeper destabilization of the cloud layer, which in turn creates a stronger secondary circulation. This modifies the dynamic nature of mesoscale processes even at the lowermost levels. Humidity profiles are given in Figure 4.21. The initial profile is characterized by a strong specific humidity gradient within the surface boundary layer and then an almost monotonically-decreasing profile with height. At 12 hr of simulation, the humidity profiles have increased with respect to the initial time at all levels up to cloud top, for CR, R1, and R2.

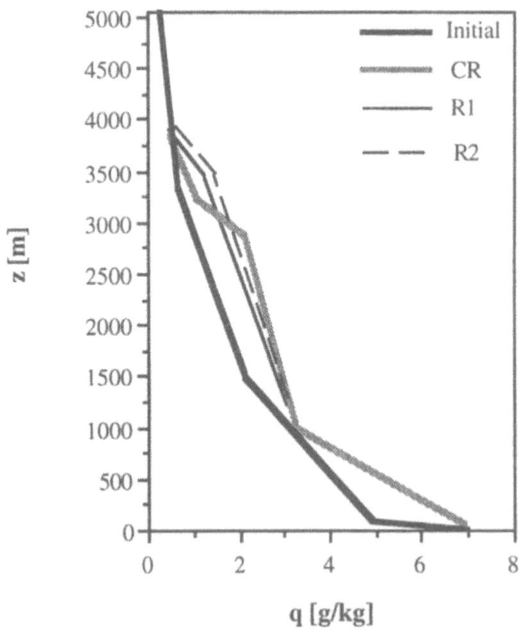

Figure 4.21: Specific humidity profile with height, initially and after 12 hr, for runs CR, R1, and R2

In the mixed layer, strong moisture fluxes have increased q by 20% on average. It is for this reason that cloud activity is still relatively strong after 12 hr; moisture availability in the lower cloud layer is high due to the very rapid transfer of moisture in the early hours of the simulation from the surface to the top of the constant flux layer. However, low humidity gradients coupled with reduced u_* values now mean that once the present moisture supply in the subcloud layer has been used up by the clouds, little further humidity will be forthcoming to maintain cloud activity. It has been observed for prolonged model simulations that the clouds do indeed dissipate due to this cloud layer/subcloud layer decoupling. In the lower cloud layer, the change in moisture is at a minimum, mainly because cloud production is a sink of ambient humidity at these levels. Higher up, detrainment in the upper cloud levels and at cloud top have considerably moistened the atmosphere as cloud droplets evaporate into the environment. There is little difference between CR and R1, R2 except near cloud top, where once again the higher clouds in R1, R2 are detraining more moisture to the environment between 3000 and 4000 m than in CR. The specific humidity variances corresponding to 12 hr are given in Figure 4.22. The strongest variances in humidity are seen to occur near cloud base, where the maximum drying of the atmosphere by condensation processes takes place.

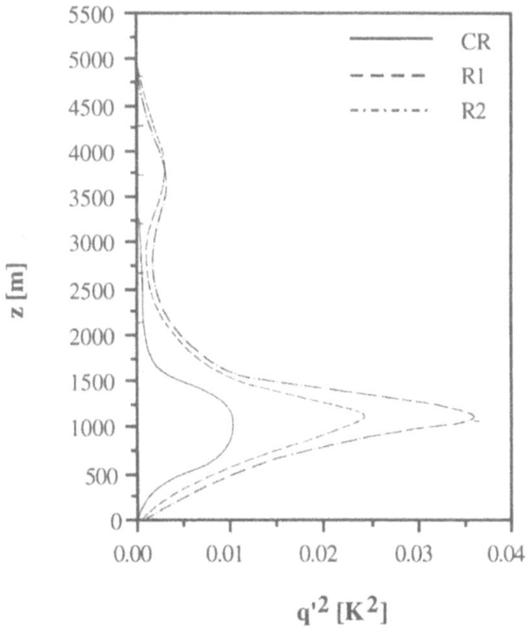

Figure 4.22: Specific humidity variance with height, initially and after 12 hr, for runs CR, R1, and R2

Additionally, subgrid-scale moisture fluxes are also felt at this level, and contribute somewhat to the q'^2 profiles. Another peak at cloud top represents the contribution of cloud top detrainment to the moistening of the atmosphere at that level. Near cloud base, R2 variances are greater than R1 variances, which in turn are significantly larger than those of CR. This is a reflection on the influence of a progressively stronger circulation induced by cloud activity on the moisture budget. Time series of rainwater specific humidity, q_R, are shown in Figure 4.23. The series shown are for the 550 m level, which is below cloud base, and consequently rainwater production processes are no longer operative here. The principal terms which act on q_R at this level are the grid and subgrid-scale transports, as well as the evaporation and fallout sinks of q_R. In this case, values of rainwater humidity rapidly grow in the first 3 hr of simulation before decreasing and reaching a quasi-steady situation. This leveling off of the q_R profile represents an equilibrium between the production of rainwater in the clouds, which is limited by cloud size, rate of condensation, and convective vertical velocities, and the sinks of rain water as the rain falls out from its saturated environment. In a similar manner, R1 and R2 time series of rain water increase rapidly until two hours, before decreasing as in CR, but in a far more oscillatory manner. This can be explained by the more intense nature of cloud activity as a result of the intricate feedbacks between radiation, cloud and mesoscale flow. Until about 8 hr of simulation, the values of rain water in R1 and R2 are about 20% higher than in CR, but this difference diminishes after 12 hr. One reason may be that after some time, grid-scale and subgrid-scale transports begin acting as strong sinks of q_R at this level, thus bringing q_R down to values comparable to CR despite the more important cloud activity in R1 and R2.

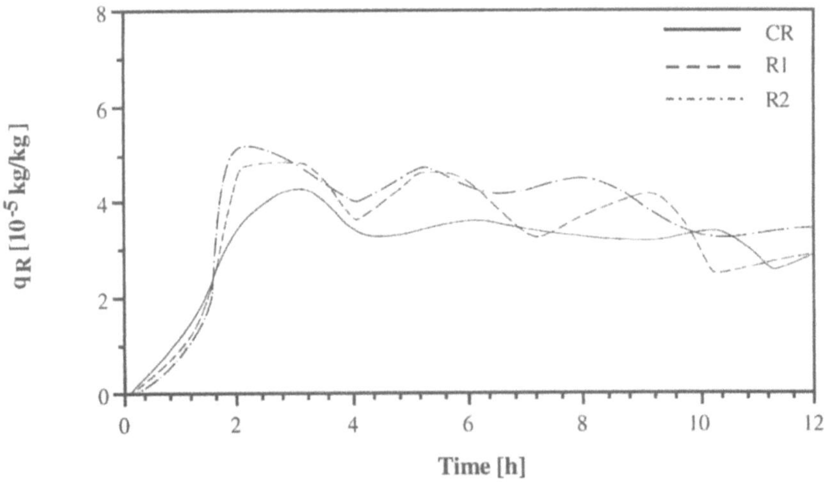

Figure 4.23: Time evolution of rainwater specific humidity, from 0-12 hr, for runs CR, R1, and R2

4.3.3.4 Influence on dynamics

Although direct coupling between convective phenomena and the dynamic fields in the mesoscale model are not explicitly considered, indirect effects are possible through disturbances in the hydrostatic pressure field as a result of convective warming or cooling. It is of interest, therefore, to see whether the radiative forcing is as strongly felt by the mesoscale dynamic fields as by the thermodynamic fields that have been discussed up to now. One variable which is interesting to examine in this context is the average mass flux generated by the model. The mass flux here is defined as $|\rho w|$, with w the grid-point vertical velocity. Hence mass flux is a measure of all the dynamic forcings in the model, from horizontal convergence or divergence of the mesoscale flow, to secondary flow effects such as cloud-induced subsidence. While it is of course not possible to separate the various contributors to the mass flux profile, it nevertheless gives an overall idea of how mesoscale dynamic processes respond to different forcings. Figure 4.24 shows time series of mass flux at model levels just above cloud base and just above cloud tops (1050 and 3550 m respectively). The mass flux at cloud base reaches a nearly steady level after 3 hr. This is true for the CR and R1, R2 runs, although mass fluxes are slightly higher in the latter cases. This is a result of the stronger secondary circulation induced by the bigger clouds in R1 and R2. Beyond 10 hr, profiles are seen to rise sharply, perhaps reflecting stronger cloud-environment interactions. However, these 'jumps' are short-lived and new steady-state levels are reached after about 13 hr of simulation.

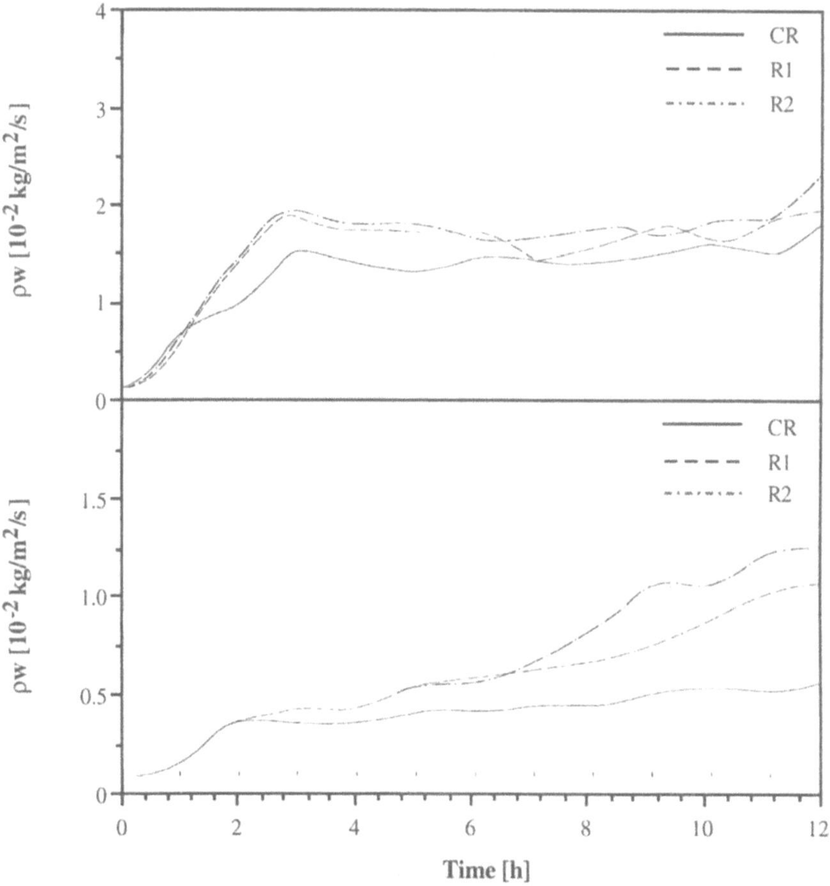

Figure 4.24: Time evolution of mass flux from 0-12 hr for all three simulations. Upper: Mass flux at 1050 m (first model level above cloud base). Lower: Mass flux at 3550 m (upper cloud regions)

Mass fluxes near cloud tops (Figure 4.25, lower) exhibit significantly different development for the three simulations. The increased buoyancy due to the infrared cloud cooling more than doubles the mass fluxes in R1, R2 with respect to CR. In addition, the smoothed radiation scheme shows an even larger mass flux than R1 because of the artificially introduced cooling at grid levels above cloud tops. Most of the mass flux at this level can be accounted for by subsidence motion, which is generated when air cooled at cloud top creates a pressure fluctuation such that a horizontal divergence of air will take place. It should be noted that mass fluxes at 3550 m are roughly 50% smaller than those at 1050 m. At the lower cloud levels, grid-scale vertical velocities are fairly high due to low-level convergence and contributions to buoyancy from the surface and condensation near cloud base. At or near cloud tops, however, upward vertical velocities are slowed down by the stability of the atmosphere, while subsiding motions are likely to be much slower, hence resulting in the much smaller mass fluxes. Above cloud tops, mesoscale mass fluxes decrease steadily through to the top of the model domain.

4.3.3.5 Influence on cloud morphology

As mentioned at the beginning of this section, it is also of interest to ascertain the extent to which radiation forcing modifies cloud activity itself. It has become apparent from the discussion that thermodynamic and dynamic fields have been significantly modified by the introduction of an explicit infrared scheme. However, one needs to see precisely what has changed in the cloud morphology. With this objective in mind, two representative parameters for cloud activity were chosen: the number of grid points where clouds occur, and the variation of cloud-top heights with time.

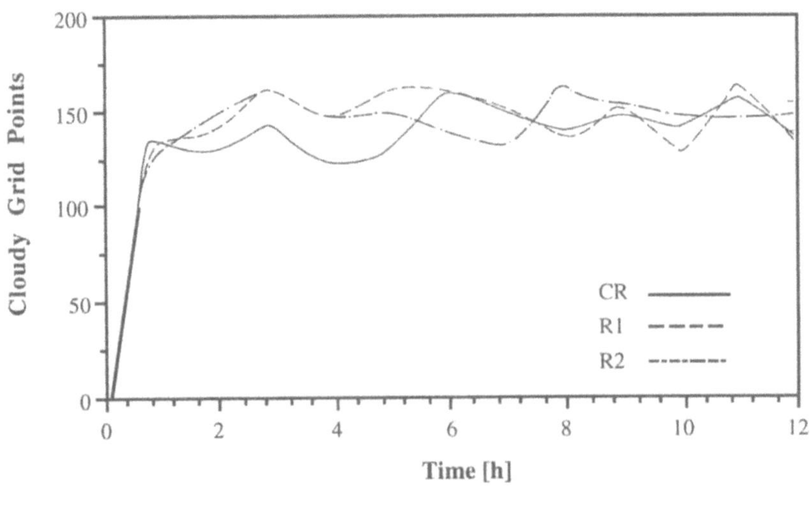

Figure 4.25: Time evolution of the number of cloudy grid-points in the DREAMS model

A DREAMS model grid point where a cloud appears represents an average value of cloud properties for the grid volume enclosing that particular grid point. Fractional cloud cover would be a more useful parameter than the number of cloudy points because it can be compared with observations. However, there is at present no way of estimating this quantity. The cloudy grid points are illustrated in Figure 4.25, where it is seen that a large number are covered by cloud in the first hour, and that afterward, the number of clouds oscillates within 10% of an average cloud number of 155 (out of a total of 400 grid points) for all three runs. The differences between CR and R1, R2 are small and, especially after 6 hours, become insignificant. It seems that the strength of convective-mesoscale interaction is not a determinant factor in the number of cloud regions which will be generated. This is perhaps a feature of the organization of convection, in particular in the case of mesoscale open cells, which probably limits the numbers of clouds which can occur in a particular model domain.

On the other hand, Figure 4.26 illustrates the direct influence of radiation on cloud development. The time series of average cloud top heights shown here supports the conclusions which have been described for dynamic and thermodynamic interactions within the model. The radiation cooling, by rapidly creating an unstable region near cloud tops, allows vertical development of cloud to occur more efficiently than in CR. After 12 hr, cloud-top heights in CR have reached only 3400 m, while in R1 and R2, they are close to 3800 m. The additional height of these clouds accounts for much of the convective influence on the mesoscale fields. Because the cloud-top radiation filtering in R2 cools the atmosphere even above the mean cloud-top level, vertical cloud development is faster than in R1, and near maximum values of cloud-top height are attained at least 3 hr before R1. During these three hours, the added strength of cloud-environment interactions in R2 have enabled even more profound modifications of the mesoscale environment than in R1, as was seen from the profiles of $\overline{\theta'^2}$, $\overline{q'^2}$, and the time series of mass fluxes.

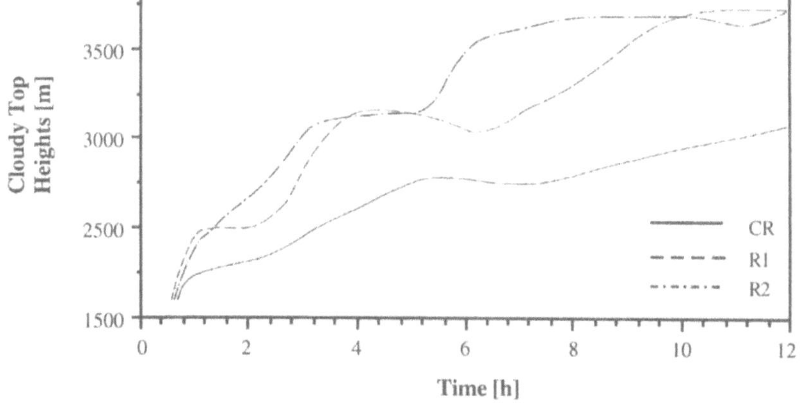

Figure 4.26: Time evolution of cloud-top heights in the DREAMS model

The relatively steady cloud-top levels after 7 hr in R2, and after 10 hr in R1, are due to the strength of the inversion induced by the cloud-radiation cooling in upper cloud layers. In order to erode this capping inversion, and consequently to allow further cloud growth, much energy would be required; it is doubtful that in the absence of new forcing such as strong surface buoyancy fluxes, the present cloud activity would be capable of increasing the depth of the cloud layer. The fact that cloud growth exhibits a step-by-step progression is a result of the model grid, which has a vertical extent of 500 m; only crude estimates can be made of cloud-top height through interpolation, so that this is reflected in the time series in Figure 4.26. Truncation error due to the coarse vertical grid could be one possible contributor to the increase in cloud-top heights, though layer destabilization by radiative processes is certainly playing the dominant role here.

4.3.3.6 Influence on heat sources

Figure 4.27 summarizes many of the points which have been made throughout this section. The figure presents the different contributions to the heating rate in the R1 run. These are namely the radiative flux divergence, boundary-layer turbulent transports, convective heat sources, and the total heating rate which combines the three parameterized sources. Clearly the most dominant features are the convection, especially in the lower layers, and the very characteristic peak of cloud-top infrared cooling. This is considerably greater than the cloud-top evaporational cooling, but of the same sign, so that the combined effects yield a large total cooling. This combination at cloud top has profound implications for mesoscale processes, and consequently cannot be neglected in a mesoscale model.

Figure 4.27 can be compared with a similar profile set given by Veyre et al. (1982). It is interesting to note that absolute magnitudes of the dry static energy are not very much larger in the present study than they were in the Veyre et al. study, which was for boundary-layer cumulus clouds. It is true that the clouds simulated by the DREAMS model are intermediate between shallow cumuli and deep convective clouds, which explains why source terms are not so different. However, one fundamental difference between the present study and the Veyre et al. (1982) study is that the cumuli in the mesoscale model have their summits well above the top of the planetary boundary layer, which is not the case in the microscale model study. This means that the turbulent heat transports were much more interactive in the latter study than in our study, where subgrid flux contributions to the heat sources vanish above 1500 m, and thus cannot balance in some way the heat sinks near cloud tops. In the case of the Veyre et al. (1982) study, turbulent fluxes near cloud tops (ca. 1200 m) were sufficiently strong so that the total heat source was considerably reduced. In the present study, the lack of this additional term means that only grid-scale vertical advection can be counted upon to counteract the combined convective-radiative cloud-top cooling.

Such a study is therefore quite significant if the present model is to be used as a prognostic tool for mesoscale convective events. It has been demonstrated that, firstly, the treatment of parameterized radiation significantly modifies the dynamics and the thermodynamics of the model with respect to a case without an explicit parametric model; and secondly, that smoothing or filtering of cooling rates as was undertaken here can also have strong and perhaps adverse effects on mesoscale processes. This conclusion is supported by Stearns (1983), who has suggested that the growth or dissipation of clouds is highly sensitive to the selection of grid points where infrared cooling is computed.

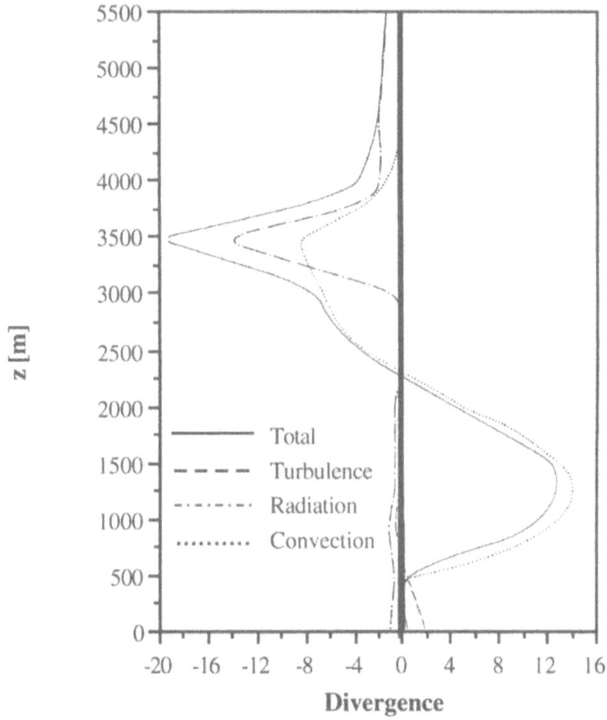

Figure 4.27: Contribution to the tendency equation for potential temperature of various heat sources, in simulation R1. Heat sources illustrated above include: turbulent subgrid-scale transports in the PBL; convective heat sources due to cloud activity; infrared cooling/warming; total heat source, combining all the individual sources

4.3.4 Mesoscale response to interactions between subgrid-scale clouds and combined infrared and solar radiation

4.3.4.1 Introduction

The use of a solar and infrared radiative transfer scheme in a numerical model allows one to study the influence of radiation on atmospheric dynamic and thermodynamic fields in a fully interactive way, including non-linear feedback mechanisms (Schmetz and Beniston, 1986). The present work is an extension of Section 4.3.3, itself a summary of work by Beniston and Schmetz (1985), hereafter referred to as BS. Two simulations, starting with the same initial conditions, have been undertaken. In the first, infrared cooling is the only radiative forcing term in the model, whereas in the second model run, both infrared and solar radiative terms are included. These two simulations will be refered to as IR and IR + SO, respectively. In this manner, the effects of solar radiation on mesoscale dynamic and thermodynamic fields can be isolated.

It was concluded in the BS study that cloud development is more vigorous than when non-interactive infrared cooling is applied to the mesoscale model. In order to save computer time, the radiation subroutines are called only each 10 min, and the net radiative cooling/warming is applied at each point over the 10 min interval; this means that during the lifetime of cumulus clouds, the radiation scheme is called several times. This is an improvement in the time resolution over the BS study, in which the infrared cooling scheme was called only each 30 min.

Model initial conditions are as in BS and the other studies discussed in this chapter. Within the first 20 minutes clouds are seen to form, and after some 2 hours of simulation, clouds have generated significant secondary circulations. From this time onward, a highly interactive regime is in operation, combining cloud, radiation, turbulent transports and mesoscale processes. In order to determine the effects of solar radiation on mesoscale fields, the radiative scheme was run with a constant solar zenith angle of 60 .

4.3.4.2 Model results

The results presented in this section pertain to a time 12 hours after starting the integration. The most significant influence of solar radiation in the model is that the solar part of the radiative parameterization will to some extent counteract the strong infrared cooling effects. This can already be seen in Figure 4.28, which illustrates the horizontally averaged vertical profiles of potential temperature at the initial time and after 12 hr of simulation for the experiments with infrared cooling only, and with combined infrared and solar contributions.

Both profiles are characterised by what was already demonstrated in BS, namely low-level warming through buoyancy fluxes in the mixed layer (50-650 m), condensation effects and upward transport of heat within the clouds (650-2000 m), and higher-level cooling by combined radiative and cloud evaporational effects (up to 3500 m). Above this level, a strong capping inversion has developed which tends to limit further upward cloud development.

Radiative processes play an insignificant role below cloud base, where warming is dominated by surface buoyancy fluxes. The differences in profiles between IR and IR + SO become apparent above cloud base: at all levels above 900 m, the θ profile for IR + SO is warmer than IR, by as much as 0.7 K around cloud tops. Above the cloud layer, IR + SO and IR profiles are both parallel to the initial temperature gradient, the effects of IR having uniformly cooled some 0.8 K in 12 hr, as opposed to 0.3-0.4 K in the same lapse of time for IR + SO. The implication of the profiles shown in Figure 4.28 is that upper cloud-layer destabilization in IR + SO is not as enhanced as in IR. If this is the case, then it must be expected that cloud-induced activity on the mesoscale fields will not be as marked as in IR where cloud activity is more vigorous due to stronger cloud-layer destabilization.

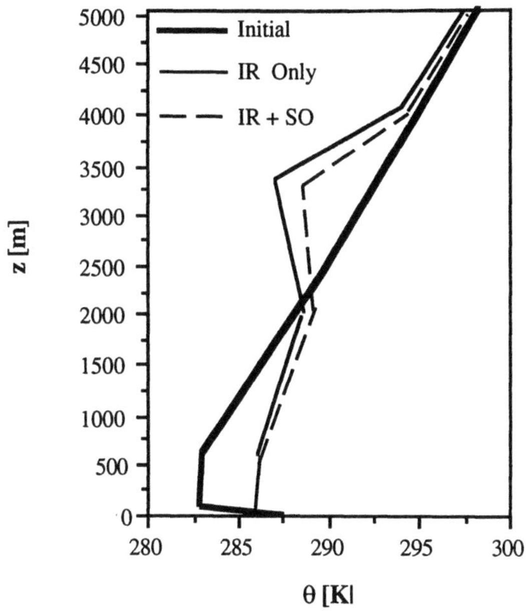

Figure 4.28: Comparisons of vertical temperature profiles for the initial time and after 12 hours for the IR and IR+SO simulations

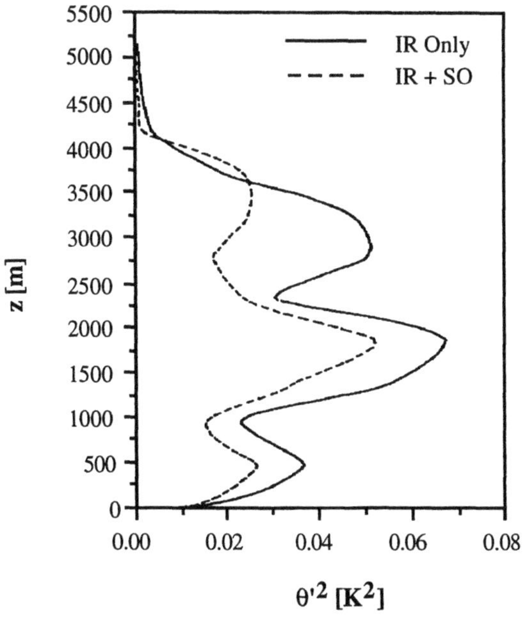

Figure 4.29: Comparisons of horizontally-averaged spatial variance profile of θ for the IR and IR+SO simulations

An analysis of the horizontal variance of θ with height, given in Figure 4.29, shows that this conclusion is indeed true. Though both profiles of θ'^2 with height show remarkably similar tendencies, the IR + SO profile has lower variance than IR. This latter profile shows three distinct peaks, characterizing mixed-layer buoyancy and cloud-base condensation effects, middle-cloud-layer heat transfer, and upper-cloud-layer cooling. Because solar radiation to some degree offsets infrared cooling, the less active nature of cloud activity is reflected in the lower variances in IR + SO. One characteristic of solar radiation is to contribute to warming at and above cloud tops; this is reflected in the rather flat peak of variance which stretches from 3000 to 4200 m, and represents not only the contributions from cloud-top cooling but also from solar warming above cloud tops. In the vicinity of cloud tops, the variance in IR + SO is only about 50 % of the variance in IR, indicating that the reduction in cloud-radiation interactive processes by solar warming is not a negligible process.

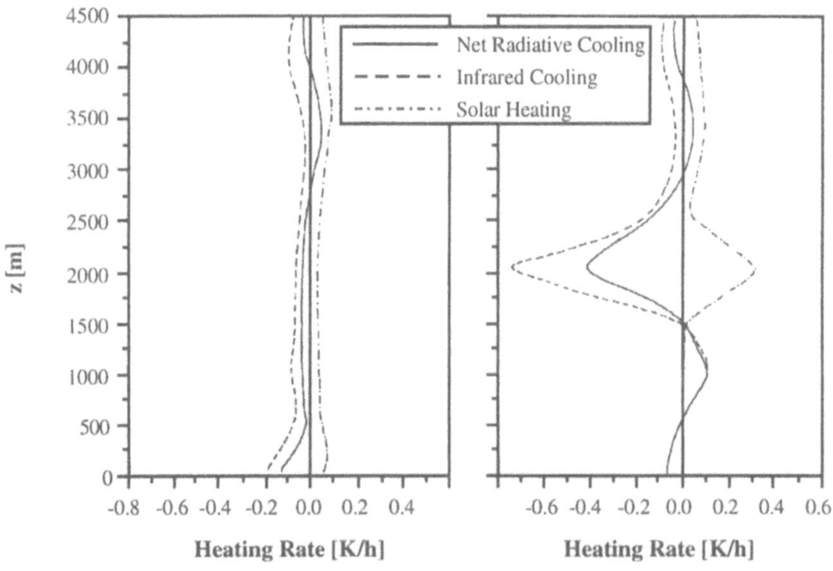

for representative clear-air (left) and cloudy (right) columns in the mesoscale model

The preceding discussion pertains only to horizontally-averaged quantities at each level of the model domain. It is of value to examine the processes taking place in given atmospheric columns. Two locations were selected in the model, representative of a cloud-free column of air and of a column containing cloud. Figure 4.30 shows the relative contributions of infrared cooling, solar warming, and the resultant net radiative heating rate for these two columns. In the clear-air case, profiles are seen to be relatively independent of height, with infrared cooling only slightly greater than solar warming except in the layer above cloud top where specific humidity jumps have a small influence on solar heating, and a small net radiative warming occurs between 3000 and 3500 m. In the cloudy column, however, it is seen that the sharp negative peak of infrared cloud-top cooling is counterbalanced up to 40% by cloud-top solar warming, thus reducing the magnitude of the net radiative cooling rate, which is the term contributing to the tendency equation for θ. Above cloud tops, a small net warming is observed as in the clear-air case, and below cloud top within the cloud layer, solar radiation is very rapidly suppressed due to the high liquid water content of the clouds. Within a cloudy column, therefore, solar heating only operates most significantly at and above cloud tops. Within an optically thin layer, solar heating could increase the long wave infrared heating at cloud base.

Figure 4.30 illustrates not only the importance of solar radiation as a modulator of net radiative cooling processes, but also indicates that it is essentially in cloudy regions that the radiative scheme plays an important role. This implies firstly that in models dealing with dry convective processes only, a constant cooling rate would be a reasonable approximation; and secondly, that cloud-radiation interactions cannot be neglected even on the mesoscale - a conclusion which was previously reached in BS.

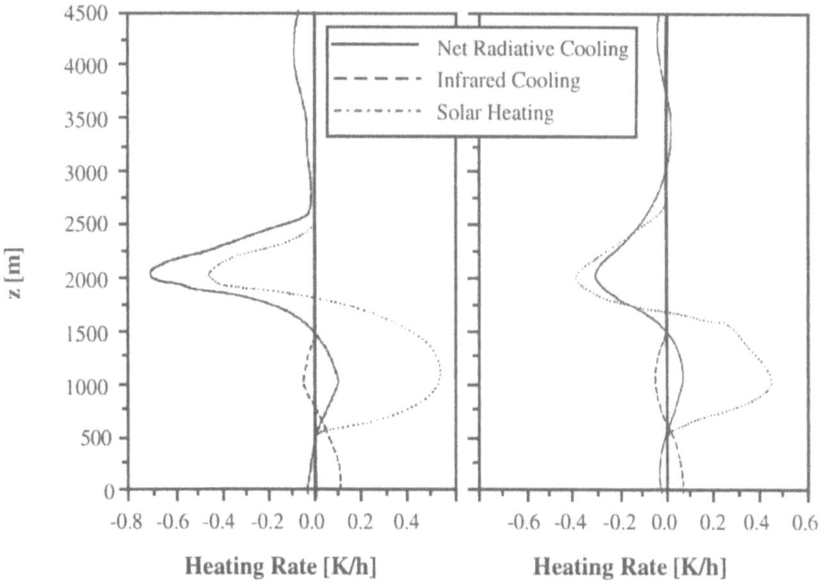

Figure 4.31: Parameterized fluxes (radiative, convective, and turbulent) for simulations containing only the infrared scheme (left) and combined infrared and solar scheme (right) for the same cloudy column as in Figure 4.30

Furthermore, while it is clear that solar radiation is a vital parameter in a model containing a solid surface, where albedo differentials resulting from cloud shading generate inhomogeneous buoyancy flux distributions, the solar part is also of importance in the moist convective layer. In the parallel simulation containing only infrared cooling, the principal feature to be seen is the strong cloud-top cooling, approximately 0.1 K/h stronger than the infrared cooling peak observed in

Figure 4.30. This stronger cooling in IR is due to the fact that, as convective activity is more important than in IR + SO, cloud liquid water content is also higher and thereby enhances the infrared cooling rate.

A further notable difference between simulations IR and IR + SO is highlighted in Figure 4.31, which shows the relative magnitudes of the parameterized heat sources, resulting from radiative, convective, and boundary-layer turbulent fluxes. While the turbulent and convective fluxes have essentially the same profiles and magnitudes in both simulations, the radiative cooling terms differ at and above cloud tops, for reasons already explained. The significant feature here is that in the IR simulation, the radiative cooling is stronger than the cloud-top evaporational cooling, leading to a very strong total cooling at this level. With the compensatory effects of solar radiation, the net radiative cooling is somewhat lower than the convective cooling, so that radiation is not the dominant term at cloud top. Though the combined cooling is still strong, around 0.85 K/h, it is nevertheless substantially lower than the 1.25 K/h cooling observed in the IR case. When such differential forcing is applied to a highly interactive and non-linear system as takes place in the mesoscale model, it is to be expected that differences in the details of mesoscale processes will occur.

As in BS, an analysis of time series of representative dynamic and thermodynamic terms yields valuable information on differences between IR and IR + SO. The time evolution of rainwater specific humidity from 0 - 12 hr in both simulations has been analyzed. Rainwater amounts are observed to be on average twice as large in IR as they are in IR + SO, indicating the more active nature of the clouds in the IR simulation. Though average cloud tops are of the same order of magnitude in both runs, the liquid water contents are significantly higher in IR because of the location of the strong capping inversion. This is a result of the stronger cloud-mesoscale interaction in IR, generated by the much higher cloud-top cooling and hence the more vigorous secondary circulations.

Such conclusions are also reflected in the dynamic terms, one particular example being the average mesoscale mass flux at 1050 m (i.e., the first model level above cloud base). After about two hours of model time , mass flux time series in IR and IR + SO begin to diverge, those in the latter simulation being somewhat lower than in the former. This is again indicative of the reduced cloud-mesoscale activity in the model due to the lower instability at cloud-layer top because of solar warming effects, as illustrated in Figure 4.32.

Similar comparisons and conclusions can be extended to other dynamic or thermodynamic parameters, such as humidity or wind terms. However, it is not the purpose of the present sub-section to enter into such detailed analyses of interactive radiative effects on cloud and mesoscale fields; many of the conclusions reached in BS are valid here also for the IR + SO simulation. The principal conclusion which this section seeks to demonstrate is that solar radiation partly compensates for the intense cloud-top infrared cooling, thus reducing the strength of cloud secondary circulations and consequently diminishing the influence of the parameterized source terms on the mesoscale dynamic and thermodynamic processes. The present study has also shown that even in a model with a relatively coarse grid, the radiative effects in IR + SO cannot be neglected when a cloudy atmosphere is considered.

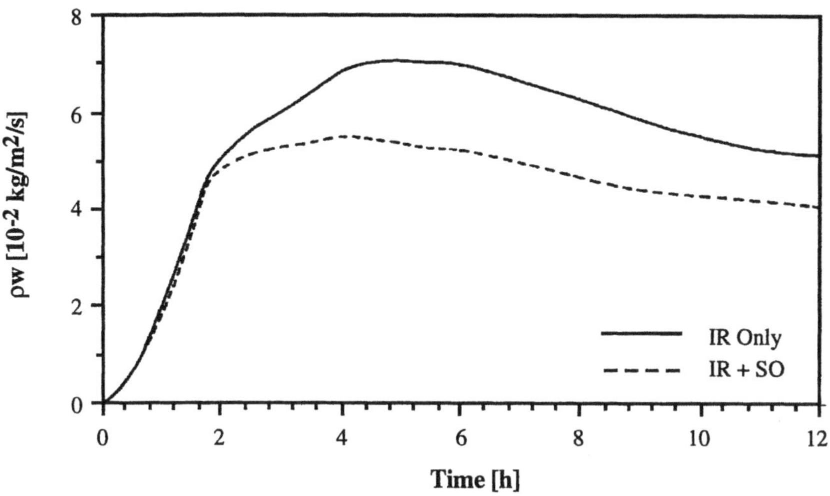

Figure 4.32: Time evolution of mass flux in the IR and IR + SO numerical experiments

4.3.4.3 Effects of a fractional cloud cover parameterization

A shortcoming of the above study is the assumption that a cloudy grid-point contains 100% cloud over the surrounding grid area (here 10 x 10 km). The type of cumuli simulated by the model is more typical of shallow medium-scale convective cloud and it is unlikely that a single cloud would cover an entire grid area. In order to simulate in a more realistic manner cloud-radiation interactions, radiative terms could be weighted with the relative magnitude of cloudy and clear air, respectively.

The prediction and diagnosis of fractional cloud amount in dynamic models is still a largely unsolved problem. A simple diagnostic cloud-cover scheme developed by Albrecht (1981) for trade-cumulus amounts has been chosen to overcome the problem to some extent. The principal assumption of Albrecht's parameterization is that cloud cover is a function of the time needed for inactive cloud material to detrain into the cloud-free environment. This assumption is based on the numerous observations concerning the active and inactive regions of cumuli; while the former may occupy under 5 % of a given region, the inactive cloud cover may be up to an order of magnitude greater. Albrecht attempts to quantify the decay time of cloud matter by estimating the rates of production of active updraft and rates of production of decaying clouds. The cloud cover N is therefore given in terms of relative humidity near the top of the cloud layer, and the initial liquid water content of the decaying elements. This leads to a formulation of the form:

$$N = \frac{\tau'}{\tau_0} \frac{(SR - 1)}{(SR - RH)} \qquad 4.39$$

where SR is the supersaturation ratio and RH is the layer-average relative humidity. τ' is a time constant for the decay process of the cloud and τ_0 is a time scale for adjustment of the cloud layer to convective processes. As these two time scales should be roughly the same, the ratio τ'/τ_0 is taken as unity. Despite the fact that the scheme was initially designed for trade-wind cumulus cloud amounts, this parameterization is nevertheless adopted here for the mesoscale model sensitivity study. The radiative cooling/heating terms are then weighted with the diagnosed cloud cover.

In the simulation, the domain-average cloud cover settled down to about 7% after an initial spin-up time of less than 2 hours. This value should not be confused with typical grid-average values of cloud cover which can exceed 35% of a grid area (i.e., 35 km^2 out of a possible 100 km^2). The results presented in the following paragraphs pertain to a simulation containing infrared and solar radiation parameterizations, along with the cloud-cover scheme. Comparisons will be made with the same run omitting the scheme, so that the outstanding differences when applying the fractional cloud parameterization can be stressed.

As a first example, Figure 4.33 illustrates the response of the mesoscale model horizontally-averaged potential temperature profile to physical processes weighted by cloud fraction and unweighted (hereafter refered to as CS-cloud cover scheme - and NC - no cloud-cover scheme). The initial temperature profile is also given in order to visualize how this has changed in the 12 hours of simulation.

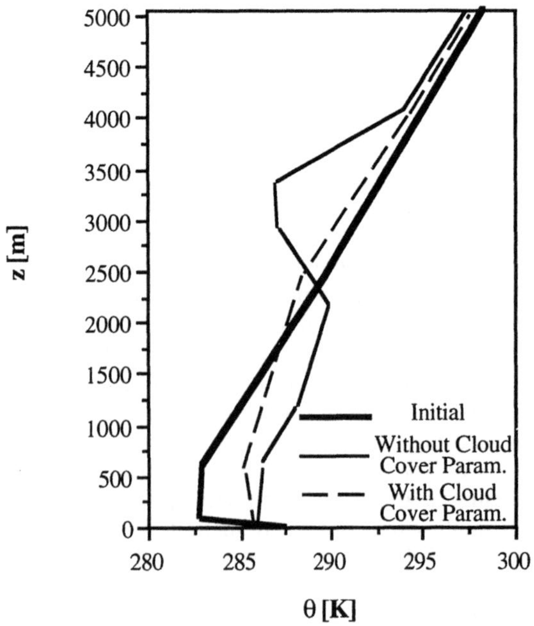

Figure 4.33: Comparison between the horizontally averaged potential temperature profiles in simulations with and without cloud-cover scheme. Initial profile is given to indicate changes which have occurred in the 12 hours of simulation

The temperature profile at 12 hr for NC is characterized by strong warming below 3400 m, due to subcloud-layer buoyancy transfers and cloud-layer condensation and transports, and cooling between 2400 and 4200 m resulting from detrainment evaporation and cloud-layer top radiative cooling. The CS profile is considerably different after 12 hr, the magnitudes of the warming and cooling being nowhere as great as those in NC except in the lowermost 500 m. In this layer, the dominant terms are buoyancy fluxes from the surface and turbulent transfer above the surface layer, which explains why the differences between NC and CS are slight. However, within the cloud layer, lower-layer warming and upper layer cooling in CS are small compared to NC.

Figure 4.34 illustrates time series of rainwater, q_R at 550 m. The NC series was already seen in Figure 4.31; the first appearance of rainwater occurs after 2 hours, with a maximum peak occurring around 5 hours before dropping to an equilibrium level after 10 hours (representative of a temporary balance between moisture availability, cloud liquid water production, and rainwater production).

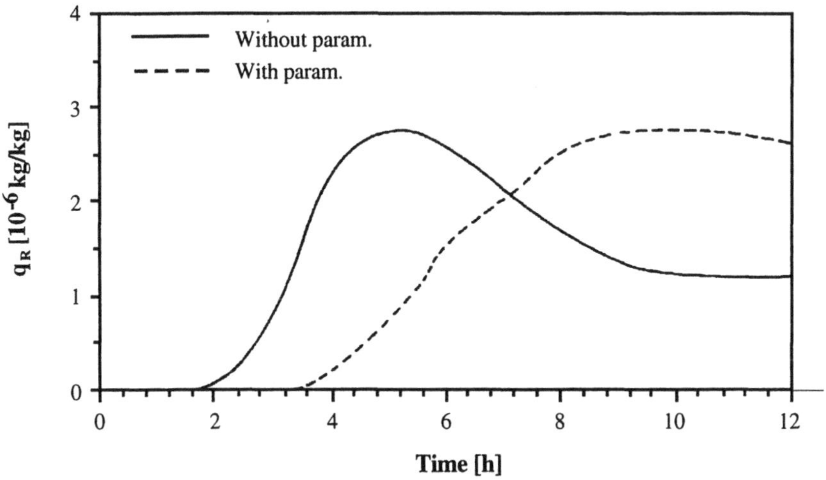

Figure 4.34: Changes in rainwater specific humidity at 550 m height as a function of time in cases with and without cloud-cover scheme

Significant amounts of rainwater occur in CS only after 4 hr of simulation, and rise steadily over the next 5 hours to reach a magnitude comparable to the maximum peak in NC. Rainwater amounts begin to decrease very slowly only after 10 hours. It is interesting to note the type of solutions a non-linear model can yield: rainwater amounts are not less in CS compared to NC, but reach similar values, and indeed after 7 hours exceed rainwater amounts in NC by a factor of 2. It is likely that the much slower development of convection leads to more moisture availabilily for condensation processes in CS, thus yielding reasonably high liquid water contents and consequently a higher supply of liquid water for the autoconversion process. It is probable that the observed slow decrease will continue in time and that an equilibrium level will be reached as in NC, but after a much longer time span.

4.3.5 Influence of a water surface on mesoscale cloud formation and induced circulations

4.3.5.1 Nature of the problem

Surface heterogeneities of momentum, heat and moisture inevitably feed into the atmospheric system through perturbations to the dynamic and thermodynamic fields. It has been shown in Sections 4.3.2 - 4.3.4 that clouds, as a subgrid-scale phenomenon in a mesoscale model (cf. definition of the subgrid-scale in Chapter 1), may have considerable influences on initially unperturbed mesoscale circulations. In this last example of a study with the DREAMS model, it will be shown how a surface inhomogeneity of moisture, namely a body of water surrounded by a land surface may in some instances generate a field of clouds, which then begin to interact with their mesoscale environment as illustrated in the previous sections of this chapter. The purpose of this investigation is to illustrate the manner in which a grid-scale feature may force subgrid phenomena, which then feed back into the resolved scales after some time. As in the sections pertaining to Rayleigh-Bénard type convection, the key process on the subgrid-scale which will be discussed here are clouds and their turbulent characteristics in a mesoscale atmosphere.

4.3.5.2 Model boundary and initial conditions

In the simulations reported here (Beniston, 1986), radiation effects are taken into account through a constant cooling rate of 1.5 K/day at all grid-points. The more complete paramaterization as given in Beniston and Schmetz (1985) was not considered necessary for the purposes of the present simulations.

Boundary conditions at the surface are constant in time at values which are specified initially. The present experiments were undertaken for a hypothetical lake occupying 4 % of the total surface area of the domain of integration (i.e., 8 x 8 km out of a total domain of 40 x 40 km). The lake is identified by a change in surface roughness (z_0 = 0.01 m over land to 0.0001 m over water) and surface moisture (values of q at the surface are 6 g/kg for the land surface and 8 g/kg for the water surface).

Upper boundary conditions are also invariant in time; this can be considered as a lid at the model top. Such a simple condition is justified when perturbations from meteorological activity at lower levels are damped with height, which is the case here. Lateral boundary conditions in the zonal direction make use of open boundaries with a Perkey and Kreizberg (1976) condition to avoid internal wave reflection as a perturbation reaches an outflow boundary. In order to identify only the moisture contributions to dynamic perturbations, temperature remains uniform over the entire surface. This implies that there is no land-water temperature difference as an additional complicating factor in the physical processes.

A control experiment was undertaken with no lake present; the 8 x 8 km area in the domain center retains only the surface roughness change in order to perturb the dynamic fields. The purpose of the control experiment, hereafter denoted by CR, is to identify how the dynamic perturbations are modified when a body of water is included, as well as for different stability conditions. CR has neutral stratification within the boundary layer from the surface to 1000 m, above which stable conditions characteristic of a normal atmosphere prevail. Four further simulations were carried out with the lake present, namely:

- R1: all conditions identical to CR but with the lake,
- R2: as Rl, but with unstable stratification in the first 250 m,
- R3: as R1, but with stable stratification in the first 250 m,
- R4: as R1, but with artificial suppression of cloud formation and activity.

In all experiments, a zonal wind of 10 m/s, constant with height from the top of the surface layer (50 m) upwards, is specified. Figure 4.35 illustrates the initial conditions of temperature for the five simulations.

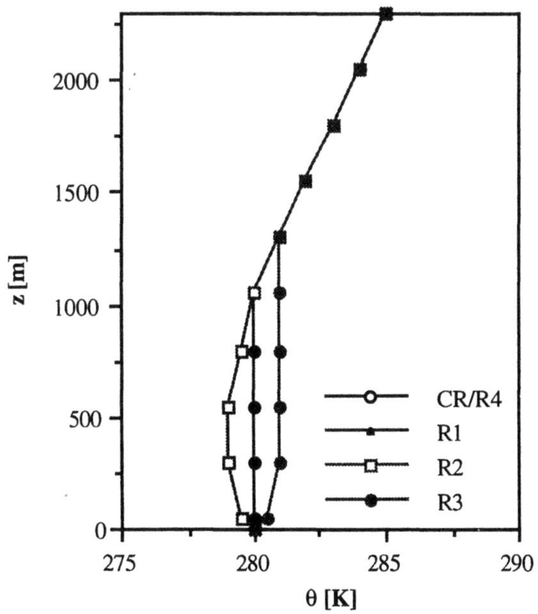

Figure 4.35: Initial profiles of potential temperature for all five experiments

4.3.5.3 Discussion of results

In order to ensure conditions of reasonable stationarity, and to have a direct comparison of results, all experiments have been conducted over a period of 6 h. By applying a different surface roughness in the 8 x 8 km 'lake' centered in the 40 x 40 km domain, the wind velocity near the ground tends to accelerate over the region of lower roughness and decelerate as the flow moves over the rougher terrain. This leads to an induction of the meridional wind component, v, through the Coriolis term, and also of the vertical velocity component, w, through divergence and convergence. A simple geometric roughness length change will, therefore, lead in time to complex three-dimensional flow patterns; furthermore, when clouds are present, these will result in cloud scale/mesoscale flow coupling, resulting in further complexity of the flow.

The foregoing discussion concerning mesoscale wind fields influenced by surface roughness change is not a new concept. Modelers and experimenters have published numerous papers in reference to this process. Early theoretical work by Estoque (1961) was followed by two-dimensional models, as well as theoretical investigations and *in situ* experiments which have underlined the importance of roughness changes at the surface. Work by Townsend (1965), Bradley (1968), Peterson (1969), and Taylor (1970) are but a few examples of research pertaining to this subject. With the advent of large computers enabling the development of three-dimensional mesoscale models, this problem has been treated extensively, in particular in Pielke's (1974) sea-breeze model.

After six hours of simulation, the solutions are independent of initial conditions and are in steady state. In such an experiment of atmospheric sensitivity to modifications in key parameters, it is necessary to examine not only the horizontally-averaged profiles of u, θ, or q, but also vertical profiles of these quantities above selected points along the flow. These were chosen along a line passing through the lake center in an east-west direction, with 'soundings' located at the lake center, and at 5, 10, and 15 km downstream of this central point. Figure 4.36 illustrates the 'soundings' above these points for the zonal wind component in all five experiments. The profiles are depicted for the first 2000 m of the atmosphere; above this level, conditions are almost unchanged since this height lies above regions where surface effects and even small cumulus effects are significant. Figure 4.36A indicates that for all experiments, wind speeds have slowed down due to surface friction. The deceleration is smallest for the CR and R4 simulations, and largest for R2 where instability-generated turbulence coupled with reverse flow induced by clouds leads to a considerable slowing down of this profile. At the 5 km sounding (i.e., 1 km downstream of the lake) (Figure 4.36B), it is seen that profiles CR and R4 are relatively unchanged whereas R2 has greatly accelerated. This paradox where the greater roughness over land apparently accelerates the flow is explained by the cloud-scale dynamics; this particular sounding is located in the region of convergent flow upstream of cloud formations. The uptake of air due to cloud activity and recirculation of air on the cloud scale largely overcomes the surface friction effects. This is illustrated schematically by the vectors in Figure 4.37.

Both the neutral (Rl) and stable (R3) cases reflect this tendency 10 km downstream of the central point (Figure 4.36C); simulations CR and R4 are still relatively independent of surface conditions (i.e., in time, the equilibrium condition is reached along this east-west section such that the velocity profile appears relatively unchanged). R2, on the other hand, exhibits a marked reduction in wind speed due essentially to the fact that this sounding is located in the segment of cloud circulation acting opposite to the mean flow. Simulations R1 and R3 conversely show that the sounding lies in accelerating air upstream of cloud activity. Finally, in Figure 4.36D, it can be seen that 15 km from the central point, the sounding is located beyond the region of direct cloud activity.

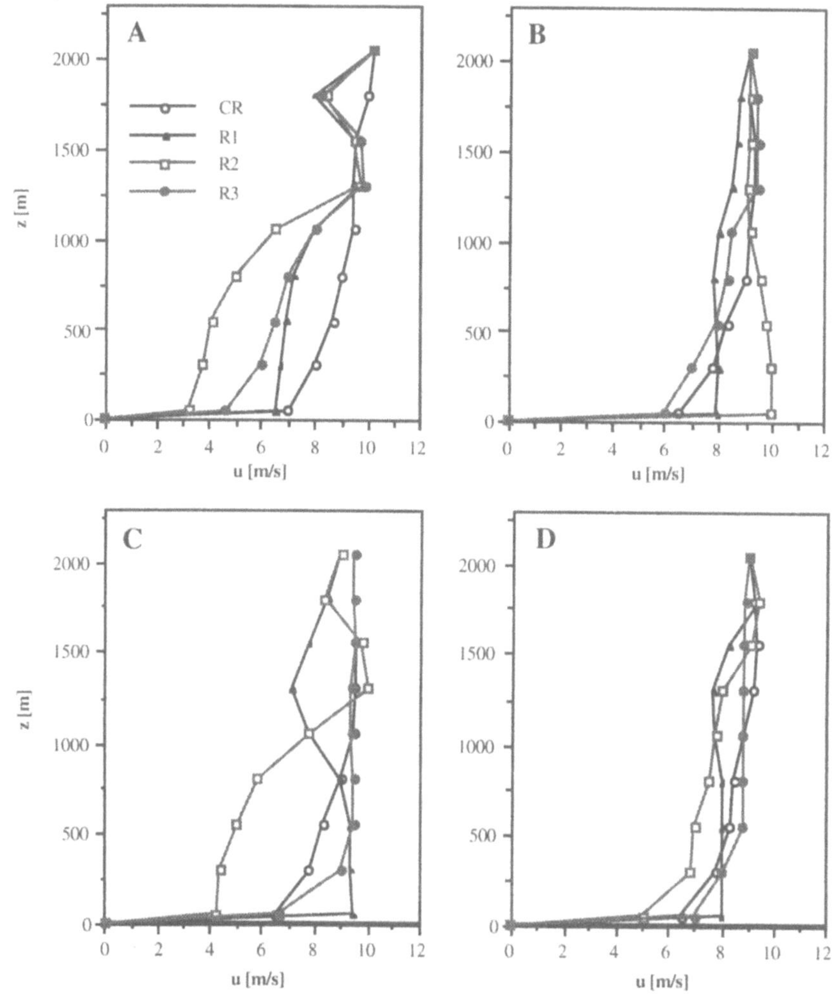

Figure 4.36: Wind profiles for all five experiments at selected points for an east-west section through the model domain center. Locations: (A) Lake center; (B) 5 km downstream of lake center (i.e., lake shore); (C) 10 km downstream; (D) 15 km downstream

Four of the simulations yield profiles which are similar within 10% at low levels, while the case with an unstable stratification at low levels (R2) has somewhat lower wind velocities; furthermore, the R2 wind profile in 4.36D has accelerated in comparison to the profile in 4.36C. This means that temperature-induced pressure gradients have overcome the surface stresses, despite the presence of strong turbulence which would normally lead to greater deceleration compared to the neutral or stable situations. In all soundings, it is seen that for the cases without lake (CR) and with lake but without clouds (R4), the profiles of u are quasi-identical. This substantiates the notion that the role of the lake is important if it is capable of initiating a field of clouds which then interacts strongly with the mesoscale atmospheric environment.

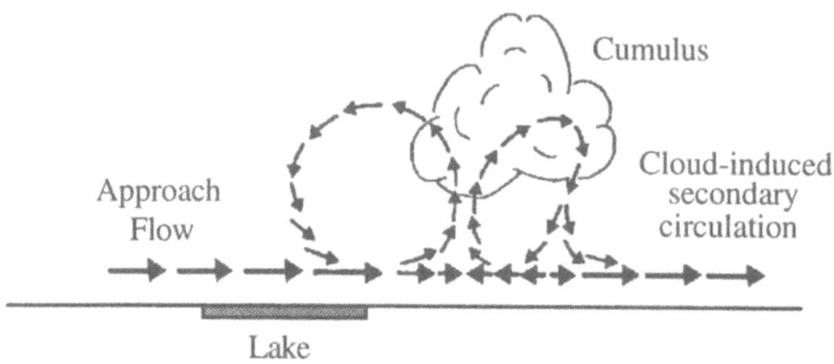

Figure 4.37: Diagram shwing flow perturbed by lake/land roughness discontinuities and cloud activity

Similar conclusions may be reached when examining the potential temperature soundings in Figures 4.38a-d. In the cases with cloud activity, the general cloudtop level is characterised by a negative peak of temperature corresponding to detrainment and evaporation of cloud droplets into the mesoscale environment; this level lies around 1500 m, giving a mean cloud-layer thickness of 900 m. Above cloud layer top, strong stability conditions are encountered, and above 2000 m little change is observed in relation to the initial conditions.

It should be noted here that cloud activity does not occur solely downwind of the lake; this is evidenced by the negative peaks in the θ profile at locations even above the lake. Such a situation is due to the presence of minor cloud activity resulting from confluence of air (i.e., the u and v components of low-level velocity upwind of, or above the lake, can result in favorable conditions for cloud formation). The schematic diagram in Figure 4.37 shows only the larger cumuli downwind of the lake for the sake of clarity, and to stress the importance of cloud-scale circulations related to these clouds.

As in the case of the dynamic fields, all significant signals from the surface conditions and cloud activity are confined within the ABL. The differences in the θ profiles from one experiment to another can reach 4 K, and reflect not only the initially different temperature conditions at each level, but also the differences in turbulent intensity. This controls both the vertical distribution of temperature through the surface flux effects, and the parameterized cloud vertical velocity which influences the magnitudes of cloud-mesoscale interactions. As can be expected, the greatest changes both in time and space are for simulation R2 where the unstable conditions lead to enhanced turbulence and cloud activity. The smallest changes are for simulation R3, where the stable conditions suppress turbulent activity and where even cloud activity is weak.

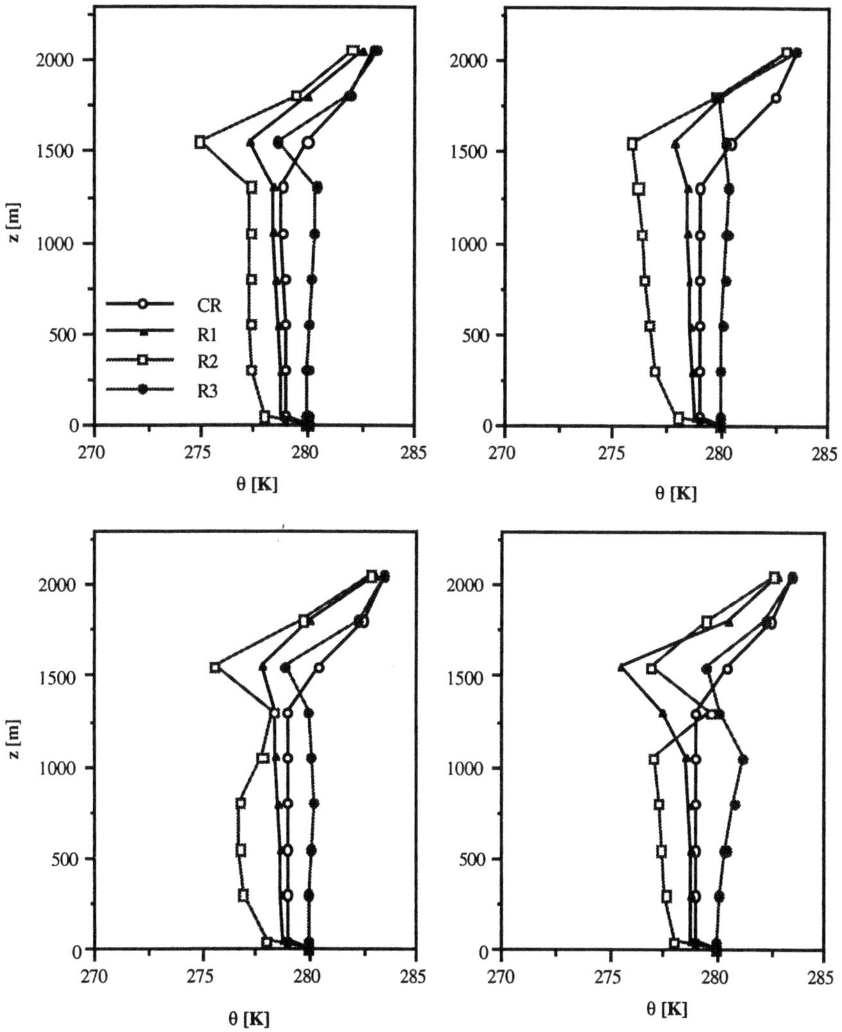

Figure 4.38: As Figure 4.36, except for potential temperature

Cases CR and R4 show identical profiles at all sounding points; again, this is to be expected since in the absence of cloud activity, specific humidity is a passive variable exerting no discernible influence on the temperature field. The combined effects of local stability and cloud activity are also important in the redistribution of humidity within the first 2000 m of the atmosphere. The characteristic cloud-top peak of moisture due to the detrainment of cloud particles back into vapor form at this level can be seen in simulations R1 - R3. Maximum humidity increases are for the unstable case R2, especially above and near the lake (Figures 4.39a and 4.39b) where turbulent transfer is most important.

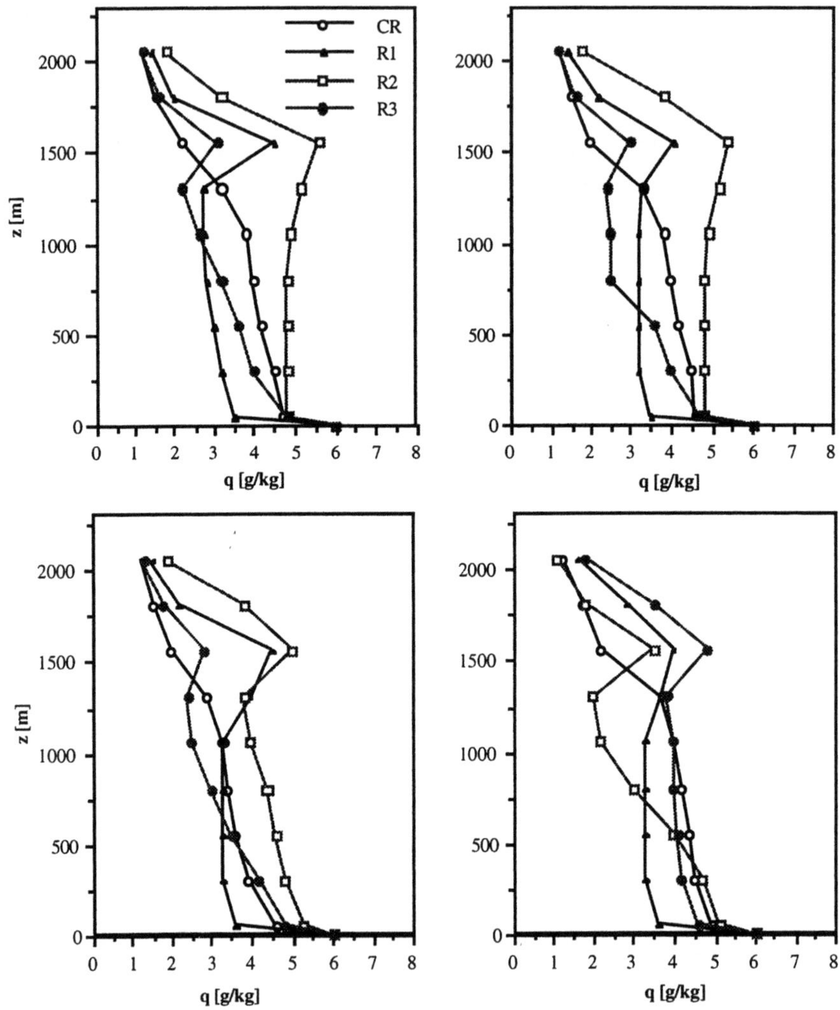

Figure 4.39: As Figure 4.36, except for specific humidity

The neutral case R1 shows a drop in moisture; this is due mainly to the uptake of low-level humidity by clouds downstream of the lake, with only a slow replacement of moisture from the lake. The stable profile R3 decreases with height to about 1300 m, unlike the Rl and R2 profiles which are quasi-independent of height through to this level (Figures 4.39a and 4.39b).

This reflects the slight changes near the surface and only moderate uptake of moisture by the weak cloud activity. Downstream of the lake (Figures 4.39c and 4.39d), profiles are controlled by cloud activity rather than by the nature of the surface, where the lake is no longer present. The cloud secondary circulations redistribute the moisture in the vertical according to uptake or re-injection of moisture between clouds and atmosphere.

In the foregoing discussion, it is clear that differences between CR and R4 are minor; this means that without cloud formations, the effect of the lake per se is minor when compared to the case of interacting cloud-mesoscale circulations. For this reason, the same symbol (open circle) is used to denote these two experiments. In Figures 4.39a - d, the open circle denotes only simulation R4, since the notable effect of the lake is to transfer moisture to low levels of the atmosphere, which is obviously not the case in CR.

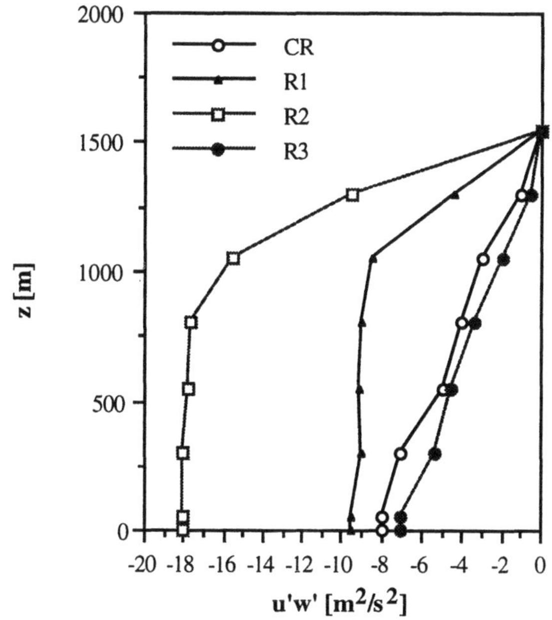

Figure 4.40: Vertical profiles of $\overline{u'w'}$

It is of interest to examine how the mean turbulent fluxes of dynamic $\overline{u'w'}$ and thermodynamic $\overline{\theta'w'}$ quantities are affected by atmospheric stability and lake-induced processes, since these influence directly the tendency terms of the prognostic equations through their vertical gradients. The fluxes are given in Figures 4.40 and 4.41, respectively. The $\overline{u'w'}$ profiles for the CR experiment (and R4 which is identical) show a monotonic decrease in absolute value from the top of the surface layer to the top of the ABL. This is a typical situation for neutral atmospheric conditions and simply illustrates the fact that downward transfer of momentum diminishes with height above the ground. The profile for R3 is parallel to CR and R4 at a lesser value, indicating the effects of turbulence suppression by local thermal stability. The R1 and R2 profiles, on the other hand, are characterised by strong fluxes in the first 800-1000 m, indicative of the more intense turbulent mixing mainly induced by cloud-scale circulations.

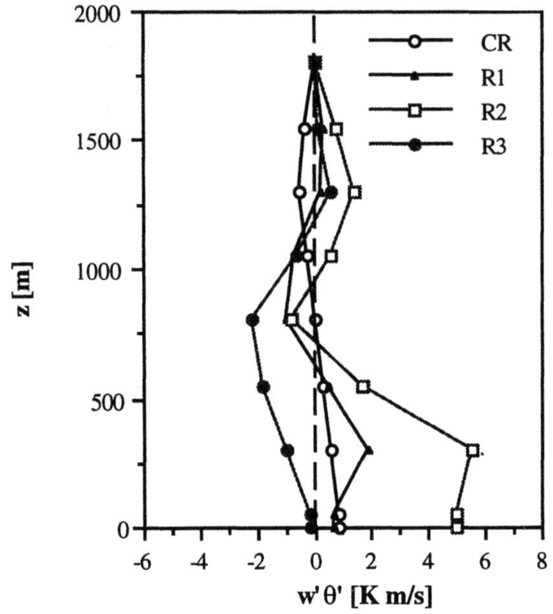

Figure 4.41: Vertical profiles of $\overline{\theta'w'}$

Similar conclusions are reached when examining the θ flux profiles in Figure 4.41. The initially neutral profiles without cloud activity (CR and R4) are characterized by slight upward fluxes through to 800 m and weak downward fluxes between 800 and 1300 m. It may at first sight appear surprising that for the cases with initially neutral stratification near the surface (simulations CR, R1, and R4), a minor positive heat flux is observed after 6 hr of simulation. This is due to the fact that the radiative cooling, fixed at 1.5 K/day has cooled the atmosphere down to the lowest levels, but not the ground which remains at the same surface temperature. Hence, a slight instability is generated at low levels, resulting in small upward heat fluxes. The effect is enhanced in simulation R1 by downward transport of air by the cloud circulation, with higher values of $\overline{\theta'w'}$; this yields the relafive peak of $\overline{\theta'w'}$ at 300 m. The initially neutral case with cloud activity (R1), and the unstable case (R3) both show a region of positive fluxes up to cloud base, negative fluxes within the lower cloud layer, and slight positive fluxes in the upper cloud layer. The irregularity of the $\overline{\theta'w'}$ profiles in R1 and R3 is directly related to the gradients of potential temperature as influenced by the cloud activity. The stable situation (R2) is seen to comprise a region of negative fluxes up to 1100 m and slight positive fluxes, again due to cloud activity.

4.4 Concluding remarks on the DREAMS mesoscale atmospheric simulations

This section has presented a number of numerical experiments to illustrate the manner in which subgrid-scale parameterized processes play a determinant role in the temporal and spatial evolution of the mesoscale atmospheric kinematic and thermodynamic fields. Clouds and the associated turbulence can profoundly modify the three-dimensional structure of the atmosphere, while being an apparently minor component in volume (and surface area) as well as an ephemeral phenomenon. It should be stressed here that the presence of cumulus activity is symptomatic of the re-stabilizing of the atmosphere in which a pre-existing instability was present; this is particularly the case of cold-air advection over a warm ocan surface, which results in the organized cloud patterns discussed in Sections 4.3.1 and 4.3.2. The absence of the unstable situation close to the surface, described in Section 4.3.2, does not result in any marked convective activity.

Enhancement or damping of cloud activity is observed when an explicit longwave and shortwave radiative flux parameterization is allowed to interact with the cloud field. Infrared cooling alone will enhance the instability at the cloud top, allowing clouds to grow higher and therefore induce stronger secondary circulations and turbulence within the ABL. The addition of shortwave solar radiation offsets to some extent the infrared cooling influences, but the general effect of a full radiation scheme is to stimulate cloud activity when compared to the case without an explicit radiation code. The examples presented in Sections 4.3.3 and 4.3.4 illustrate the fundamental influences which small-scale processes can ultimately exert on local to regional scales.

Modifications to the lower boundary conditions through the presence of a small lake, associated with different stability profiles and surface roughness lengths, have also contributed to illustrate the concept of small-scale perturbations feeding into a larger-scale atmosphere, ultimately dominating the processes therein. The "lake" introduced in Section 4.3.5 only occupies 4% of the surface area of the model, but is a sufficiently significant feature to generate moderate to strong turbulence and redistribution of momentum, heat and moisture in the entire model domain.

The different illustrative examples using the DREAMS mesoscale model point to the fact that if small-scale processes feed into the mesoscale in an important manner, then mesoscale features will ultimately feed into the largest scales of the atmosphere. Since it is difficult to simulate the actions of the very small scale on the very large scale, the interactions between scales can only be studied through the linking of a hierarchy of models. Exceptions to this statement will be given in Chapter 6, where the DREAMS model will drive small-scale pollution which does not interact directly with the mesoscale atmospheric fields, and in Chapter 7, where global atmospheric fields are used to drive a Limited Area Model in which there are no feedbacks to the climatological scales.

5. Simulations of Climate and Climate Change

5.1 The climate system and the greenhouse effect

The difference between "weather" and "climate" is often a common source of misconception and not always clearly defined. A typical definition is that given by the Webster dictionnary (1994), which states that climate is "*the average course or condition of the weather at a place usually over a period of years as exhibited by temperature, wind velocity, and precipitation*". The IPCC (1996) defines climate in its simplest term as "*the average weather*". Both these definitions are insufficient to truly differentiate between "weather" and "climate", however, because the notion of temporal scale is not sufficiently emphasized.

Climate is more accurately described as the long-term effect of the sun's radiation on the rotating earth's varied surface and atmosphere. It can be understood most easily in terms of annual or seasonal averages of temperature and precipitation. Day-by-day variations in a given area constitute the *weather*, whereas *climate* is the long-term synthesis of such variations. A simple, long-term summary of weather changes, however, is still not a true picture of climate. To obtain this requires the analysis of daily, monthly, and yearly patterns. Investigation of climate changes over geologic time is the province of *paleoclimatology*, which requires the tools and methods of geological research. Spatial scale is an additional consideration in the definition of climate, and the broad atmospheric scale separations illustrated in Figure 1.2 can be applied to macroclimates, mesoclimates, and microclimates, bearing in mind that the long-term statistics of atmospheric processes at each scale are being considered here. A description of the climate over a period of years to centuries involves the averages of various elements of weather (e.g., pressure, temperature, wind velocity, moisture and precipitation) over that period together with the statistical variations of those elements.

Fluctuations of climate occur on many scales as a result of natural processes, and this is referred to as climate variability. The definition of *climate change*, which has attracted much media and public attention over the last decade, may differ from climate variability according to the scientific or political forum it is aimed at. The IPCC (1995) refers to climatic changes as those which occur as a result of natural and anthropogenic influences; the UN Framework Convention on Climate Change (UN FCCC), on the other hand, makes reference only to the forcing of the climate system over the next century as a result of human activities. This is a much more constraining view of the issue of climatic change, and in scientific terms is probably an oversimplified definition since it is by no means trivial to disaggregate natural and anthropogenic forcings, particularly because there are a large number of counter or interactive effects between these two distinct types of forcing. The FCCC has adopted its own definition, however, because the aims of the Convention are to mitigate, by political, economic, and other means, the effects of *anthropogenically-induced* climate change. The most recent assessment report of IPCC Working Group I (1996) have concluded that "*the balance of evidence suggests that there is a discernible human influence on global climate*", suggesting that despite the numerous uncertainties associated with this statement, it may now be possible to identify the anthropogenic signal on the climate system and attribute some causes; the statement may thereby in time provide valuable scientific support to the FCCC definition of climate change.

Figure 5.1 provides a schematic illustration of the climate system and its different components. While most of what will be discussed in this section concerns variables essentially associated with the atmosphere, a full description of climate requires addressing issues related to other components of the system. The atmosphere is certainly the most rapidly reacting element of the climate system to internal and external forcing, but in terms of long-term variability, the oceans are probably the major element of the system. Oceans act as a regulator of atmospheric temperature and gas concentrations; the complex surface and deep ocean circulations transport and store large quantities of heat and greenhouse gases, through exchange mechanisms which are still incompletely understood. Recent findings (IPCC, 1992; 1995) indicate that carbon uptake by the ocean is perhaps less important than previously estimated, so that probably a vital climate control resides in the biosphere. Living organisms, particularly forests and plants, play a key role in atmospheric heat, moisture, and energy budgets close to the surface. However, knowledge of two-way interactions between the biosphere and the atmosphere are too poorly known for their effects to be adequately included in present-day climate modeling systems. Other components of the climate system include the cryosphere - continental ice caps and floating sea ice - which contributes to instabilities in the atmospheric general circulation as a result of temperature differences between the Poles and the Equator. In addition, the surface energy balance is significantly influenced by the high reflectivity, or albedo, of the ice. The seasonal fluctuations of sea ice, and its possible overall spatial reduction in the Arctic Ocean in a warmer atmosphere, could have significant repercussions on accelerated global warming. Any abrupt changes in the forcing of one or more subsets of the climate system would inevitably have impacts on the entire system.

Natural variability in the climate system includes seasonal changes readily observed in mid-latitudes, or more irregular features acting on different time scales such as the prolonged droughts which different regions of the world periodically experience (e.g., the Sahelian zone, southern Africa, California, SE Australia) or extreme flooding episodes (China, central US, etc.). Direct observations and the decrypting of various biological and geological records generically referred to as proxy data have highlighted the natural variability of climate on scales ranging from a few years to several thousand years. During the past millenium, for example, a warm period between the years 1100 and 1500 (the "Climatic Optimum") led to agricultural settlements in Greenland and bountiful harvests in Europe. Between the 16th and 19th Centuries, the tendency was reversed during the so-called "Little Ice Age", where glaciers advanced significantly in the Alps; harvests frequently failed and resulted in severe social consequences. Average global temperature changes of less than 1°C were sufficient to entail a dramatic change from one regime to another. Further back in the past, records stored in the ice caps of Antarctica and in marine sediments have brought to light a range of periodicities associated with climate change. These are linked to the so-called Milankovitch cycles, where major shifts in climate - in particular the glacial and interglacial periods - are correlated with changes in the astronomical parameters of the Earth's orbit. These changes modify the amount of solar energy - the main driving force of the climate system - which is received at the Earth's surface. These major cycles are identified to have periods of approximately 22,000, 40,000, and 100,000 years. Climate, therefore, has changed naturally in the past and will do so in the future.

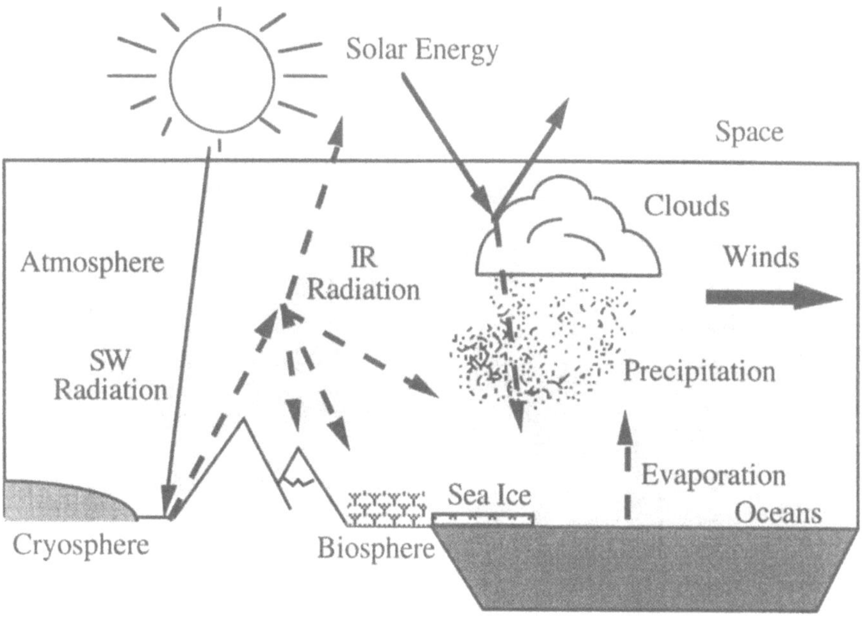

Figure 5.1: Schematic illustration of the climate system

The principal external driving force for the climate system is provided by solar energy, and Figure 5.2a illustrates the manner in which solar radiation is distributed as it enters the atmosphere; much of the atmosphere is transparent to visible wavelengths, where most of the incoming energy is located as illustrated schematically in the solar energy spectrum in the small insert to the right of the figure. Clouds reflect about 25% of the incoming energy back to space, with about 23% reaching the ground directly. Scattering by air molecules accounts for 7% of the energy returned to space and 5% transmitted towards the surface. About 17% is absorbed by the atmosphere through aerosols, clouds, and other opaque material, while scattered diffuse radiation ("sky radiation") accounts for about 21% of energy reaching the surface. Thus, on average, about half the direct solar radiation intercepted by the atmosphere reaches the ground.

The Earth re-emits much of this intercepted radiation, which takes place with a shift in the wavelengths compared to the incoming solar energy, as seen in Figure 5.2b. Peak energy is about 30% of that of solar radiation at the top of the atmosphere and takes place around the 10 mm wavelength. The Earth radiation loss therefore occurs essentially in the infrared (sensible heat) wavebands. Because of the presence of a number of absorbing gases, such as water vapor, carbon dioxide, or methane, the nature of heat loss to space is by no means trivial. If the atmosphere were totally transparent to infrared radiation, then surface temperatures would be some 35° C cooler than today. The absorption and re-emission by various trace gases maintains a far warmer temperature, about 15° C as a global average. The Earth surface is always losing energy to space, but the atmosphere can emit either to space or back to the surface according to the influence of greenhouse gases and their action on a specific part of the spectrum. On average, there is an imbalance in net radiation (the difference between incoming and outgoing radiation) at the surface polewards of about 40° latitude. In order to maintain *globally* an energy equilibrium, heat needs to be transfered from the low to the high latitudes. This mechanism is one of the fundamental driving forces of the climate system.

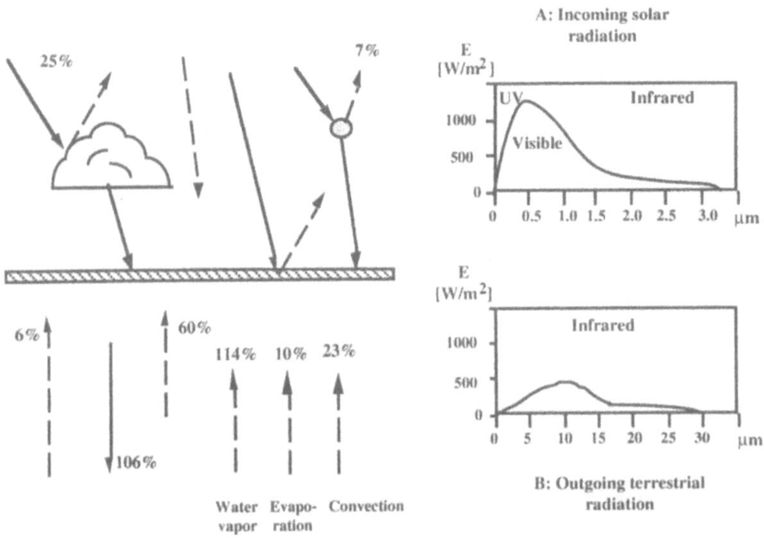

Figure 5.2: Distribution of incoming solar (A) and outgoing terrestrial (B) radiation

Figure 5.3 illustrates very approximately the typical absorption spectrum for the range of wavelengths occuring in the atmosphere, from the short-wave UV bands to the long-wave IR spectra. In the UV wavelengths, most of this energy is absorbed by ozone in the stratosphere, while the transparent visible wavelengths stand out very clearly. In the infrared, numerous peaks of absorption are to be found at various wavelengths or wave bands of the spectrum, and these are due to the radiative properties of water vapor and carbon dioxyde. Other trace gases such as methane, nitrous oxide, CFCs also generate signals in the spectrum and act to warm the atmosphere. These are the so-called greenhouse gases (GHG), and it should be noted that it is the substantial increase of these gases which is generating concern among climatologists as to their effects on climate and their potential for producing global warming of the atmosphere at an unprecedented scale.

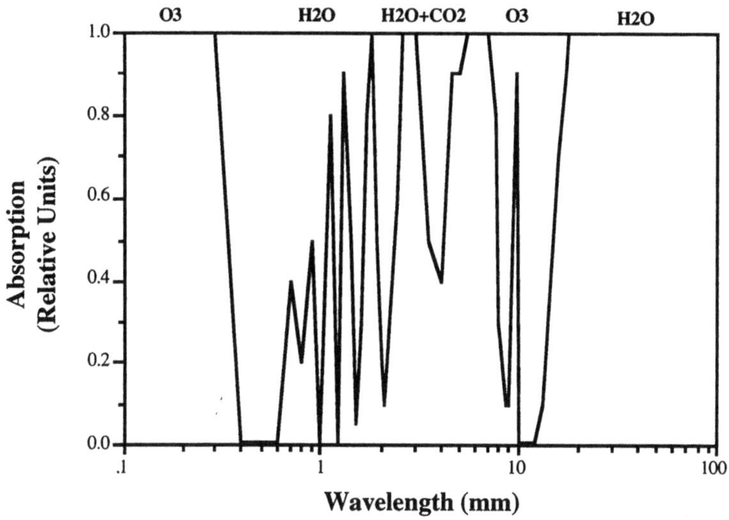

Figure 5.3: Schematic illustration of some absorption bands in the IR spectrum

The Intergovernmental Panel on Climate Change (IPCC) published its first conclusions in 1990, based essentially on mathematical models of the climate system. One of the most significant results is that, if nothing is undertaken to halt anthropogenic sources of GHGs, then the global atmosphere may warm on average by 1.5 to 4.5°C by the end of the next century. While this may appear to be a minor warming - after all, diurnal or seasonal amplitudes of temperature often exceed those figures - it should be noted that for an average over the Earth, this is a rate of warming unprecedented in the last 10,000 years. The consequences of this accelerated atmospheric warming may be catastrophic for the natural and human environments.

Figure 5.4 gives an overview of the contribution of the various trace gases of anthropogenic origin to the radiative forcing the atmosphere. CO_2 accounts for 55% of this effect, while methane, nitrous oxydes, and CFCs account for the rest. CO_2 is widely mentioned to be the principal GHG; however, molecule for molecule, it is by no means the most efficient in terms of warming potential; methane, for example, has a warming potential some 60 times greater than CO_2, nitrous oxide on the order of 250 times, and certain CFCs up to 6000 times. At present, atmospheric concentrations of all but CO_2 are still very low; however, significant increases in methane would result in greater efficiency of atmospheric warming, since methane concentrations 50 - 60 times lower than CO_2 would in fact produce the same radiative forcing. The biggest single GHG, however, is water vapor; increases in global air temperatures will allow the atmosphere to hold more water vapor, thereby further enhancing the greenhouse effect.

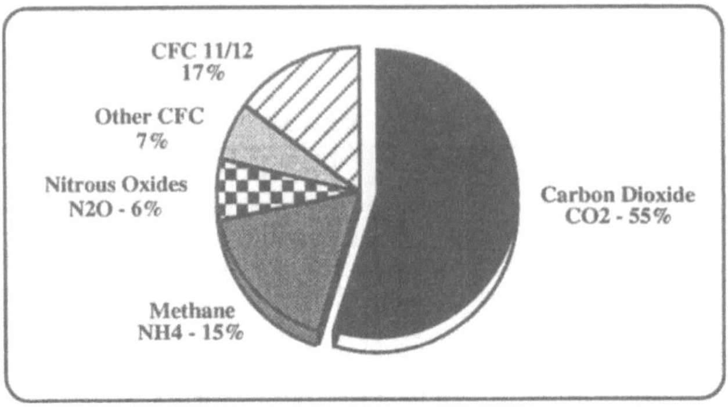

Figure 5.4: Contribution of various greenhouse gases to global warming

Atmospheric CO_2 concentrations measured at Mauna Loa, Hawaii, have increased significantly since the beginning of the industrial era, while estimated fossil fuel consumption has increased less rapidly since the major oil-price rises in 1974. Observed CO_2 increases since 1976 could be a result of accelerated deforestation, or a lower capacity of the oceans or the biosphere to absorb this CO_2 excess.

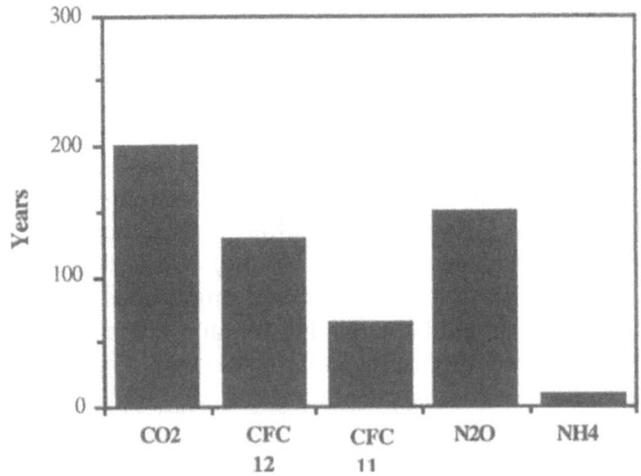

Figure 5.5: Average lifetimes of greenhouse gases in the atmosphere

Industrial activity, energy production, and transportation are the essential sources of man-made GHGs, while agriculture is a source of methane and nitrous oxides. In a world where population is rising and where pressures on food and energy supply are correspondingly increasing, the concentrations of these gases in the atmosphere will inevitably augment if no political and economic measures are taken to reverse this trend. Observed increases in major GHGs in the last 250 years, i.e., from before the industrial revolution until today, show that there is an accelerating trend which is directly related to industrialization, in particular since 1950.

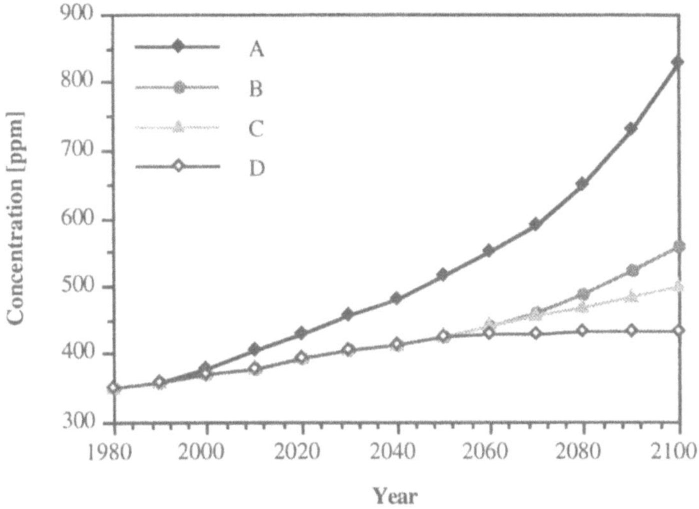

Figure 5.6: Global atmospheric CO_2 increases based on IPCC scenarios

Figure 5.5 shows the residence times of certain GHGs in the atmosphere. Gases such as CO_2 have life-times on the order of a century. Methane is relatively short-lived, on the order of a decade or so. This figure implies that even if all GHG emissions were halted today, their influence will continue to be felt for a century or more. The IPCC developed a number of scenarios in 1990, from Scenario A, "Business as Usual," in which little attempt is made to control GHG emissions, to Scenario D where energy supply shifts to low carbon fuels and renewable sources of energy, deforestation is reversed, CFCs are phased out, and stringent controls are imposed on agricultural emissions. These scenarios have since been refined (IPCC, 1992), but the conclusions still hold here.

The implications of these scenarios for global increases in carbon dioxide concentrations, and corresponding average atmospheric temperature rise are summarized graphically in Figures 5.6 and 5.7. In Figure 5.6, it is seen that only with the strict measures of Scenario D can carbon dioxide concentrations be stabilized from the middle of the 21st century, at levels some 50% higher than in 1985. In Figure 5.7, the corresponding temperature increase is seen to be on average 1.5°C globally, significantly less than the 4.5°C expected from the "Business-as-Usual" Scenario A. Even with the very constraining "Scenario D", temperatures would continue to rise throughout the next century, mainly because of the "commitment" of current emissions to future climate change, but also because the stabilization of *emissions* of GHGs does not imply a stabilization of *concentrations* of these gases in the 21st Century. Stabilization of concentrations could be achieved through a drastic reduction of emissions well below current levels; in its 1994 interim report, the IPCC (1994) computed the pathways to stabilization for a number of arbitrary concentration levels (Figure 5.8). Even at very high concentration levels of 750 ppmv of CO_2 at some time in the future (i.e., a three-fold increase in atmospheric concentrations compared to pre-industrial levels), the

IPCC concludes that emissions would need to be reduced below current levels within a relatively short time-frame.

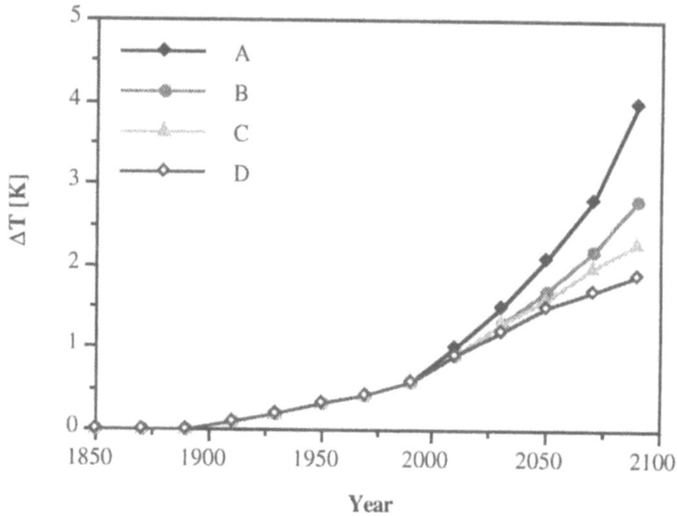

Figure 5.7: Global atmospheric temperature increases based on IPCC scenarios

The challenges faced by world population regarding the impacts of climate change on the natural and social environments would be immense. Among the possible consequences of Global Change are increased desertification, sea-level rise, reduced water resources, threat to global food supply, irreversible ecosystem damage and loss of bio-diversivity, drop in energy availability, etc. In the context of a rapidly changing world social context, the abrupt climate change predicted for the next century needs to be prepared for by adaptation and/or mitigation.

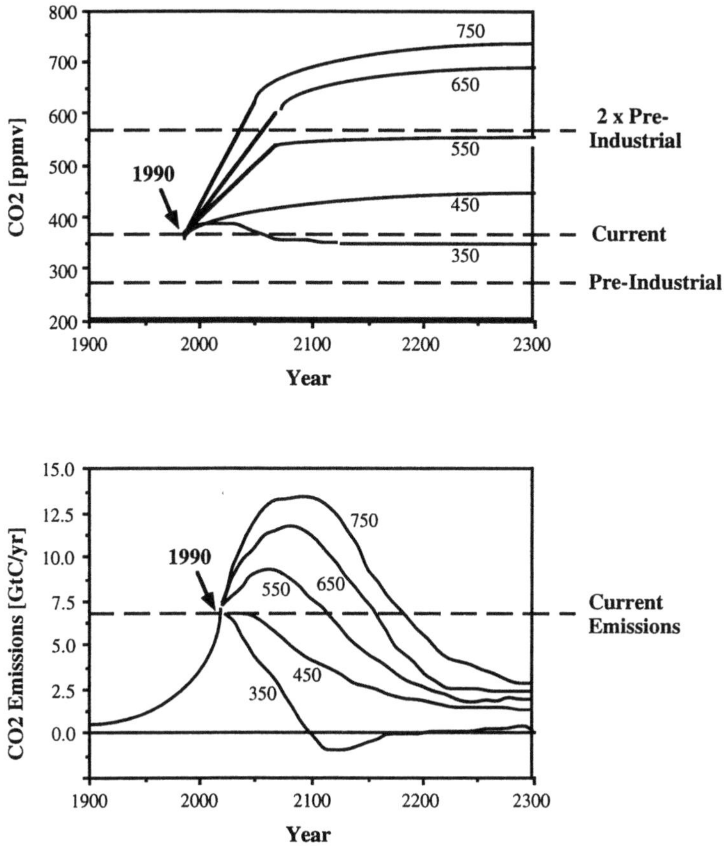

Figure 5.8: Arbitrary concentration levels (upper) and possible pathways to stabilization at the selected levels (lower). Note that in all cases, emissions need to be lower than current levels at some time in the future

5.1.1 The impacts of climate change

In its Second Assessment Report, the IPCC (1996) has summarized its main conclusions on the potential impacts of climatic change on a range of natural and socio-economic systems. While there are many adverse effects of climate change, some may be beneficial; furthermore, in a world under numerous social, economic, and environmental pressures, climate change represents one additional stressor which, for certain systems, may not necessarily be the dominant one. Because of these interacting stresses and the complexity of these systems, it is unlikely that ecological and socio-economic impacts can be attributed unequivocally to climate change.

The most vulnerable systems are those whose sensitivity to changes in climate is greatest and for which adaptation options are limited, for example unmanaged ecosystems and the cryosphere. While there will be some beneficial effects of climate change, there will be many adverse effects, with some being potentially irreversible. The net impact will be determined by the degree of success in implementing adaptation options and the ability to move towards sustainable management practices.

The efficacy and cost-effectiveness of adaptation strategies will depend upon management, institutional, legal and regulatory practices. Adaptation options for managed systems such as agriculture and water supply are generally improving because of technological advances. However, access to these technologies is not evenly distributed around the globe. Moreover, societies are becoming more vulnerable to extreme climatic events as population density increases in river basins, coastal plains, hill slopes, and marginal lands.

Measures to assist adaptation in unmanaged ecosystems, such as design of migration corridors, are not well developed. Human activities fragment most landscapes, thus limiting adaptation potential. Ecological systems disrupted by climate change will take decades to centuries to recover, if they can recover at all. Species extinction can never be reversed.

5.1.1.1 Impacts on water resources

Poor irrigation concepts often lead to increases in soil salinity and, in very arid areas, high evaporation rates which make ever greater demands on water resources. Industrial requirements are often critical in areas which are sensitive to fluctuations in hydrological regimes. In addition, industrial waste often contaminates regional water supply. Domestic sources of water are in many parts of the world plagued by high levels of contamination. Intensive agricultural use of fertilizers overloads water sources with undesirable chemical compounds. Politically also, the sharing of water resources poses immense problems: major hydrological basins are in general shared by several different countries. It becomes exceedingly difficult to manage the waters of a river which crosses several international boundaries. These problems will become exacerbated by climate change in sensitive regions, especially in the context of increased demands by an ever-growing world population and in the face of increased drought episodes in already sensitive areas.

Natural storage capacity (essentially in ground water) is a function of yearly input by the hydrological system. Any reduction in hydrological budgets would lead to significant cutbacks in storage capacity of fresh water. In general, water consumption is far too high for the available resources; this is especially true in highly developed societies. Interannual variability of precipitation regimes may lead to critical reactions within different hydrological basins, thereby disrupting continuous water supply. Increased extreme events, both droughts and floods, will have a severe impact on the capability of exploiting existing sources.

Over-exploitation of water resources associated with increased drought episodes would lead to higher salinity of the unsaturated zone (the layer between the surface and the water table at depth), which in turn would have consequences for natural and managed ecosystems (plants, forests, crops, etc.). In coastal zones, the lowering of water tables by over-use would draw salt into the ground water, leading to water unfit for domestic and agricultural use. This situation has already occurred in many parts of the world, for example in the Nile Delta of Egypt, or the Colorado Delta in Baja California, Mexico. Changes in river and lake levels would require major adjustments in the urbanization of settlements located on rivers and lakes;

in some cases, the costs of such adjustments may be too high for certain less privileged nations. A critical example is that of the Aral Sea in the ex-Soviet Union.

5.1.1.2 Impacts on agriculture

Changes in global climate will involve shifts in the latitudinal belts which are today optimal for certain varieties of crops. Global warming scenarios predicted by many general circulation models indicate that, for example, the climate of Finland will become similar to that of present-day northern Germany, and that of Iceland will experience conditions close to those occuring in Scotland. This does not mean, however, that crop productivity will simply follow the geographical shifts in climate belts. Other conditions may prove to be more constraining than favorable climates. Among these, soil types may be totally inadequate to support intensive agriculture as practiced today in the main producer countries. Even if, for example, sub-Arctic Canada experiences climatic condition in future similar to those of its grain-producing regions, soil composition may be too poor to sustain crop growth.

Photosynthesis would in principle be stimulated by higher CO_2 atmospheric concentrations; laboratory experiments confirm that because of increased carbon uptaken by plants, they grow bigger and more quickly. However, other factors will certainly counter-act this positive effect. Model results tend to indicate that present grain-producing areas, in particular the Great Plains of the US, will experience more frequent periods of drought and heat waves. Different plants have different threshold values for resistance to drought or heat, and extended periods of extreme conditions would destroy crops, negating completely the potentially greater productivity by CO_2 fertilization. During the extended drought of 1988 in the US corn belt regions, corn yields dropped by 40%; for the first time since 1930, US grain consumption exceeded production. If one considers that the corn-belt is one of the world's major producing region for crops, then future prospects for global food supply in the face of growing population demand and increased frequency of extreme climate episodes are likely to be poor.

Two schools of thought exist today in the scientific community concerning possible impacts of global change on agriculture. One school believes that increased productivity in an atmosphere containing more CO_2 will also reinforce the plant's capacity to resist extreme events; it is also believed that shifts in climate belts may lead to comparative regional advantages for intensive crop cultivation. Another school of thought, however, suggest that the "available evidence in support of this is inconclusive and is not based on the complex dynamics of the interactions between agricultural production processes and the environment. For example while longer frost-free periods in mid-latitudes could, by themselves increase crop yields... the most predictable effect of climate change is that it will cause significant increases in pest populations... advantages of present temperate regions may be cancelled by increased temperature and uncertain precipitations" (Source: S.H. Schneider: *Global Warming*, Sierra Press, page 175).

The primary source of two major GHGs, methane and nitrous oxide, is a consequence of agricultural practices. Rice cultivation, the burning of biomass, and the fermentation of organic matter in the digestive system of cattle and other domestic animals, inject annually into the atmosphere between 150-450 million tons of methane. Certain studies indicate that up to 80% of methane emissions are of recent biological origin and not the results of fossil-fuel burning. Nitrous oxide is emitted from soils through nitrification and denitrification processes, both heavily dependant on the use of fertilizers. Total emissions represent between 4.4 - 10.5 million tons per year which is small compared to the estimated 5.7 billions tons of CO_2. But it must be recalled that nitrous oxide has a warming potential some 250 times greater than CO_2, thereby placing this gas, even at low concentrations, as a significant contributor to the greenhouse effect.

5.1.1.3 Impacts on forests and natural ecosystems

Species will react in various manners to global climate change; increases can be expected for some species while others will undergo marked decreases. Under conditions of rapid warming, it is quite likely that certain ecosystems may not adapt quickly enough to respond to abrupt climate changes; this could be the case of forests which may not be able to migrate fast enough, or of species with very local ranges which have nowhere to migrate to. The possibilities of migration of species to higher / different altitudes or latitudes, in order to find similar conditions as today's, will probably be limited by other factors such as soil types, water availability, and the human barriers to migration such as settlements and highways. Analogies with migrations of the past are hazardous at best because of human perturbations to the environment which were not present millenia ago, and because the expected climatic change will occur at a rate unprecedented in the last 10,000 years.

Mankind has changed the planet's terrestrial ecosystems in a very short period of geological time, so that true natural ecosystems are to be found only in very limited or remote areas of the globe. Overexploitation of ecological resources is contributing to the enhanced greenhouse effect, through the release of methane and nitrous oxide by agricultural activities, and large amounts of carbon dioxide through deforestation. Deforestation has a number of adverse effects on the climate: it releases carbon to the atmosphere, which was previously sequestered in the trunks and leaves, either directly by burning or indirectly by the decomposition of logs; it changes the nature of the surface over vast areas, thereby perturbing the radiation budget through albedo (reflectivity) changes and the hydrological budget through reduced evapotranspiration wihin the tropical rain forests in particular. Soils once covered by trees will become poor in nutrients, more saline through leaching of tropical soils, and will lack the water retention capacity and, hence, water availability essential for the agricultural productivity which has provided one of the justifications of massive deforestation.

In the northern latitudes, vast amounts of organic material are at present trapped in permafrost, i.e., permanent layers of ice just beneath the thin surface of the soil which insulates the ice during brief periods of warmth in the summer months. Some specialists fear that atmospheric warming, predicted to be of greater amplitude at high latitudes, will melt significant amounts of permafrost. This in turn would release large quantities of methane which up till now have been sequestered. Rapid increases of methane concentrations, a gas 60 times more efficient than CO_2 in terms of warming potential, may lead to a spiraling effect in which atmospheric temperature increases would be even more rapid than predicted.

Because ocean circulations and mixing of surface waters with deeper layers may be affected by climate change, the nutrient availability to phytoplankton and their access to solar radiation for photosynthesis would be affected, too. These nutrients control the primary biological production in marine environments, and thereby the rest of the food chain. Marine ecosystems will be affected in direct proportion to the extent of physical perturbations in ocean circulations by climate change. In this context, it is feared that stratospheric ozone depletion, resulting in significant increases in UV radiation, may cause irreversible biological damage to minute organisms such as krill which form the lowermost level of the oceanic food chain.

5.1.1.4 Impacts on sea-level rise

Fluctuations of levels of the world oceans are correlated to those of climate. Despite difficulties in observations, it seems that sea-levels have risen by 1 - 2 mm/year in the last 100 years. It is even possible that acceleration of sea-level changes have occurred this century, though the evidence is open to controversy.

Two major factors can explain changes in sea-levels: thermal expansion of the oceans consecutive to global warming, and fresh-water discharge by continental ice caps and mountain glaciers. The Intergovernmental Panel on Climate Change (IPCC) has concluded that for its "Business-as-Usual" Scenario A, where no specific measures are taken to abate greenhouse-gas emissions, predicted rise is 20 cm by 2030 (expected time of doubled CO_2 concentrations) and 65 cm by the end of the next Century. While thermal expansion and melt-water from Greenland will contribute to this rise, the ice cap of Antarctica is expected to modulate overall sea-level fluctuations. This is because model predictions indicate increased snow accumulation in Antarctica associated with higher precipitation consecutive to global warming. Nevertheless, the expected rate of change is 3 - 6 times faster than over the last 100 years. Sea-level rise will not be distributed evenly over the globe. The figures given by the IPCC may appear small, as in the case of average global warming. However, a half-meter rise in sea level will create irreversible problems for low-lying island communities (Maldives, Pacific atolls). Major harbor areas will experience frequent flooding during high tide or storm surge episodes, and civil engineering costs for harbor defence projects to abate flooding are likely to be prohibitive. Countries already prone to devastating floods would be submitted to higher frequencies of such episodes; low-lying Bangladesh would be particularly affected. In Indonesia, which has one of the longest coastlines of any country in the world, agricultural settlements in marshland areas close to the coast would be particularly sensitive to small shifts in ocean levels; these communities would once more have to be displaced. Salinity of ground water close to coastal regions would increase, and so would coastal erosion.

5.1.1.5 Impacts on human infrastructure

Small islands and low-lying coastal areas are particularly vulnerable to climate-change effects such as rising sea levels and increases in flooding, coastal erosion, and storm frequency or intensity. Many low-lying areas, such as parts of the Maldives, Egypt, and Bangladesh would be inundated and made uninhabitable by a 50 cm sea level rise. Integrated Coastal Zone Management could help both developing and industrialized countries to achieve the twin objectives of addressing immediate socio-economic concerns while developing responses to longer-term events, such as climate change. However coastal zone management costs would be very large as a proportion of GNP for many small island and developing countries.

In general, the sensitivity of the energy, industry and transportation sectors is relatively low, while the capacity for autonomous adaptation is expected to be high, if climate change is relatively gradual. The sub-sectors most sensitive to climate change include agro-industry, renewable energy production including hydroelectric generation, construction, offshore oil and gas production, and manufacturing heavily dependent on water supplies. The most vulnerable human settlements are located in damage-prone areas of the developing world that do not have the resources to cope with impacts.

Property insurance is vulnerable to extreme climate events. The insurance industry is currently stressed from a series of "billion dollar" storm events every year since 1987, resulting in drops in capacity and significant increases in losses. If the frequency or severity of extreme events were to increase as a result of climate change, insurance in vulnerable areas could be withdrawn, with implications for other economic sectors, including banking. The cost of such incidents could increase insurance rates world-wide.

5.1.1.6 Impacts on human health

Direct health effects include increased exposure to very hot weather events and more frequent weather hazards (droughts, floods and severe storms), and are projected to increase the loss of life. Indirect effects will include increases in the potential transmission of vector-borne diseases (malaria, dengue and yellow fever, and schistosomiasis) via extension of the geographic distribution of disease vectors; increased transmission of contagious diseases (e.g. cholera); and potential impairments in nutritional status. Quantifying the predicted impacts is difficult because the seriousness of climate-induced health disorders depends on many other factors including environmental and socio-economic circumstances, nutritional and immune status, water purity, population density, and access to and quality of health care services.

5.1.1.7 Options to control the sources and enhance the sinks of GHGs

An extensive array of technologies and policy measures capable of mitigating greenhouse gas emissions is available. There is significant controversy regarding ease and cost of implementation of these technologies and measures, however, due to potential social, institutional, financial, market and legislative barriers to their application and implementation. Without effective policy intervention, the actual contribution of these options to mitigation of emissions may be limited.

Sustainably managed agricultural lands, rangelands, and forests can play a key role in reducing current emissions of carbon dioxide, methane, and nitrous oxide. Emissions can be reduced, or carbon sequestered, through improved management of agricultural soils, restoration of degraded agricultural lands, establishment of plantations, agroforestry, forest regeneration, and slowing of deforestation. There are significant uncertainties associated with estimating the amount of carbon that can be conserved and/or sequestered.

Energy production from fossil fuels is responsible for the largest portion of greenhouse gas emissions. In 1990, commercial primary energy use was approximately 350 EJ, resulting in the emission of about 6 Gt of Carbon. It grows to approximately 540-840 EJ by 2025 and 540-2500 EJ by 2100 (IPCC IS92 Scenario). The mix of fuels and the geographic sources of emissions are expected to change substantially over this time period. While the likely rate of demand growth is less certain, advanced technologies can significantly reduce both energy demand and resulting emissions.

Over the course of the next century, the world's energy infrastructure will be replaced completely at least twice, offering opportunities to change the energy system in step with the corresponding investment processes. Advanced technologies can significantly reduce energy demand in all sectors. The most promising options in the industrial sector are more efficient motors, control systems, co-generation, energy cascading, and materials recycling. Changes in vehicle design, propulsion systems and fuels, as well as measures to influence driving behavior, could reduce emissions by 40 percent in 2025 relative to a high emission scenario. Projected growth in energy demand in commercial and residential structures could be cut by one-half over the next 35 years with more efficient heating, cooling, lighting, and appliances

The most promising technology-based approaches for reducing greenhouse gas emissions from energy supply include gas turbine technologies, coal and biomass gasification technologies, production of transportation fuels from biomass, approaches to handling intermittent generation of electricity, wind energy utilization, electricity generation with photovoltaics and solar thermal electric technologies, fuel cells for power generation, and hydrogen as a major new energy carrier, produced first from natural gas and later from biomass, coal, and electrolysis.

Renewable energy sources could provide a substantial portion of foreseeable energy needs of the world over the next century; potentially most promising is biofuels for electricity generation and transportation fuel. Considerable uncertainty exists regarding the costs and rate of market penetration without major policy interventions.

5.2 General circulation models: three-dimensional modeling of the climate system

There are numerous approaches to simulating climate or components of the system. Models can be 1, 2, or 3-dimensional, and be applied to a single physical feature of climatic relevance, or may contain fully interactive, three-dimensional processes. In order to gain insight into the functioning of climate, the most complex models are not necessarily the most explicit; the type of model to be selected is therefore closely linked to the nature of the problem to be investigated. There are three broad types of climate models, namely the quasi one-dimensional Energy Balance Models (EBM), the two-dimensional zonally-averaged Statistical-Dynamical Models (SDM), and the full three-dimensional General Circulation Models (GCM). It is this latter type of model which this section will focus on. Descriptions of the other types of model are to be found in a number of textbooks dealing with modeling approaches to the climate system (cf. Henderson-Sellers and McGufie, 1987, for example).

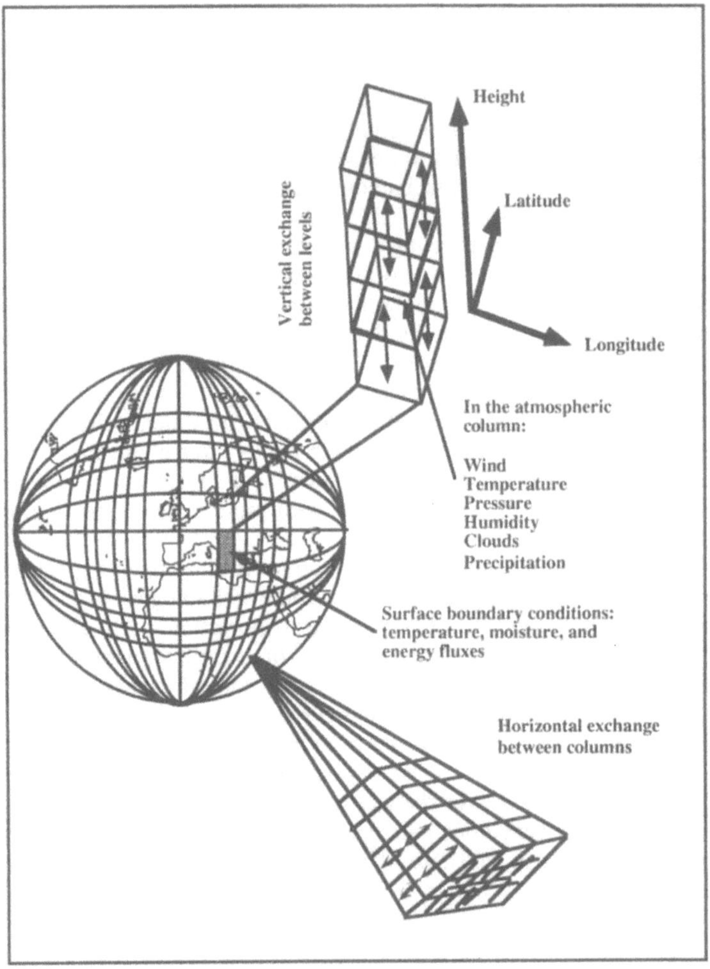

Figure 5.9: GCM structure and physical mechanisms

GCMs are essentially weather-forecasting models which operate at lower spatial resolution in order to integrate further ahead in time; they attempt to incorporate as many elements of climate as possible, i.e., not only the atmosphere, but also the oceans, the cryosphere, and the biosphere. GCMs are among the largest and most demanding operational applications in terms of computing resources. They typically solve large sets of equations at up to several hundred thousand grid-points, and these computations must be repeated 50 or more times per simulated day in order to represent the temporal evolution of the system with sufficient accuracy. Because the GCMs are producing results over a network of grid points distributed in three dimensions over the globe, computer time and space requirements are extremely large, and much of the physics representing feedbacks within the climate system needs to be parameterized, often in an oversimplified manner. As an example of computing resource requirements, an advanced model such as the

Max-Planck-Institut T-106 ECHAM3 / ECHAM4 spectral models (wave number truncation 106, corresponding to a spatial resolution of approximately 120 km, or 1.1° latitude/longitude, over the globe) requires 130 hours of supercomputer time to simulate a one-year atmospheric cycle (NEC - SX/3 vector machine of the Swiss National Scientific Computing Center, Centro Svizzero di Calcolo Scientifico / CSCS in Lugano-Manno, Switzerland). The data generated for this one-year period exceeds 16 Gbytes. Despite the impressive computational resources used for climate modeling, GCMs have numerous problems associated with them, in particular a resolution to crude to provide adequate climatological information on the regional scale; the "regional" experiments undertaken during the IPCC process (Houghton et al., 1990) were disappointing at best. Chapter 8 will return to this question and outline some solutions to the problem of regional climate scenarios of use to climate-impact modelers and other potential end-users.

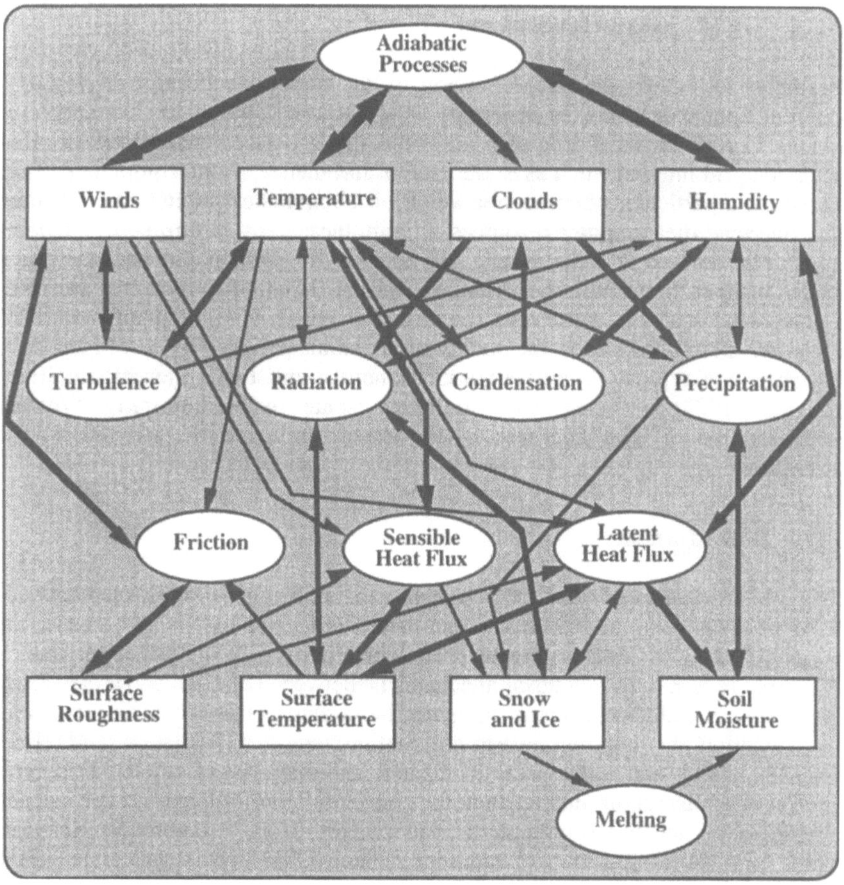

Figure 5.10: Typical physical parameterizations and their interactions in a GCM

General circulation models are based on the physical laws of the atmosphere which were introduced in Chapter 2, which describe the redistribution of heat, momentum, and moisture resulting from atmospheric motion. The governing equations are non-linear partial differential equations whose solution can only be achieved by numerical methods. Figure 5.9 illustrates the typical structure of a GCM and the principal processes which are included in such a model.

The numerical methods consist in subdivision of the atmosphere vertically into discrete layers, and horizontally either into discrete grid-points or by a finite number of mathematical functions (spectral models, as reviewed in Section 2.3). The values of the predicted variables, such as surface pressure, wind, temperature, humidity, and rainfall are integrated at each grid point (or for each spectral function) with respect to time, in order to obtain a prediction of the future state of these variables.

5.2.1 GCM parameterizations

Parameterization, or the simplification of the equations through physically-coherent approximations, is necessary in a numerical model for a number of reasons. Certain physical processes may be acting at a scale smaller than the characteristic grid interval, such as in the case of turbulence, or the complete physics describing a particular phenomenon would, if computed explicitely at each time step, overload the computer resources; in both these cases, a *parametric scheme* needs to be devised in order to take into account the phenomenon in some meaningful manner. In general, the parametric scheme attempts to relate the variables at unresolved scales to those resolved at model grid-points; the quality of GCM results are directly related to the quality of the parameterized physics. Figure 5.10 illustrates the typical parameterization schemes and their interactions (from Houghton, 1984) for the ECMWF (European Center for Medium-range Weather Forecasts) model. The thickness of the arrows indicates the strength of the interactions.

5.2.1.1 Radiation

The parameterization of radiation is probably one of the most crucial elements in a GCM which is being applied to the man-made climate change issue, because the radiative forcing of GHGs is the key term which needs to be treated in the model. This leads in turn to the feedback mechanisms between radiation and water vapor and clouds. A radiation scheme computes the radiative balance of the incoming solar radiation, the outgoing terrestrial radiation, long-wave radiation, and reflection. Absorption and re-emission of infrared radiation are calculated in several spectral bands by taking into account the concentration of different absorbers and emitters, in particular carbon dioxyde, water vapor, ozone, and aerosols. Seasonal cycles form an integral part of radiation schemes, but diurnal cycles are often neglected for reasons of computer economy. Cloud-radiation feedbacks are extremely sensitive for long-term climate simulations, and parameterizations of this effect are based either on estimates of cloud amount (dependent on the relative humidity) within a grid volume, or on the variation of cloud optical properties through the water content in clouds. There is still much controversy on the net radiative effect of clouds on climate: it can consists in a positive feedback (net warming) through the trapping and re-emisision of outgoing infrared radiation, or in a negative feedback through the reflection of intercepted solar radiation.

More detailed information on the radiation parameterization used in the ECHAM3 and ECHAM4 models will be forthcoming in Section 5.3.1.

5.2.1.2 Clouds and precipitation

The spatial resolution of GCMs is generally too crude to "see" clouds, except for stratus-type clouds which may have a considerable horizontal extent, and yet it is known that clouds exert a major signal on climate dynamics. Cumulus-type convection is one of the main heat-producing mechanisms at subgrid scales in a vertically unstable atmosphere. The technique known as *moist convective adjustment* is one of the primary parametric schemes for cumulus convection. The procedure consists in adjusting the temperature and water vapor profiles to a conditionally stable state. Large-scale clouds are allowed to form when the relative humidity within a given grid volume exceeds a certain threshold value of relative humidity (between 80 and 100% according to the GCM), which sets the cloud cover at between 95 and 100%.

It should be noted that cumulus-type clouds have a small areal extent but penetrate deep vertical layers, thereby profoundly modifying the dynamic and thermodynamic structure of a large vertical segment of the troposphere, whereas the contrary is true for stratiform clouds. In the case of these clouds, radiative feedback mechanisms, and not direct dynamic or thermal effects, will be the dominant feature.

Large-scale rainfall is usually based upon saturation vapor pressure in each model layer, and moisture is allowed to condense out of supersaturated air. If the temperature of the lowest layer is beneath the freezing point, snowfall occurs and snow depth at the surface is incremented accordingly. In the case of surface temperatures exceeding 273° K, precipitation falls as rain and soil moisture content is incremented.

5.2.1.3 Subgrid-scale transports

The principle driving force for the climate system is the heterogeneous distribution of heat at the surface, resulting from differential absorption of solar energy by the different types of surface. The energy which drives the dynamics of the general circulation of the atmosphere becomes available after it has been transfered through the Atmospheric Boundary Layer. ABL processes are characterized by small-scale (subgrid-scale) turbulent exchanges of heat, moisture, and momentum. Most of the momentum dissipation, which slows down or at least balances the other forces responsible for the general circulation, takes place within the ABL.

In most GCMs, the turbulent fluxes of heat, moisture, and momentum at the surface are calculated from empirical bulk aerodynamical formulae. Fluxes at the top of the ABL, which is generally fixed in height, are either neglected or based on mixing-layer theory. More detailed GCMs which resolve the ABL use eddy-diffusion methods, such as the one described in Section 4.2.1.

5.2.1.4 *Land-surface processes*

It is important in climate models to parameterize the heat and water transfer within the soil, such as the balance of surface hydrology processes, which include precipitation, snow melt, ground-water storage, and runoff. This is extremerly valuable in research on man-made climate change, because these processes will determine how local climates may change from humid to dry according to significant shifts in the atmospheric general circulation. Changes in surface vegetation and land-use, if well parameterized, should also produce a signal in the climate system.

Soil moisture can be parameterized in the form of a thick soil layer acting as a single reservoir in which moisture increases through precipitation, or decreases through evaporation. When all the moisture is used up, evaporation stops. More sophisticated techniques use two or more soil layers, the uppermost layer responding most rapidly to precipitation. If the upper layer dries out, some moisture is available for evaporation from the deeper, reservoir, layer. If the model incorporates vegetation as a land-surface characteristic, then moisture in the reservoir layer is taken up by the vegetation during dry periods for evapotranspiration.

At any model grid-point over land, a balance between precipitation, evaporation, runoff, and soil moisture accumulation is evaluated. If precipitation exceeds evaporation, then accumulation will occur locally until saturation is achieved. Excess runoff occuring under saturated soil conditions is used for fresh-water input into the ocean, which can have significant local effects in ocean models.

5.2.1.5 *Gravity wave drag*

The transfer of momentum between the atmosphere and the earth occurs at all scales. At the small end of the scale spectrum, dissipation of energy occurs through friction with the surface. At larger scales, additional processes lead to momentum dissipation, and these are linked essentially to the presence of orography. Mountains which are present over large horizontal distances and with significant heights may generate internal gravity waves which results in gravity wave drag, as sketched in Figure 5.11. The dissipation of energy by gravity waves is different from that occuring in the ABL.

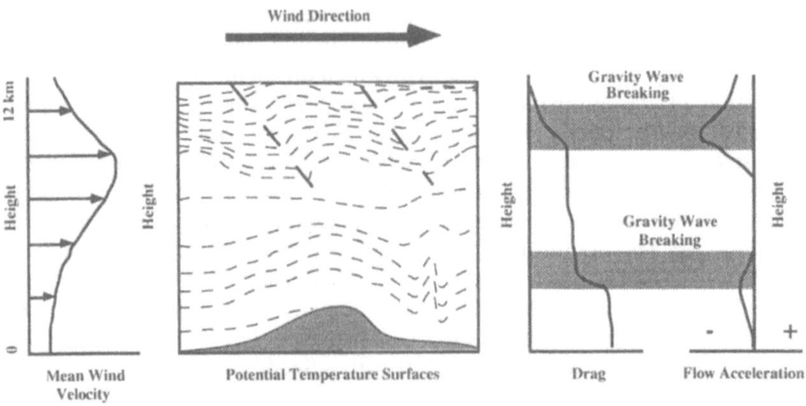

Figure 5.11: Sketch of the dynamic effects of a mountain barrier on the large-scale flow

Large scale models will in general resolve rather poorly even important mountain chains, so that mountains need to be included in some parameterized form, such as "envelope topography" which smooths the real orography over continental areas. Large-scale orographic forcing is then be derived by filtering the orography at smaller scales and ABL effects. Gravity waves generated by the presence of underlying ororaphy on the atmosphere are capable of breaking in a similar manner to ocean waves at the seashore, and in doing so transfer substantial quantities of momentum from the large-scale to the small-scale flows. They also have an influence on the formation of clouds and precipitation close to mountains, which are in turn indirect mechanisms of heat and moisture transfer in the vertical.

5.2.1.6 The cryosphere

The cryosphere is more than a particular case of surface processes. A global model needs to incorporate snow and ice features because of their determining effect on radiation, as well as dynamic mechanisms through Pole-to-Equator temperature gradients. In the summertime, the dominant component of the surface energy balance is solar radiation; small changes in albedo can have large impacts on the response of climate, so that surface albedo requires accurate specification or prediction. In particular, models need to distinguish the sea ice and continental ice surfaces, which have significantly different reflectivities; sea ice albedo is often modeled as a function of latitude and thickness, the latter being determined by the hydrological balance. It is obvious that if a model has difficulty in predicting other variables such as precipitation, then this may reflect upon the accuracy of computations of features such as sea ice.

Similarly, the albedo of surfaces which are snow-covered will be higly different if vegetation is taken into account in a model or not. Snow falling in the Tundra will raise the surface albedo from 0.2 to 0.8, whereas in a coniferous forest, practically no albedo change will be observed as a result of the forest canopy. In a model with no description of vegetation, a sharp rise in albedo would be observed in the region which in reality is covered by forests, thereby feeding into the climate system in a different manner.

5.2.1.7 The biosphere

Biospheric processes are a key factor in surface-atmosphere feedback mechanisms; the complexity of biological systems precludes anything more than a very crude parameterization of the presence of different vegetation types and their interactions with dynamic, thermodynamic, and chemical processes. In this latter respect, vegetation on land controls the magnitude of the fluxes of several GHGs, including CO_2 and CH_4. Any changes in gaseous composition will change the response of plant types and the redistribution of GHGs within the atmosphere.

Terrestrial ecosystems are also important in their role on the surface fluxes of heat and moisture, as well as determining to a large extent surface albedo, both areally and seasonally. Evapotranspiration releases vast quantities of moisture which can be taken up within the hydrological cycle, thereby redistributing heat and moisture in the atmosphere. Furthermore, the presence of vegetation exerts a dynamic influence through its roughness characteristics. Flow of air over a tundra region, which consists of sparse and low vegetation will be significantly different from flow over large forested areas, as a result of the different characteristic roughness lengths of these surfaces.

Parameterization of the biosphere is being attempted in some GCMs, using a combined multilayer soil-vegetation surface. As described above, soils have at least two layers, one acting as a reservoir for plant moisture availability, while the vegetation is parameterized by a canopy layer, surface and root layers, and variable albedo and hydrological characteristics. These latter depend essentially on the leaf-area index (LAI) of the plants, which determine the ratio of transpiring to non-transpiring canopy surface, and the area over which dew formation and precipitation interception occur. More water can be retained by a surface which has a greater LAI, while albedo is determined by the seasonal distribution of this parameter (e.g., LAI will undergo a strong seasonal cycle for deciduous trees in temperate regions, but will be less marked or even insignificant for tropical forests or coniferous trees). In the debate over the influence of massive deforestation in the tropical regions on climate change, the necessity of sophisticated vegetation parameterization techniques becomes obvious, though research into the incorporation of the biosphere in GCMs is still in its infancy.

5.2.2 Model validation

Abstract theories and models need to be validated, in order to have some degree of confidence in their projections and analyses. As previously mentioned, the limits of our understanding reduce the predictive ability of the modeling systems, so that the purpose of validation is not so much to prove the absolute exactitude of a model as to define the degree of uncertainty of its results and limits of applicability. By comparing as much as possible model results with measurements and observations, improvements in model design or recommendations for developing new models can be expected.

Theories and models are normally designed to address classes of similar issues. Before they can be used to answer specific questions, they must be adapted to the particular conditions of the problem to be analyzed. This is achieved through the specification of initial and boundary conditions. Model initialization occurs when the state of the system is specified at the beginning of the computation, by setting each of the dependent variables to some reference value obtained from observation or hypothesis. In a GCM, model initialization is not trivial, because the information used to initialize a simulation must not only be processed in a manner suited to the internal model structure, but it must also be consistent with the laws and relations which define the model.

Boundary conditions are also a necessary part of a model simulation. These describe constraints on the evolution of the system, or interactions between the system and its environment which are not described explicitly by the model. Much of the atmospheric data used for model initialization and verification is standardized by the World Meteorological Organization in Geneva, based on a worldwide network of meteorological stations. These provide not only ground-based weather information but also upper-air soundings based on simultaneous releases of weather sondes around the world. For the last 15 years of the climatological record, the European Center for Medium-range Weather Forecasts (ECMWF, Reading, UK) now has available a global gridded data base of use for GCM initialization and intercomparison purposes. A similar data set is also available from the US National Meteorological Center (NMC, Washington, DC).

Remote sensing techniques have provided an additional and powerful method of observing the climate system, in particular through remarkable progress in satellite technology, radiation measuring instruments and optics, and computerized data post-processing and graphics. Sensors located on satellite platforms constitute the only global and repetitive source of data concerning the state and evolution of the Earth system. Among the spaceborne instruments currently used for terrestrial observations, SPOT HRV, Landsat TM, Meteosat, GOES, and AVHRR are the most widely used. The first two are used for high-resolution regional studies, while GOES and Meteosat have an essentially meteorological orientation. AVHRR has been used for monitoring continental ecosystems and provides the bulk of data for large-scale and long-term studies.

It should be noted that remote sensing techniques are intrinsically limited to variables that directly or indirectly affect the transfer of radiation through the environment. This means that much of the data needs to be processed, as the sensor is not measuring directly a variable such as temperature or moisture, but the radiative transfer through the atmosphere affected by that temperature or humidity. Conversion of radiation data to the value of the desired variable is not trivial and may be a source of error. It should also be noted that in coming years, the capability of satellite observations will be so large that they will be generating enormous quantities of data, on the order of one hundred or more GBytes of data *per day*, which is far more than any human operator can look at. Even today, only a fraction of the information received from satellites is processed and used, due to lack of personnel, the cost of data acquisition, and the lack of appropriate software tools. This problem will need to be solved if a meaningful use of satellite technology for climate research is to be made in coming years.

There are several manners in which GCM results can be examined to evaluate the performance of a model, though none will prove beyond doubt the validity of the model. The most obvious approach is to test overall model skill in reproducing today's climate in order to check the consistency of the model. In particular, the ability of a GCM to reproduce seasonal cycles of temperature is one of the standard control experiments. However, simulation of seasonal changes does not mean that a model would be capable of reproducing long-term changes induced by numerous factors, such as changes in radiative forcing, sea surface temperatures or cryospheric characteristics. Another validation method is to test separately physical sub-components of the model directly against real data, or against results from high-resolution models. Validation of one isolated parameterization does not guarantee, however, that all interactive physics has been correctly treated. Other validation tests include comparisons of model-generated statistics at selected grid-points with observed variability at the same locations.

The differences in observed and simulated data can help determine in what part of the model's physics or numerics improvements would need to be made. Testing of the sensitivity of results to model resolution can yield information on the performance of feedback mechanisms between different elements of the climate system in a GCM. A further rigorous test of GCMs can take the form of simulating very different climates which have occured in the past on Earth, and for which substantial data sets exist for intercomparison purposes; simulations of the atmospheres on other planets have also been attempted as a test of the robustness of a model to vastly different initial and boundary conditions from those prevailing on

Earth. Whatever methods are chosen, it should be noted that none will be entirely satisfactory or sufficient; no data sets exist which allows direct comparisons between models and reality at the spatial and temporal resolution typical of a GCM. The sparseness of available data is a general problem for model validation, whatever the scale considered.

5.2.3 Coupled model systems

Modelers have recognized the need to incorporate as many elements of the climate system as possible within one model, though in practical terms this is an exceedingly difficult task due to the computational resources required and the fact that an integrated model would be attempting to simulate processes acting on vastly varying time scales, as illustrated in Figure 5.12.

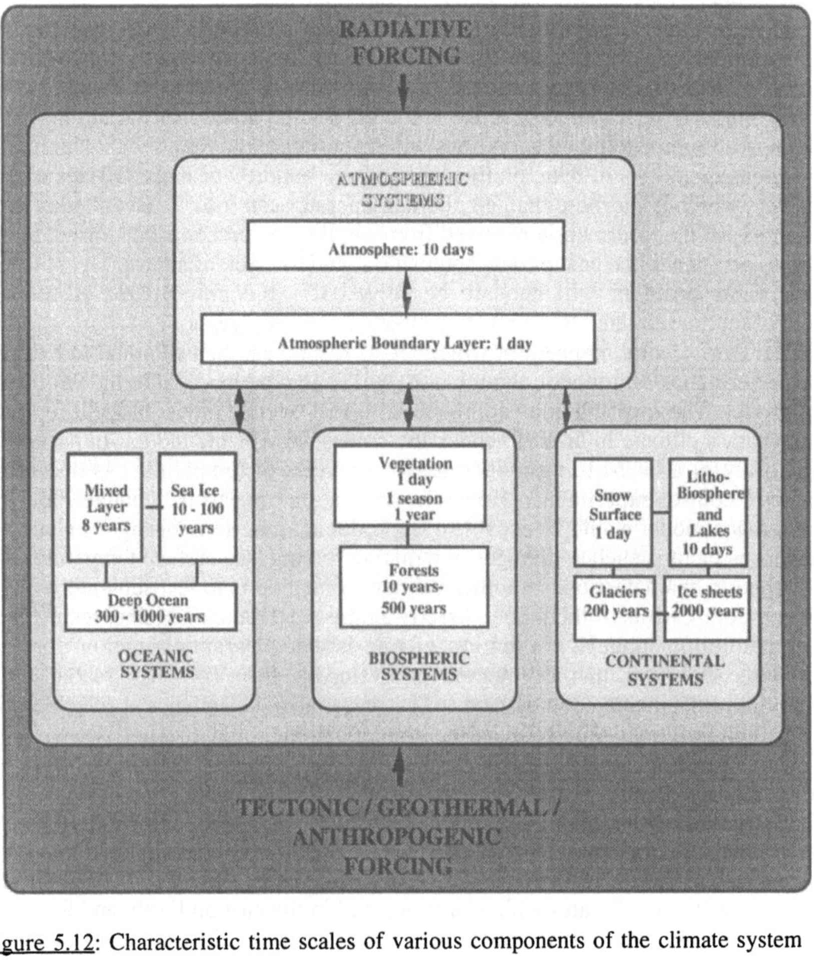

Figure 5.12: Characteristic time scales of various components of the climate system

It can be seen that, according to the component of climate considered, response time to a particular forcing can vary from a few hours to several centuries. The direct coupling of different elements of the system in an integrated climate model is not facilitated in view of this situation, and even the expected increase in computing power in coming years will not alleviate this problem to any significant extent.

Because of its dominant role in climate processes, and the large volume occupied on the planet, the oceans have been modeled in an analogous manner to the atmosphere and, up to a few years ago, separately from atmospheric models. Until recently, the ocean interactions with the atmosphere have been modeled either by prescribing observed sea-surface temperatures (SST) or by specifying the meridional energy transport of the oceans. This has obvious constraints for any long-term climatic simulations, and the necessity of taking into account not only the ocean surface processes but also those acting at depth in the deep ocean circulation have lead to experiments in ocean-atmosphere model coupling on a global scale.

As seen in Figure 5.12, the oceans have vastly different response times from those of the atmosphere, and the ocean subsystem spans a greater range not only in time but also in space. An oceanic GCM requires high spatial resolution to capture eddy processes which are key features of ocean dynamic exchange, and also bottom topography and basin geometry. High-resolution ocean models are therefore exceedingly costly in computer time. The link between an ocean model and an atmospheric GCM is difficult because of the discrepancies in response times illustrated in the Figure 5.12. One approach is to synchronize coupling between the two model types at different time scales for the atmosphere and for the oceans; in some long-term, low-resolution coupled experiments investigating long-term climate change, the atmospheric component is synchronized each 4 years and the oceanic component each 1,200 years. This involves managing a major computerized data base in which the results of the ocean model and the atmospheric GCM can be stored and mutually accessed in a synchronous manner.

For decadal to century timescale simulations with higher-resolution GCMs, particularly those investigating climate change and variability in the 21st Century, the initialization of the ocean and atmospheric models poses a particular problem often referred to as the "cold-start problem". An atmospheric model requires sea-surface temperatures (SST) as one of the lower boundary conditions, while an ocean requires surface fluxes of energy, salinity and momentum as upper boundary conditions. This requires a separate "spin-up" of the atmosphere and ocean models so that their boundary conditions eventually reach a state representative of "present-day" conditions, after which time the models are effectively coupled. During the spin-up phase of the ocean model, if forced by specified fluxes, the ocean will attain an unrealistic state since the feedbacks of the atmosphere to the ocean, in particular the increase in GHG forcing prior to the "present-day" conditions, are absent. The ocean-atmosphere system will therefore be colder than reality, hence the term "cold-start".

There are a number of solutions which can help overcome the cold-start problem, which are essential since inconsistencies in surface fluxes of both the ocean and atmosphere models during spin-up inevitably lead to climate drift when the models are coupled. More recent GCM experiments, however, have taken the initial state not from the latter part of the 20th Century but from the middle of the last century, when GHG values were close to their pre-industrial values. For example, the Max-Planck Intitute for Meteorology in Hamburg is currently undertaking a series of coupled ocean-atmosphere simulations at T-42 spectral resolution, spanning the period 1860 - 2100 (Bengtsson, personnal communication).

The atmosphere and the ocean interact via fluxes of heat, momentum and fresh water, which are generally on the order of several hundred W/m^2. Oceans are warmed in summer and cooled in winter, so that the fluxes cancel out on an annual average. Coupled models may generate residual fluxes which lead to a drift in model results away from the observed climate; these residuals may be on the order of $\pm$ 5 W/m^2, which is small compared to the absolute values of the fluxes at the ocean-atmosphere interface, but would be sufficient to lead in time to very large errors if not corrected. Many atmospheric models therefore need to adjust certain fluxes before they are passed on to the ocean component; the latest generation of coupled ocean-atmosphere GCMs, however, tend to overcome the problem of flux correction through improved intialization and model formulation techniques. Coupling of the ocean to the atmosphere has been achieved with a hierarchy of ocean models, many of which are oversimplified representations of reality. The simplest "ocean" model is known as a "swamp model" because the ocean is represented as a water surface with only prescribed SSTs and no heat storage capacity. A more complex variant to the swamp model is the fixed-depth ocean which allows for changes in heat capacity but where no dynamics occurs. The mixed-layer ocean model is one in which dynamic processes are simulated in response to atmospheric temperature and dynamic forcing; indeed, two-way interactions between the oceans and the atmosphere can be taken into account in a simplified manner.

In view of the significant influence of the oceans in terms of heat storage and GHG absorption, long-term simulations of climate requires a full three-dimensional ocean model, where features such as the formation of deep water are simulated. Changes in the intensity - and location - of deep water formation can, and have had, profound effects on the atmosphere. In the past, changes in the thermohaline circulation of the oceans have resulted in major atmospheric responses, such as the cold "Younger-Dryas" period which affected Europe after the end of the last glaciation. Furthermore, if only SSTs are prescribed for the ocean component, climate predictabilty in the long-term become doubtful because features such as the periodic El-Niño/Southern Oscillation (ENSO), which perturb the climate system and depend on dynamic interactions between the atmosphere and the oceans, are not predicted. Coupled ocean-atmosphere systems have been the first attempt at integrated climate modeling; however, further integration of of other climate system component models, especially the cryosphere and the biosphere, are in the long term necessary in order to obtain more realistic simulations of climate on decadal to secular or longer time scales. Because of the vastly different nature of these systems, and due to the limits imposed by even the most powerful supercomputers available today, a fully interactive three-dimensional climate modeling system is unlikely to be operational in the near future.

5.2.4 A brief review of current climate simulations

There are some 30 GCMs which have been developed worldwide for climate research; many of these have been used in the IPCC context for assessing model capability in reproducing present-day climate, and for providing an insight into the future, GHG-enhanced climate of the 21st Century. Much of the IPCC Second

Assessment Report which focuses on current climate simulations with GCMs is based on the AMIP (Atmospheric Model Intercomparison Project; Gates, 1992) experiment which has analyzed model output for 30 GCMs. Other intercomparisons between GCM results have taken place in the past, such as the very detailed examination of the interactions between clouds and climate based on 19 GCMs and reported by Cess et al. (1990).

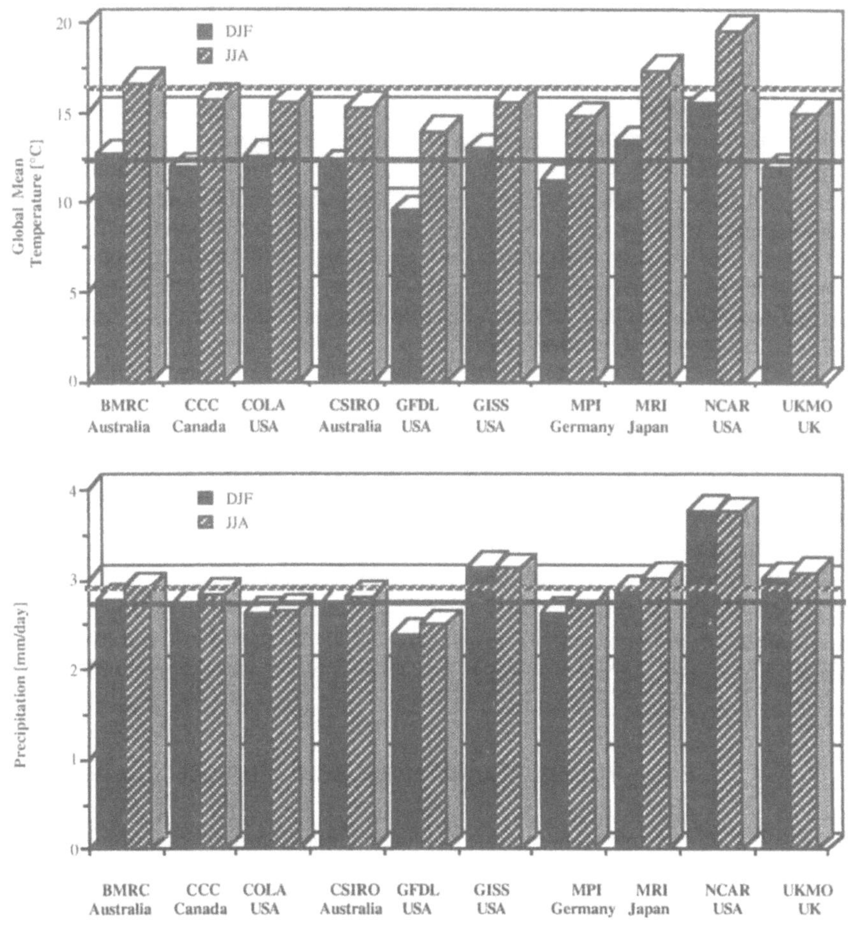

Figure 5.13: Globally-averaged winter (DJF) and summer (JJA) temperature (upper) and precipitation (lower) simulated by 10 different general circulation models, compared to observations (continuous lines with appropriate color coding)

In terms of current climate simulations, coupled ocean-atmosphere now begin to capture many elements of the observed climate with a reasonable degree of skill, particularly when initialized with the observed sea-surface temperature distribution and its recent evolution. Some regional detail in interannual climate variation is possible with higher-resolution GCMs, such as the T-106 version of the Max-

Planck Institute model (to be discussed further on). However, the computational costs associated with running a high-resolution GCM are such that they preclude numerous experiments which would be required to increase confidence in the model results.

Figure 5.13 illustrates the performance of a number of independently-developed, coupled ocean-atmosphere GCMs in reproducing observed summer (JJA: June, July and August averages) and winter (DJF: December, January and February means) globally-averaged temperatures and precipitation. It is seen that the models successfully simulate the large-scale seasonal distribution of temperature. Differences beween one model and another are the result of differing physical parameterizations, the manner in which processes such as seasonal sea-ice and the strength of the temperature/albedo feedbacks are taken into account, and the simulation of interactions between different elements of the climate system.

While the ability of GCMs to simulate component elements of the atmosphere, ocean and land surface have clearly improved in the 1990s, major problems remain, in particular the radiative effect of clouds. Land surface parametric schemes have been significantly ameliorated in recent years, with the inclusion in some models (the MPI ECHAM4, for example) of up to 16 biome types as opposed to a crude vegetation surface; however, the simulation of soil moisture is still uncertain because of a dearth of observations.

A reasonable simulation of the precipitation rate, as a measure of the intensity of the hydrological cycle, is especially important in coupled model systems because of possible changes in the ocean thermohaline circulation through freshwater fluxes. As seen in the lower part of Figure 5.13, the GCMs capture the broad-scale structure of global precipitation, to the degree that the precipitation itself is not particularly well known on a global basis.

Other key atmospheric variables, such as mean sea-level pressure, are generally well represented in current-generation GCMs; pressure is dependent on atmospheric dynamics and thermodynamics, as well as the surface stresses that drive the ocean. Most GCMs have problems with an accurate description of sea-level pressure over mountain regions, perhaps as a result of gravity-wave generation.

5.3 GCM experiments for climate and climate change

Since 1992, GCM experimentation has been undertaken for the first time in Switzerland, through a joint research collaboration with the Max-Planck-Institute for Meteorology (MPI) in Hamburg, Germany, the National Center for Atmospheric Research (NCAR) in Boulder, Colorado, USA, and the Swiss Scientific Computing Center (CSCS: Centro Svizzero di Calcolo Scientifico) in Lugano-Manno, Switzerland. This section will report on the broad findings of the GCM results for current and future climate. The ultimate focus of these climate investigations were to determine the extent to which climate in mountain regions, particularly the Alps, can be simulated and predicted through advanced numerical modeling techniques. This topic, which involves coupling of a GCM and a Limited Area Model (LAM) will be discussed in further detail in Chapter 7.

5.3.1 The MPI general circulation model, ECHAM3 (upgraded in 1994 to ECHAM4)

The ECHAM3 model is a spectral transform model which includes 19 vertical levels; the highest level is located at 31 km (or the 10 hPa level). The model is based on the ECMWF (European Center for Medium-range Weather Forecasts, Reading, UK) weather-forecasting model but has been significantly modified in terms of its physical parametric schemes in order to apply it to climate-relevant research. It includes advanced parameterizations for radiation, convection, cloud and precipitation formation, planetary boundary layer and surface processes (DKRZ, 1992). In particular, the model incorporates an explicit moisture formulation which includes an equation for water vapor and cloud water and ice with conversion/production terms based on the Kessler (1969) parameterizations.

The cumulus parameterization is that of Tiedke (1989). The scheme accounts for penetrative, shallow and mid-level convection, with inclusion of updraft and downdraft mass fluxes. Penetrative convection is initiated by local convective instability and moisture convergence. In the present simulation, sea surface temperatures (SSTs) and sea ice cover are prescribed daily by linear interpolation between monthly mean climatologies.

	Spectral Interval (μm)	Absorber
ECHAM3	0.215-0.685	O_3
	0.685-0.891	H_2O
	0.891-1.273	H_20
	1.273-3.580	H_20, CO_2
ECHAM4	0.25-0.68	O_3
	0.68-4.0	H_2O, CO_2

Table 5.1: Spectral intervals and absorber in the solar part of the ECHAM3 (Hense et al. 1982) and ECHAM4 (Morcrette 1991) radiation schemes

The radiation scheme, where joint developments between MPI and the Climate Modeling Group of ETH in Zurich have taken place, is based on a two-stream approximation described by Kerschgens et al. (1978) and Zdunkowski et al. (1980) with delta-Eddington approximation for clouds. The longwave spectrum is divided into 6 spectral intervals taking into account absorption due to water vapor, carbon dioxide, ozone and aerosol. An inversion procedure is employed to define the relevant optical properties by matching the two-stream solutions to more accurate model solutions obtained by a line-by-line model (Hense et al. 1982), based on spectroscopic data of LOWTRAN3 (Selby and McClatchey 1975).

For cloud droplet absorption an emissivity formulation is used (Stephens 1978). Scattering of longwave radiation is neglected. The shortwave spectrum is divided into four intervals ranging from 0.215 μm - 3.58 μm, with the same gaseous absorbers as above (Table 5.1). The optical thickness for gas absorption is a function of the effective absorber amount and is determined similarly to the longwave part. Rayleigh scattering is included via a parametric expression of optical thickness. Scattering and absorption coefficients of stratospheric, urban, and maritime aerosols are taken from a dataset provided by Shettle and Fenn (1975). Cloud optical depth and single scattering albedo are derived from the cloud water path (Stephens 1978).

The ECHAM model has been used in climate simulation studies and in climate change investigations both at MPI and at ETH. Studies have also been undertaken where the model has been coupled to different ocean models developed at MPI. In such coupled experiments, investigations have been undertaken to assess the effect of the change in the atmospheric concentration of greenhouse gases on climate. These climate investigations have been extended over very long periods, up to a few hundred years. For computational reasons it has been necessary to undertake these experiments at quite low resolutions equivalent to grid distances of around 500 km.

Experience in medium-range weather prediction has brought to light the key importance of spatial resolution. In order to resolve intense weather systems such as polar fronts, cyclones and lee-wave cyclogenesis, the horizontal resolution must be significantly improved and increased to at least the order of 100 km. At such resolution climate models are in a position to resolve synoptic weather systems with reasonable accuracy, and also to better capture the influence of orography on weather systems. They are also able to investigate the feedbacks of the smaller scales on global climate, through interactions between surface features and large-scale flow.

The latest version of the German climate model which has been used for recent investigations at ETH-Zürich, ECHAM4, is one of the most advanced atmospheric GCMs currently used for climate simulations, using either prescribed SSTs or coupled to an ocean model. It uses the spectral transform technique for the advection of vorticity, divergence, temperature and surface pressure while water vapour and cloud water are treated in a semi-Lagrangian manner. Horizontal diffusion of the form is applied to the vorticity, divergence and temperature tendencies. Specification of the e-folding time at the largest total wavenumber retained in the spectral truncation determines the diffusion constant. Moisture terms are now parameterized through a semi-Lagrangian scheme, which ameliorates the simulation of precipitation intensity and its location when comapred to observed climatology.

A further parameterization which takes into account the dynamic influence of mountains on the general circulation, particularly on atmospheric blocking , is the gravity-wave-drag scheme. The effect of orographically excited gravity waves on the momentum budget is parameterized following Palmer (1986). The surface stress may propagate up the 19 hybrid vertical levels unless it is absorbed at some layer where the local wave Richardson number falls below a critical value.

ECHAM4 incorporates the most recent improvements in a number of key climatic processes. One of the most fundamental changes between the ECHAM3 and the ECHAM4 model is the inclusion of a completely new radiation code. This code is based on the scheme developpped by J.J. Morcrette for the ECMWF numerical weather prediction model, Cycle 43 (Morcrette, 1991).

This scheme uses a two-stream method based on Fouquart and Bonell (1980) for the solar part and Morcrette et al. (1986) for the longwave part. The solar spectrum is divided only into two bands (cf. Table 5.1). Coefficients for the gaseous absorption are calculated from the 1982 version of the Air Force Geophysics Laboratory (AFGL) line parameters compilation (Rothman et al. 1983). Rayleigh scattering is included via a parametric expression of optical thickness.

Aerosol scattering and absorption coefficients are based on the work of Shettle and Fenn (1975) as in ECHAM3. The cloud shortwave radiative parameters optical thickness and single-scattering albedo are a function not only of the cloud water path as in ECHAM3, but also of the effective radius of the cloud particles. The longwave spectrum is divided into six bands with gas absorption coefficients fitted from AFGL 1982. The water vapour continuum parameterization is based on Roberts et al. (1976). As in ECHAM3, clouds are treated as grey bodies with a longwave emissivity depending on the cloud water path (Stephens 1978).

The Morcrette code has undergone a number of changes at MPI. These include additional greenhouse gases (CH_4, N_20, CFCs, and the 14.6 µm O_3 band) (R. v. Dorland, presonal communication), parameterizations of effective radii and single-scattering properties of cloud water droplets and ice-cristals according to Rockel et al. (1991), and modifications in the water vapor continuum (Giorgetta and Wild, 1995).

For other aspects of this new version of the German climate model the reader is refered to the documentation by Roeckner (1995). The model currently installed at the Swiss National Computing Center (Centro Svizzero di Calcolo Scientifico, CSCS, in Lugano-Manno) has a resolution of about 120 km (spectral truncation T-106), which is currently the most detailed resolution used for climate simulations.

5.3.2 Overview of GCM results

The Climate Modeling Group at the Institute of Geography, ETH - Zurich, focused upon two distinct features of the ECHAM3 T - 106 simulations: the parameterization of readiative fluxes and the surface energy balance, and the computations of persistance and intensity of blocking episodes. The following is therefore a summary of work undertaken by the Climate Modeling Group, some of which has recently appeared in the peer-reviewed literature. Because the credibility of a GCM needs to be high, especially when they are extensively used to address scientifically and politically sensitive issues such as anthropogenically-induced climate change and its potential impacts, much of the work reviewed here is aimed at assessing the capability of the MPI General Circulation Model in reproducing certain climatically-relevant processes of current-day climate

The treatment of radiation by a GCM is a crucial part of any climate change investigations, because significant errors in the model-computed surface energy balance will lead to erroneous conclusions on future climate trends forced by radiatively-active trace gases. In addition, the Department of Geography has at its disposal a very comprehesive data set (GEBA: Global Energy Balance Archive) which has been compiled in the context of the World Climate Program of WMO, which allowed this group to carry out advanced intercomparison analyses. In the latter case, the model investigations of blocking high pressure episodes, particularly in the Northern Hemisphere and over Europe, allow an insight into some of the causal mechanisms and can help shed some light onto some of the possible reasons for the significant increase in the persistence of high pressure over the Alpine region in the 1980s (cf. Chapter 7).

Two five-year time windows were simulated with ECHAM3, for current climatic conditions and for equivalent doubling of atmospheric CO_2 concentrations (expected to occur around the decade 2040). Exhaustive analyses of ECHAM3 results have been carried out (Wild et al., 1995a; 1995b) to determine the model's capability in simulating certain features of observed climate and how these may change in future; indeed, much of the work has focused on present-day climate studies where much data is available for intercomparison purposes.

5.3.2.1 *Radiative fluxes and surface energy budget at the global scale*

Surface radiative fluxes calculated with the ECHAM3 GCM at T21, T42 and T106 spectral resolution have been compared with the observed data in the Global Energy Balance Archive (GEBA). GEBA is a new comprehensive compilation of surface energy balance measurements which have been performed worldwide over the past decades. This observational dataset has proved to be a valuable tool for the validation of surface fluxes in GCMs.

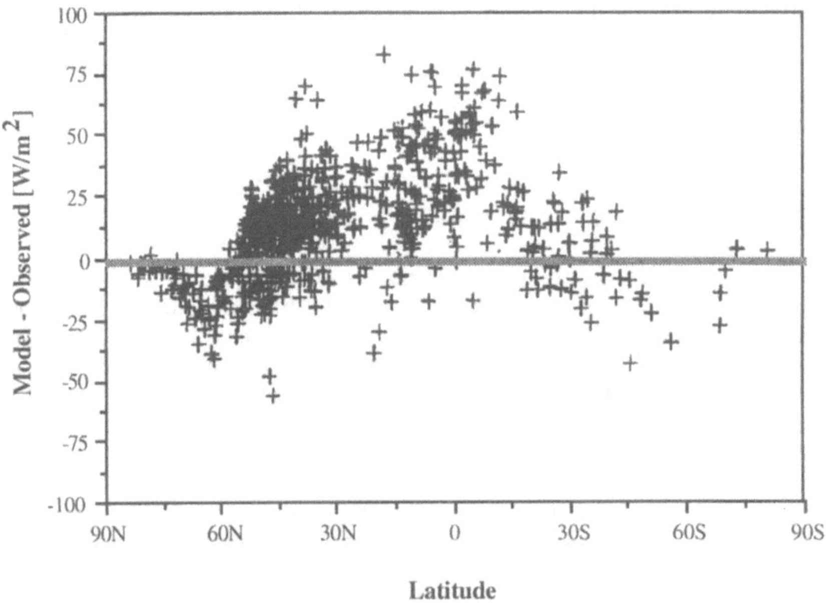

Figure 5.14: Difference between model-calculated and observed annual mean incoming radiation

On the global scale the model atmospheres have been shown to absorb a too small quantity of solar radiation, which results in an overestimation of the incoming shortwave radiation at the surface. The absorbed global mean shortwave radiation of 164 -167 Wm^{-2} in the ECHAM3 model is estimated to be in excess over observations by about 10 - 15 Wm^{-2}. Since the absorbed shortwave radiation in ECHAM3 is already comparatively small with respect to other GCMs, similar or higher overestimates are likely to be found in many current GCMs. Figure 5.14 illustrates the geographical distribution of diescrepancies in annual mean insolation for the T-106 version of ECHAM3.

One hypothesis to explain the discrepancy in fluxes is related to the inadequate performance of the radiation scheme under clear sky conditions. A stand alone validation of the shortwave scheme showed daily maximum overestimates up to 50 Wm^{-2} for clear sky cases, which may be related to the underestimation in the water vapor absorption. As a further source an insufficient shortwave absorption by the model clouds has been suggested.

The incoming longwave radiation in the ECHAM3 models at the three spectral resolutions (330 - 335 Wm^{-2}) is underestimated by 10 - 20 Wm^{-2}, an amount similar to the shortwave overestimate. This was concluded from comparisons with the measurements available in GEBA, from the analysis of the low cloud distribution which showed a lack of calculated low clouds and from an isolated validation of the radiation scheme, which showed an underestimation of the incoming longwave radiation under cloud-free conditions on the order of 10 Wm^{-2}. Underestimates of the global mean incoming longwave radiation are present in other GCMs as well.

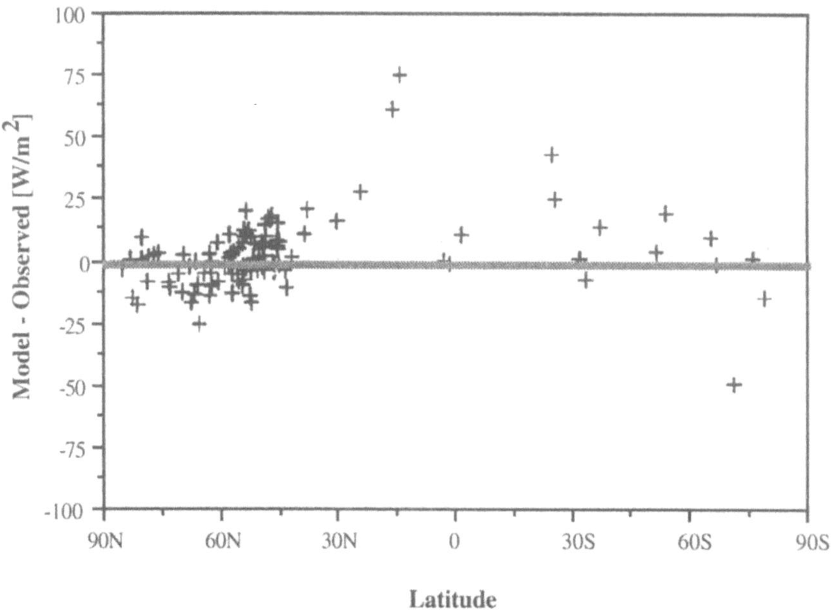

Figure 5.15: As Figure 5.14, except for mean annual net radiation fluxes

The net radiation measurements in GEBA suggest that the global mean surface net radiation is unlikely to be higher than the 102 Wm^{-2} calculated in ECHAM3. A value near 102 Wm^{-2} is further supported from the analysis of the longwave and shortwave components of the net radiation and from independent considerations on the global hydrological cycle. Figure 5.14 provides an overview of the differences in annual mean net radiation fluxes between the model and surface observations. A common feature of ECHAM3 and other GCMs is a superficially realistic simulation of surface net radiation as the overestimate in the shortwave balance is compensated by an underestimate in the longwave balance.

From a zonal and seasonal perspective, the incoming shortwave radiation in ECHAM3 is overestimated in low latitudes year round and in mid-latitudes in summer, but underestimated in high latitudes year round and in mid-latitudes in winter. This results in a too strong meridional gradient of surface solar insolation in midlatitudes. This behavior is found in all simulations independently of horizontal resolution.

The same tendency is found in the net radiation and temperature, so that too much net energy and too high surface temperatures are found in latitudes with excessive solar radiation, while a too small amount of energy is available for the non-radiative fluxes of the surface energy balance at higher latitudes. This may have major consequences on the model's dynamics and hydrological cycle. The latitudinal discrepancies in the incoming shortwave radiation are consistent with differences between simulated cloud amount and the surface-based cloud climatology of Warren et al. (1986). The large scale simulation of radiative surface fluxes is largely independent of changes in horizontal resolution. The systematic model deviations seem to dominate the effects of an increase in resolution.

5.3.2.2 *Radiative fluxes and surface energy budget at the continental scale*

Surface energy fluxes of a GCM with high horizontal resolution (T106) have been analysed over the European region. The observations have been retrieved from a new dataset, which contains the worldwide direct measured energy fluxes (Global Energy Balance Archive, GEBA). The available data allow a seperate assessment of the simulated shortwave, longwave and net radiative fluxes over the European area and provide a unique basis for the surface flux validation of high resolution GCMs and regional climate models.

A comparison of the ECHAM3 T106 calculated surface radiative fields over Europe with the observations from GEBA has revealed a number of systematic errors. An overestimation of insolation in southern and central Europe and and an underestimate in northern Europe has been identified in the model, leading to a too strong annual mean meridional insolation gradient. The available data allowed the identification of a similar overestimated gradient over North America and Asia. In summer the transition zone from under- to overestimation is shifted northward leading to an overestimation of insolation over most of Europe, while in winter the southward shift of the transition zone results in an underestimation of insolation over northern and mid Europe. This can be related to deficiencies in the simulation of the cloud amount and water vapour distribution over Europe, and to the radiation scheme itself.

The underestimated cloud amounts and atmospheric humidity in the European summer together with the deficiencies in the longwave part of the radiation code lead to an underestimated incoming longwave radiation consistently apparent at all European stations with longwave measurements. The ratio of incoming to outgoing longwave radiation over mid Europe, a rather conservative quantity, should be close to the observed ratio of 0.90 rather than the simulated 0.85 and indicates a too strong cooling in the net longwave balance. The simulated surface net radiation comes close to the observed values in the annual mean over Central Europe, but this is mainly a result of the cancellation of the excessive incoming shortwave radiation and the too low incoming longwave radiation. The overestimate of net radiation in summer and the underestimate in winter is closely linked to the deficiencies in the simulation of the annual cycle of insolation. Therefore too little energy is available in winter and too much in summer for the nonradiative components of the surface energy balance.

The T106 horizontal model resolution allows to capture topographic features previously unresolved in lower resolution GCMs. Therefore the assessment of vertical and horizontal gradients in the Alpine mountain chain has become feasible. Compared to the observations, although substantial differences exist in the absolute values of the fluxes, the model captures some of these gradients in a realistic way, particularly the difference in insolation on the northern and southern slope of the alps, the transition from an increase of insolation with altitude in winter to a decrease in summer and the annual mean decrease of incoming longwave radiation and net radiation with altitude. Quantitatively the calculated increase of insolation with altitude in winter is too small and the decrease in summer too large, so that in high Alpine areas the insolation overestimate is reduced and no longer compensates for the underestimated incoming longwave flux. This leads to a too strong vertical decrease of surface temperature and too cold high Alpine areas.

Partly on the basis of this analysis, a new radiation code has been incorporated into the latest version of the GCM (ECHAM4); preliminary analyses of a 10-year time-window for current climate simulations indicate substantial improvements in the accuracy of the radiative flux components contributing to the surface energy balance; these improvements have consequences for the improved simulation of climate close to the fround, in particular processes related to soil moisture which in ECHAM3 led to a too rapid drying of the soil in summer and a corresponding erroneous increase in low-level air temperatures.

5.3.2.3 Blocking

Current ideas on blocking mechanisms suggest that the interactions between small-scale transient disturbances and the planetary-scale waves play a key role in maintaining a blocking anticyclone which has developed in the diffluent jet exit region over the eastern Atlantic or Pacific oceans. The high horizontal resolution may thus have a positive effect on the simulation of blocking episodes, since the mid-latitude transients are expected to be captured more realistically at T-106 truncation than at T-42 resolution. Several methods of identifying blocking situations, from the classical criteria proposed by Rex (1950) to the more objective methods of searching positive persistent anomalies in large digitized data sets (e.g. Dole and Gordon, 1983), have been used by a number of meteorologists with slightly different notions of the phenomenon in mind. Here, the simple height criterion proposed by Lejenäs and Økland (1983) is adopted. From (half-) daily 500hPa height data, a longitude-dependent index is formed by taking the difference between 40 and 60° north. Figure 5.16 illustrates the observed winter 500 hPa field based on ECMWF analyses and the simulated geopotential height field as simulated by the ECHAM3 GCM.

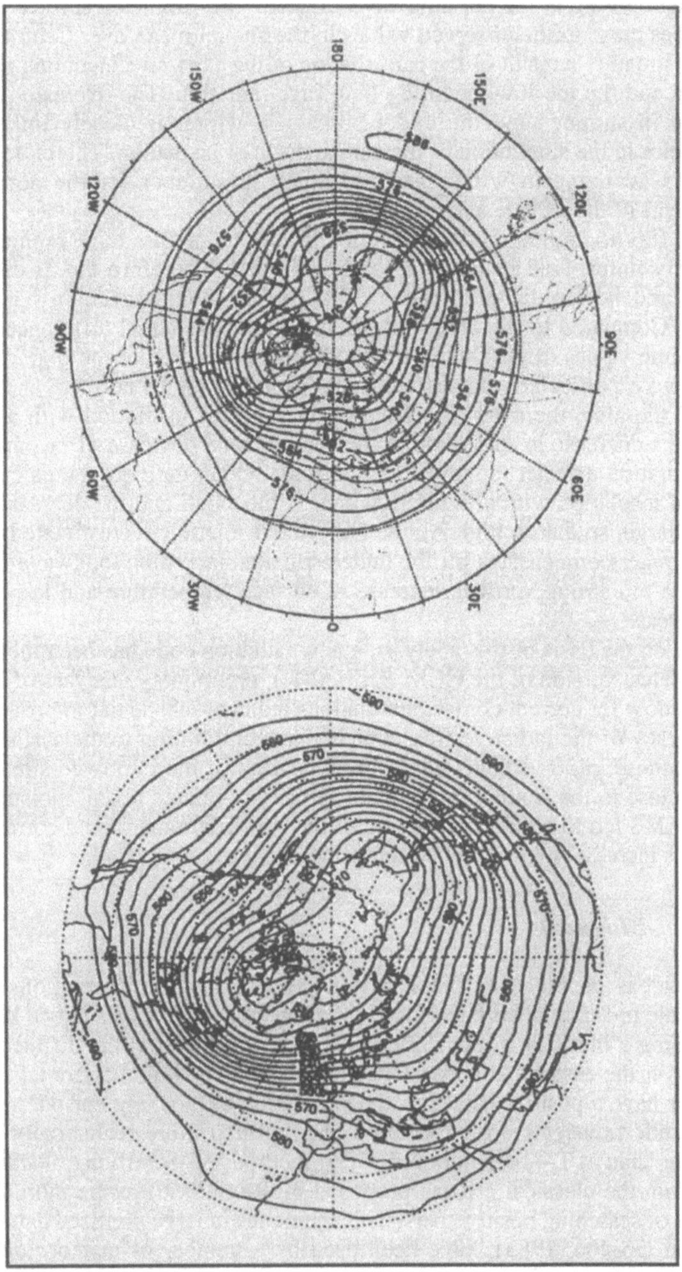

Figure 5.16: Northern hemisphere map of 500 hPa geopotential height for the winter season. Upper: observations from Bluestein (1993) based on ECMWF analyses; lower: ECHAM3 T106 simulation

Usually, this geopotential height gradient is positive, but when a high pressure cell establishes itself around 60° it may become negative. If this is the case, averaged over 30° longitude, the episode is defined as being "blocked" at the longitude in the center of the 30 degrees. While quite efficient, the method does not necessarily capture all possible blocking configurations. In the Euro-Atlantic sector, the simple height gradient method is adequate and allows comparison of blocking frequencies simulated by the ECHAM3 model at different resolutions, and observed frequencies of the 30-year period 1950-1979 as calculated by Lejenäs and Økland (1983) from NMC (US National Meteorological Center) data.

Composites of the blocking episodes lasting for more than 5 days were assembled to emphasize the typical blocking pattern of the two major centers. For the Euro-Atlantic composite only blocks within 30°W and 30°E were taken into the averaging procedures, while the sector from 170°E to 130°W was chosen for the Pacific region. The block with the largest persistence, however, occured eastwards of 30°E and lasted for 12 days. This and other cases are studied by inspecting consecutive five-day means.

In Figure 5.17, the blocked half-days of one particular model year are shown as a Hovmöller diagram to illustrate the rather irregular occurence of blocked days. The lack of blocking activity for all simulations in the Pacific area, especially in January, is particularly striking. In the Euro-Atlantic sector around the 0° meridian, the model captures the annual cycle and especially the sharp decrease of blocking occurence from May to June with great skill. A slight eastward shift of the T-106 model blocking frequencies can be noted, too. From the compositing method, 8 cases totaling 60 days (3*6+7+8+3*9) could be extracted from the 1800 simulated days for the Euro-Atlantic pattern.

For the Pacific composite the corresponding numbers include 4 cases with a total of 26 days (3*6+8). To have a larger composite, and also because all cases but one occured in winter or spring, the averaging was not separated according to season. The ECHAM3 GCM at T106 resolution simulated a distinctive split-flow type block in the Euro-Atlantic sector that extends over 60° longitude, with the anti-cyclone centered over the North Sea at 55° latitude. The Pacific pattern consists of a strong ridge which tilts backwards, and accompagnying strong westerly flow extending into Western Europe may be observed. Not much cancellation occured as a result of the averaging, indicating that all the cases developed at more or less the same location within the prescribed 60° longitude sector.

Analysis of model data at high (T-42) and very high (T-106) resolution climate integrations with ECHAM3 and prescribed climatological SSTs generally exhibit high interannual variability; both the T-42 and T-106 versions generate blocking-like structures with approximately the same frequency. In the Euro-Atlantic sector, the values compare favorably with observations, while they are much too low in the Pacific area. This discrepancy is partly due to the method used to detect blocking sutuations, which is very sensitive to the latitude where the difference in geopotential height is taken, and partly attributable to shortcomings in the simulated jet structure. The eastward shift of some European blocks can also be explained by the Atlantic jet which extends too far towards Europe. Despite these shortcomings, which show up in the climatological distribution of the 500hPa geopotential height, the model was able to simulate the annual cycle of the observed blocking activity and the two primary patterns with considerable skill.

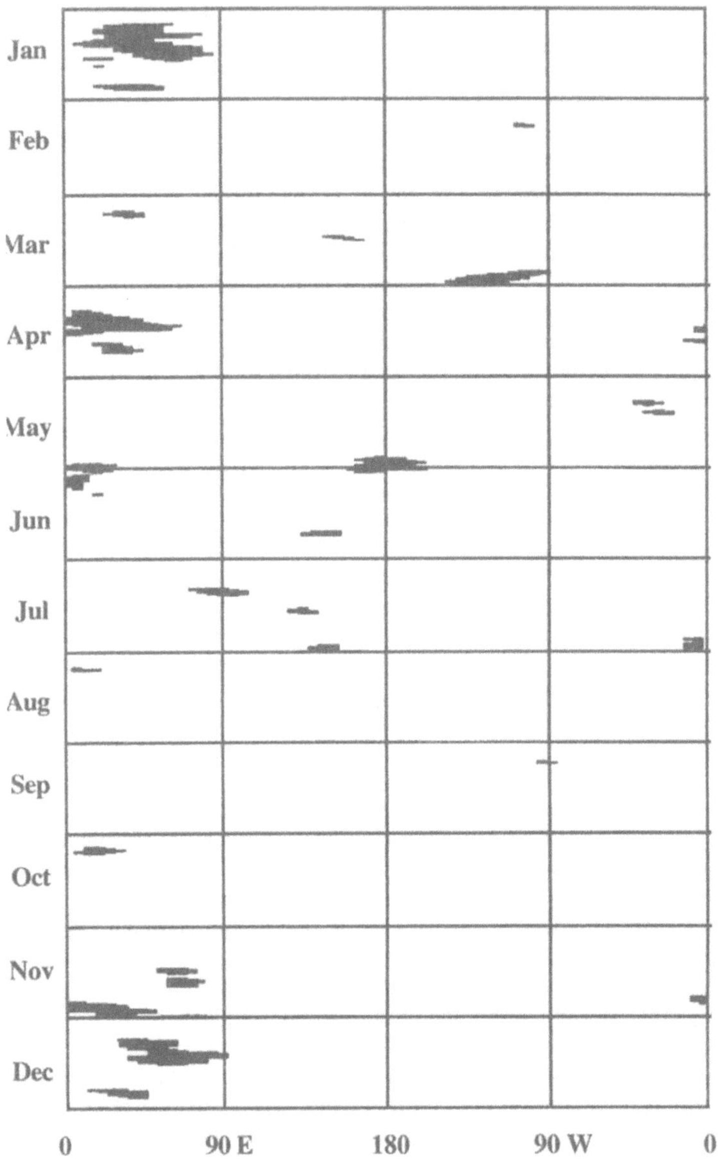

Figure 5.17: Hovmöller diagram illustrating periods and geographical distribution of blocking episodes

5.3.2.4 Differences between current climate and "double-CO_2" climate as simulated by the ECHAM3 GCM

This section will briefly report on a further application of the GCM to the problem of the enhanced greenhouse effect. Most of the initial work on climate modeling in Switzerland has focused upon the simulation of current climate. However, it has been worthwhile attempting these preliminary studies of a 2 x CO_2 simulation using the nested modeling approach because it is a topic which receives much attention from the media, the general public, and policy makers, and therefore needs to be addressed; it is also through early experiments of this kind in the 1970s and 1980s that the possible consequences of the enhanced greenhouse effect were first realized, leading to the setting up of the Intergovernmental Panel on Climate Change (IPCC) in 1988, the publishing of the IPCC reports (1990, 1992, 1995), and ultimately the preparation and ratification of the UN Framework Convention on Climate Change (FCCC).

In addition to these considerations, many of the exhaustive analyses of the model results were sufficiently encouraging to justify the simulations of future climate in an enhanced greenhouse atmosphere with these models. The objectives of fundamental research and science as a tool for decision making converge in this type of analysis.

Variable	Global	Oceans	Land
2m Temperature (°C)	+1.5	+1.4	+2.3
Precipitation (mm/day)	-0.02	-0.07	+0.09
Evaporation (mm/day)	-0.02	-0.04	+0.03
10m Wind speed (m/sec)	-0.06	-0.09	-0.01
Cloud cover (%)	+1.0	+1.0	+1.0
Integrated water vapor (kg/m^2)	+3.3	+3.4	+3.0
LW downward radiation (Wm^{-2})	+10.3	+9.1	+13.4

Table 5.2: Global annual mean changes of different variables as a result of doubled CO_2 concentrations

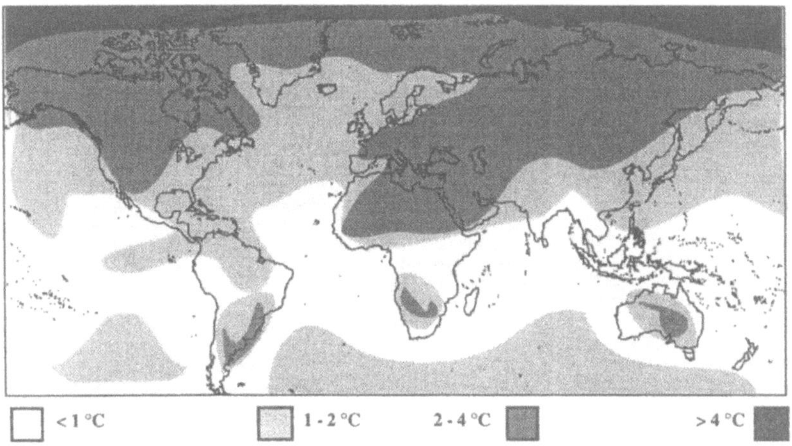

Figure 5.18: Changes in annual mean global temperatures between current climate and 2 x CO_2 climate

It is unfortunately not feasible even with current supercomputer resources to use the high resolution T-106 GCM coupled to an ocean model over decades as required in a transient scenario run for the next century. Therefore a "time window" experiment has been performed. In this experiment, sea surface temperatures and sea ice have been taken from the 100 year scenario simulation with a low resolution (T21) GCM coupled to a dynamic ocean model previously performed at the Max-Planck-Institute for Meteorology in Hamburg (Cubasch et al., 1992).

These serve as boundary conditions for the T106 simulation at the time of CO_2 doubling, which corresponds to the year 2045 according to the IPCC scenario IS92A. The integration with doubled CO_2 concentration extends over 5 model years, and a few results of this experiment are discussed in a succinct manner in the present subsection; Table 5.2 provides an overview of the changes in the global annual mean values.

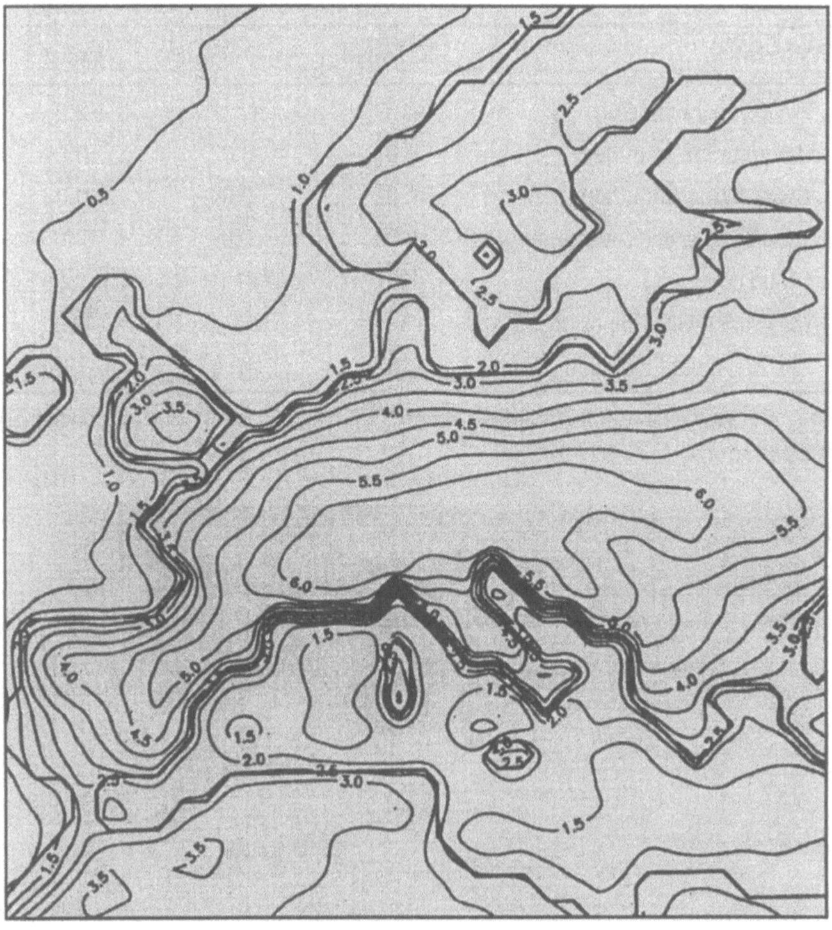

Figure 5.19: Summer surface temperature changes in Europe between current climate and 2 x CO_2 climate

The global mean temperature increase in this experiment with respect to the control simulation is 1.5°C. From Figure 5.18 it can be seen that the increase is largest over the continents and at high latitudes. The land surface is responding faster to the greenhouse forcing than the ocean due to its large thermal inertia, as shown in Table 5.2. The retreating snow cover and sea ice, and the associated positive feedbacks lead to the amplification of the greenhouse forcing at higher latitudes. The increase in the Arctic is strongest during autumn; this can be explained by delayed and reduced sea-ice formation.
Over Western Europe and in the Alps, the temperature increase is larger in summer than in winter (Figures 5.19 and 5.20 illustrate summer and winter temperature changes over Europe as simulated by the T-106 GCM). A strong drying of the model soil contributes to the high temperature increase in summer with up to 6°C over the Alps. This is a problem already identified in the control simulation (Wild et al., 1995a) which is further amplified under the CO_2 doubling conditions.

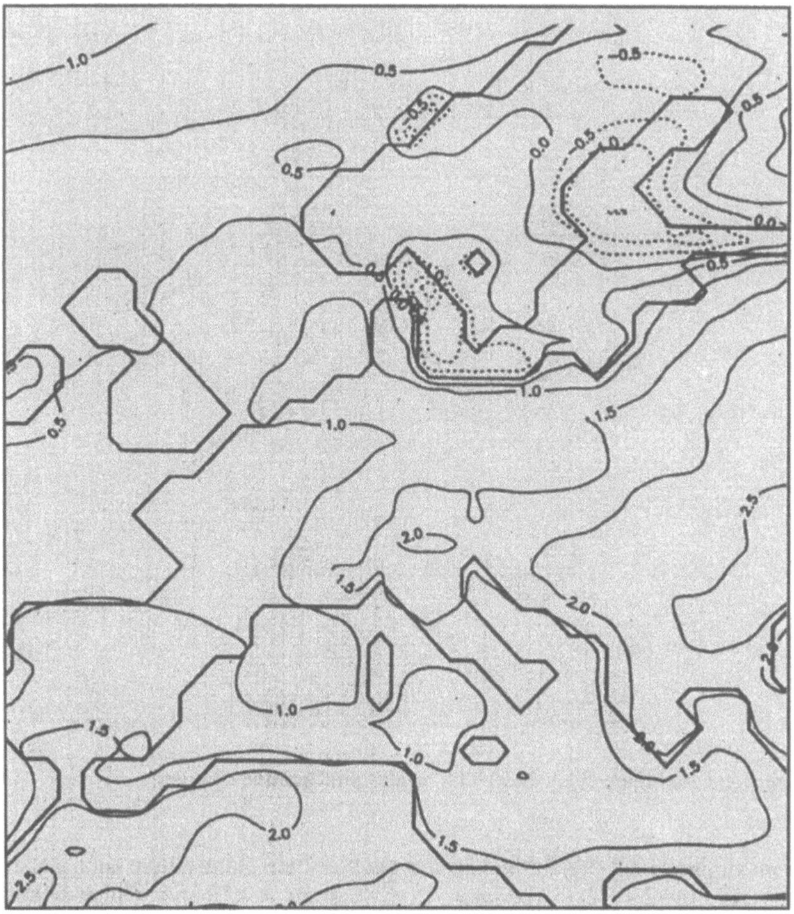

Figure 5.20: Winter surface temperature changes in Europe between current climate and 2 x CO_2 climate

The manner in which the mid-latitude weather systems will behave under these competing effects needs further investigation. Since the water vapor content increases in the warmer troposphere (cf. Table 5.2), the effectiveness of the heat transport by the transient eddies is enhanced. It is possible that eddy activity may be reduced in a warmer global climate through the expected decrease in the equator-to-pole latitudinal gradient of temperature.

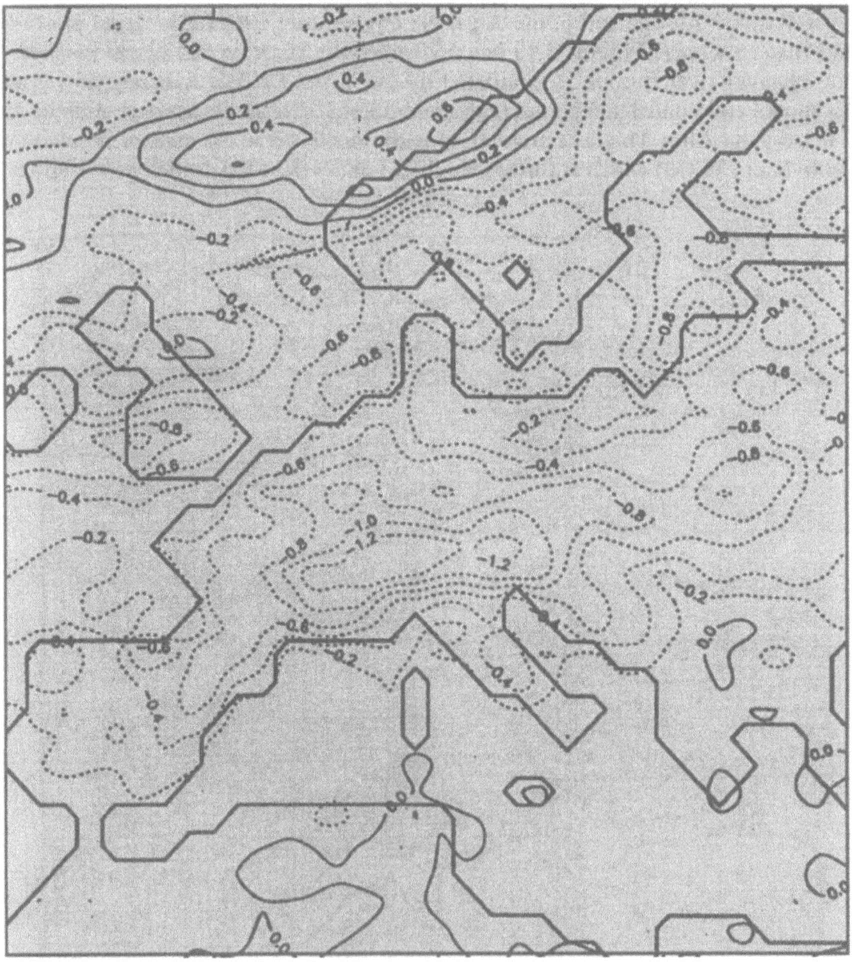

Figure 5.21: As Figure 5.19, except for winter precipitation changes

The maximum zonal wind at the axis of the Northern Hemisphere subtropical jet increases in the 2 x CO_2 experiment from 41.8 m/s to 45.1 m/s, which is consistent with the temperature changes aloft. In addition to this strengthening, the jet is shifted upwards, but not northwards as reported in the study of Hall et al.(1994).

An interesting feature of this experiment is the slight reduction of the intensity of the global hydrological cycle with increased CO_2 concentrations (cf. the decreasing precipitation and evaporation in Table 5.2). This is contrary to most other GCM experiments of this type and to the same GCM experiment with the T42 and T21 versions of ECHAM3. Further work is needed to understand how the high horizontal resolution, which has not been used before in studies related to the enhanced greenhouse effect, is contributing to this reduction. So far it has become apparent that areas of decreased evaporation are often related to areas of reduced windspeed. It can be seen from Table 5.2 that global mean wind speed is slightly reduced in the 2 x CO_2 experiment. and this may be contributing to the evaporation decrease.

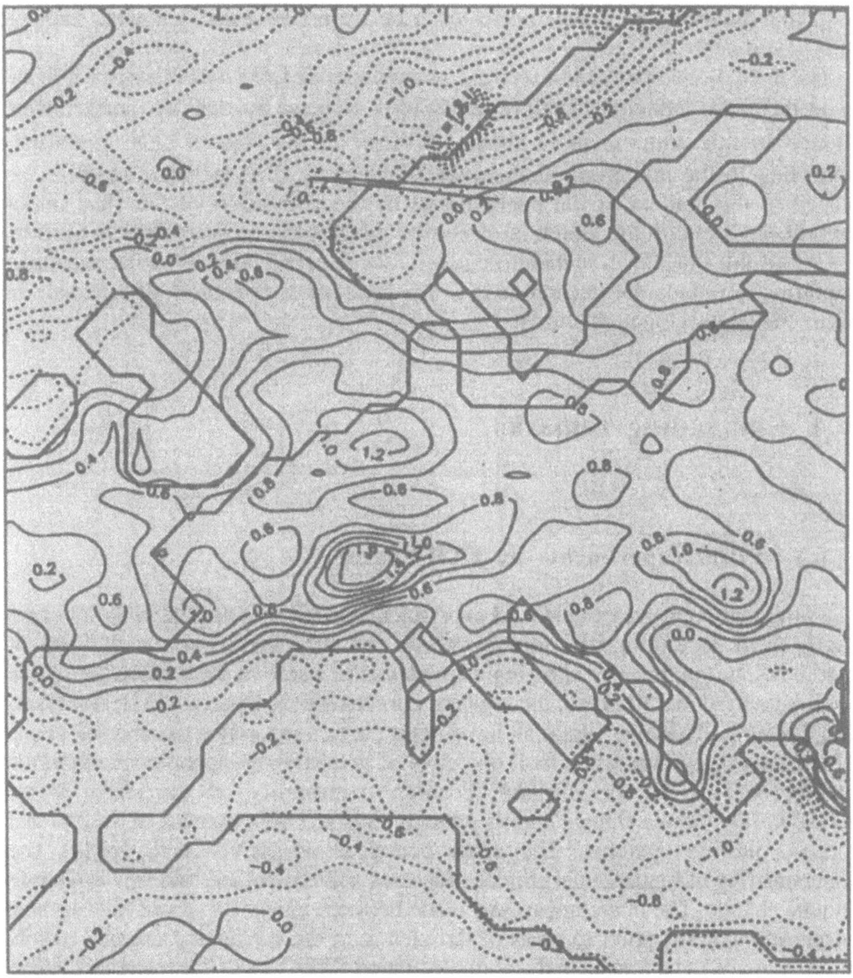

Figure 5.22: As Figure 5.19, except for summer precipitation changes

The zonal-mean temperature changes of the atmosphere show the characteristic greenhouse signal with warming in the troposphere and cooling in the stratosphere. In particular, the marked warming in the tropical upper troposphere, together with a cooling in the extratropics at the same level (200 hPa) implies an enhanced meridional temperature gradient aloft. At the ground, the Equator-to-Pole temperature contrast and thus the low-level baroclinicity is reduced.
No pronounced trend can be noted in the zonal distribution of the precipitation changes. The distribution of precipitation change over western Europe and the Alps is shown in Figure 5.21 for summer and Figure 5.22 for winter conditions. A strong reduction of precipitation is to be noted in summer. In winter, over Western Europe and the Alps, an increase of precipitation can be observed, while the Mediteranean region is drying up. However it should be emphasized here that due to the limited sampling period of five years and the large interannual variability of precipitation, the results presented here are not necessarily always statistically significant.

The most direct effect of an increase in atmospheric CO_2 concentrations is seen in an increased amount of longwave radiation directed towards the surface. The change in this component is about 10 W/m^2 at the time of CO_2 doubling, according to the model simulations (cf. Table 5.2). This increase needs to be placed in the context of the uncertainties in the simulation of this flux under present day conditions, which is of similar magnitude, as described in Section 5.3.2.2 of this chapter. It is therefore crucial for the credibility of global warming scenarios to reduce the uncertainties in the simulation of present day climate in future model development efforts.

5.4 Concluding remarks

5.4.1 T-106 experiments at ETH-Zürich

The principal objective of the GCM experiments undertaken to date in Switzerland has been to develop a methodology which can help overcome the problem of spatial resolution, thereby allowing estimates to be made of the regional response of climate to global forcings; the regional climatological aspects will be discussed in Chapter 8. The investigations have been partitioned into a number of steps, which converge towards the final objective of providing an operational modeling system for use to the climate research community, in particular those investigating climate change and its potential impacts on a number of natural and socio-economic systems. The other principal objective is to further the understanding of fundamental climate processes and control mechanisms leading to climate change. The intercomparisons made between model-simulated features and observations helps point to areas of deficiency in the model and thereby enable improvements to be formulated and implemented. The steps of the research leading to regional climate change assessments are as follows:

1. Simulation of current climate with the high-resolution General Circulation Model and exhaustive analysis of the model performance, in particular in the following domains:
- The radiation budget (surface energy balance and individual SW and LW components); intercomparison with Global Energy Balance Archive database of the Geography Department, ETH
- Comparison with T - 42 and T - 21GCM simulations to check influence of resolution on solutions
- Strength and persistence of blocking high pressure episodes

2. Simulation of future climate (climate change scenario based on a doubling of equivalent CO_2 concentrations in the atmosphere compared to 1990 values) with the General Circulation Model. Analyses of model solutions, in particular:
- Changes in general circulation patterns between current and 2 x CO_2 climate
- Changes in temperature fields
- Changes in precipitation amount, distribution, and intensity
- Changes in the surface energy budget, and in the SW and LW radiative fluxes

3. Investigation of climate data in Switzerland (to be elaborated upon in Chapter 7). This first step involving the observational record had the following objectives:
- To evaluate the nature and amplitude of climate change this century
- To establish links between global-scale signals and the regional response in the Alpine zone
- To set up a data set for intercomparison purposes between model results and the observational record.

4. Simulation of current climate (January and July daily climatologies) with the high-resolution Limited Area Model and exhaustive analysis of the model performance, in particular in the following domains:
- Boundary-layer parametric scheme in complex terrain
- Precipitation, in particular orographically-forced and convective
- Intercomparison with Swiss climatological observations

5. Simulation of future climate with the Limited Area Model. Analyses of model solutions, in particular:
- Changes in the behavior of precipitation in the Alpine area between current and 2 x CO_2 climate
- Changes in temperature fields
- Changes in wind distribution

Steps 1 and 2 have been described in this Chapter. Steps 3 - 5 will be discussed in Chapters 7 and 8.

5.4.2 Other GCM scenario experiments

In its 1995 Second Assessment Report, the IPCC (1996) has extensively reviewed the existing literature on future climate simulations by the global climate modeling community; indeed, specific GCM experiments were conducted for the IPCC. There is a certain consistency between model results for future climate simulations, and the range of possible global temperature increases projected to the year 2100 for the IS92A Scenario lies between 1.5 and 4.5 °C, with the best estimate set at 2.5°C, for a climate forced by greenhouse gases alone.

In contrast to the 1990 assessments, the 1995 IPCC reports have made use of results from a number of GCMs which have included the forcing effect of aerosols for the first time in a systematic manner. Incorporating the potential negative forcing of future changes in anthropogenic aerosol concentrations leads to a lower global temperature range of between 1 and 3.5°C, which is nonetheless a significant warming of climate. On a continental to regional scale, the influence of aerosols is seen in some instances to even counteract GHG warming, particularly over regions of high industrial activity such as parts of North America, Europe, and the emerging industrial nations such as China and India.

A summary of the state of knowledge on current and future climate, the later based exclusively on GCM results, is given below.

5.4.2.1 What is certain

- There exists a natural greenhouse effect which maintains the global atmosphere at a temperature level about 33 °C warmer than otherwise, thereby allowing various life forms to be sustainable
- The absorption of radiated energy will raise atmospheric temperatures because of the physico-chemical properties of the greenhouse gases; the extent of warming, however, will depend on feedback mechanisms which can either damp (negative feedback) or amplify the anthropogenic climate signal (positive feedback)
- The amounts of radiatively active trace gases are significantly higher than their pre-industrial values, and it will take hundreds of years to return to these levels, assuming economic and policy measures are taken to reduce greenhouse-gas emissions
- Globally-averaged surface air temperature has risen by about 0.6°C in the last 100 years

5.4.2.2 What is probable

- Lower tropospheric temperatures will continue to rise through at least the end of the 21st century.
- Land areas in high latitudes will experience amplified wintertime warming. Oceans will warm less than continents
- Higher temperatures will enhance the hydrological cycle. Summer dryness will increase in mid-latitudes

- Global warming will cause sea level rise and reduce sea ice volume in the Northern Hemisphere
- Natural forcings, whatever their sign, will be smaller in the next 50 - 100 years than greenhouse-gas forcing. This is illustrated in Figure 5.23, where the sign and amplitude of natural forcings are compared to the expected forcing by the four original IPCC scenarios
- Stratospheric temperatures will significantly cool, in part due to stratospheric ozone depletion

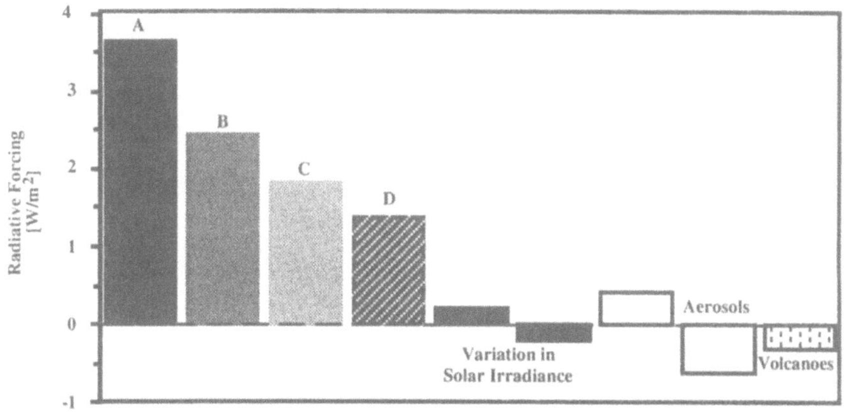

Figure 5.23: Comparison of anthropogenic forcings (IPCC 1990 Scenarios A - D) compared to natural forcings of the climate system

5.4.2.3 What is uncertain

- There may be changes in climate variability, though a warmer climate may not necessarily be accompanied everywhere by increases in variability
- Changes in regional-scale climate will be different from the global average. Observational evidence has shown that certain regions of the world have extremely different responses to global changes, such as the Alps for example which have experienced significant warming tendencies of up to 2 °C this century, compared to the global temperature increase of about 0.6 °C (Beniston et al., 1994; Auer and Boehm, 1994)
- The intensity of tropical storms may increase; this is based on the fact that the extent of regions where conditions are favorable will increase in a warmer world. However, analysis of the ECHAM3 T-106 data for the 2 X CO_2 experiment has shown that the number of tropical cyclones diminishes in the Indian and Pacific Oceans north of the Equator (Bengtsson et al., 1995). The changes in intensity of the storms was also analyzed, but the evidence of a significant change was inconclusive because of the short sampling period of the simulated time window

5.4.2.4 How can confidence in GCMs increase?

There has been substantial improvement in the performance and predictive capability of GCMs, particularly as a result of the coupling of atmospheric models to ocean models, and the inclusion of aerosols as a counter-forcing factor which is important on regional to continental scales. Numerous problems remain, however, which are currently being addressed to enhance our confidence in GCM results. These include improvements in our understanding and parameterization of:

- Clouds and their radiative effects
- Hydrological balance over land, including soil moisture
- Heat flux at the ocean surface
- Component models of the climate system and their coupling
- Global data coverage, in particular observational data for model verification, the enhancement of existing data sets (re-analysis), and information on heat fluxes at the ocean surface

6. Coupling the Mesoscale to the Microscale: Air Pollution Simulations

6.1 Introduction

Air pollution is an extremely complex phenomenon, and this complexity makes its understanding, and therefore its control more elusive. This complex behavior reveals itself not only in the physical terms of diffusion and transport of emitted pollutants, but also chemically, through the many reactions occurring in the atmosphere. It is therefore important to correlate emission and resulting pollutant concentration and to characterize the impact of different sources of pollution (industry, vehicles, domestic heating) on the environment.

The study of the dynamics of air pollution requires a good evaluation of the three-dimensional air-flow characteristics in a particular region. It is this air flow, and the associated turbulence, which will transport and disperse the pollutants as they are carried from their emission point (or source) to a region of particular interest where knowledge of concentrations (or immissions) is necessary. Such computations are of use when attempting to understand air pollution in general, but also have predictive value when determining minimal-impact locations of a future industry or highway. An additional value of modeling in this context is that accurate reproduction of particular pollution episodes can help in disaggregating the different sources of pollution contributing to immission in a given region, thereby allowing an identification of the most responsible sources.

The transport and dispersion of atmospheric pollution is commonly governed by processes acting on a wide range of scales, in particular:

- *The microscale.* Sources of air pollution are very often localized points or lines in space (e.g., industrial chimneys, traffic along highways); larger-scale or regional air pollution, such as in an industrialized urban environment, represents the combined effect of numerous individual sources. Examples of truly regional sources of atmospheric pollution are, for example, biomass burning related to deforestation in tropical areas, a subject which will not be discussed in this work. The behavior of air pollution as it is transfered from its source area to its receptor area is governed by microscale processes, namely dispersion and dilution through atmospheric turbulence.
- *The mesoscale.* In many instances, transport of air pollution by the mean flow will occur over relatively long distances, spreading the gaseous or particulate matter over large regions. Mesoscale circulations will largely determine the regions which are likely to be the most severely affected by pollution, i.e.,

those which are poorly ventilated; knowledge of regional air flow patterns, such as sea breezes or mountain and valley winds, and the influence of thermal stratification on these flows, are therefore essential for understanding the behavior of atmospheric pollution.
- *The macroscale.* Ultimately, air pollution which does not remain at ground level or close to its source will be incorporated into the planetary-scale atmosphere through different mixing and transport mechanisms. Acid rain resulting from sulfate emissions is one such example of long-distance pollution effects which act on a continental or hemispheric scale. The concerns of enhanced greenhouse-gas forcing of climate, or stratospheric ozone depletion, are related to the global influence of trace gases originating in very specific regions of the world.

Clearly, atmospheric pollution and its regional to large-scale transports cannot be treated explicitely in a single model because of the range of scales which characterize pollution. Scale separation is therefore an important element in the study of air pollution; the coupling of models representative of different scales is one approach to integrating the microscale elements of pollution embedded in larger-scale flows.

The examples discussed in this chapter will make reference to two models developed for the purposes of investigating atmospheric pollution, the DREAMS-1 mesoscale model (described in Chapter 4) and the DREAMS-2 pollution transport and dispersion module, to be discussed in Section 6.2. The models have been primarily applied to Switzerland, where the mesoscale wind-field is primarily determined by local mountain and valley systems. Other factors, however, can exert significant influences on regional-scale flows. Among these, one may identify geographical or geophysical features such as surface roughness heterogeneities (human settlements, forests, fields, or lakes); albedo changes; differential distribution of surface temperature, both in space and in time (diurnal heat cycle); and surface moisture characteristics (presence of bodies of water, etc.).

In the following text, the term immision is defined as the concentration of a particular pollutant at a given location in space. In most circumstances, immission is used to describe ground-level pollutant concentrations, although strictly speaking, it describes the three-dimensional nature of pollution. Units are generally mass per unit volume (mg/m^3) or parts per volume (ppm, ppb, etc.).

The text in this chapter draws on material from the following publicatons: Beniston (1987; 1989; 1990), Beniston and Pielke (1987), Beniston et al., (1989, 1990), and Ehinger and Beniston (1993).

6.2 The microscale atmospheric pollution module DREAMS-2

A typical simulation over a given geographical area with the DREAMS-1 mesoscale atmospheric model (see Chapter 4) will compute through to steady state (or to a particular time of a transient simulation) a set of dynamic and thermodynamic fields. These will adjust to surface conditions, local thermal stratification, and if the meteorological situation provides the necessary conditions, to cloud activity, precipitation, and cloud-induced secondary circulations. Once the desired conditions have been attained, data can be extracted for a much finer grid required for pollutant transport and dispersion simulations. Dynamic, thermodynamic and orographic information is transfered onto the grid of the microscale pollution transport and dispersion model DREAMS-2. The domain of the microscale model can either identical to that of DREAMS-1, or represent a sub-area of DREAMS-1, i.e., a numerical "zooming-in" on a region of particular interest. Interpolation of the necessary meteorological and orographic fields is achieved in this case through a 3-D interpolation technique, as will be discussed below.

The problem of determining concentrations of a particular pollutant at a given location by numerical analysis is exceedingly complex. The Eulerian technique is poorly adapted to point sources, since the source becomes specified over an entire grid volume which, even in the case of a fine mesh, is an invalid representation of the point source. The most commonly used Gaussian dispersion models yield elegant analytical solutions to the emission-transport-immision problem, but contain so many restrictive assumptions that they are unable to produce correct solutions to dispersion in complex terrain under real meteorological conditions. More recently, researchers have focused on Lagrangian solutions to the dispersion problem. The advantage of the Lagrangian approach is that individual trajectories of particles (which represent a particular pollutant) are independent of grid size. Furthermore, a Lagrangian particle technique is the only method by which a complex flow situation, in particular strong wind shear or flow reversals, may be adequately taken into account. Both continuous and instantaneous releases can be considered. Among the first workers in this field were McNider (1981), whose "conditioned particle technique" has been widely used by Pielke and co-workers (Pielke, 1984). Lagrangian particles as tracers for pollution dispersion studies have also been designed by Legg and Raupach (1982), and adapted to heavy gas dispersion by Schorling (1987), i.e., here the particles are no longer passive elements of the flow but have their own mass.

The DREAMS-2 contains schemes allowing the study of the physical behavior of pollution with either the Eulerian or the Lagrangian techniques.

6.2.1 The Eulerian formulation

This segment of DREAMS-2 solves a conservation equation for a pollutant species χ, given by the following relationship:

$$\frac{\partial \chi}{\partial t} = -u\frac{\partial \chi}{\partial x} - v\frac{\partial \chi}{\partial y} - w\frac{\partial \chi}{\partial z} + F_\chi + Q_\chi \qquad 6.1$$

where F is the turbulent diffusion or dispersion of the pollutant as it moves in the flow, and Q is the emission rate of the pollutant at its source location. The F term in Equation 6.1 can be written as:

$$F_\chi = K_X\left[\frac{\partial^2 \chi}{\partial x^2} + \frac{\partial^2 \chi}{\partial y^2}\right] + \frac{\partial^2}{\partial z}K_z\frac{\partial \chi}{\partial z} \qquad 6.2$$

with K_X and K_Z representing the horizontal and vertical diffusion coefficients, respectively. These are recomputed at the higher spatial resolution of the DREAMS-2 code according to the dynamic and thermal stability information provided by the DREAMS-1 computations.

6.2.2 The Lagrangian formulation

In view of the success of this methodology, the "conditioned particle technique", described at length by McNider (1981) and Pielke (1984), has been implemented in DREAMS-2. The idea behind this approach is relatively simple and is illustrated schematically in Figure 6.1. A large number of numerical "particles" represent the mass of pollutants emitted by a particular source, which can be either a point source such as a chimney, or a linear source used to simulate pollution due to traffic along a highway. In both cases, the source can be instantaneous (puff) or continuous (plume). As soon as the particles are injected into the atmosphere, they are transported along by the mean flow and dispersed according to local turbulence conditions. By counting the number of particles, which have a given mass, in a so-called receptor volume, the concentration of any non-reacting pollutant can be evaluated to a certain degree of accuracy, as discused by Beniston et al. (1990).

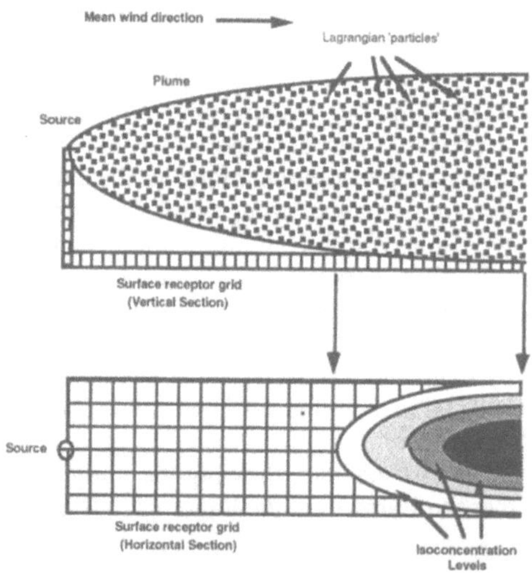

Figure 6.1: Schematic illustration of the principles of the Lagrangian formulation

The position in 3-D space of a given particle is described by:

$$x_i\,(t+\Delta t) = x_i(t) + [\, u_i(t) + u'_i(t)\,]\,\Delta t \qquad 6.3$$

where u_i is the grid-resolved wind vector and u'_i is the subgrid scale velocity component. This latter component may be written as:

$$u'_i(t) = u'_i(t-\Delta t)\, Ru_i(\Delta t) + u''_i(t-\Delta t) \qquad 6.4$$

where Ru_i are the Lagrangian autocorrelation functions for each velocity component in terms of the separation time Δt; the frozen turbulence hypothesis is used in this case. The u_i'' terms in Equation 6.4 are random fluctuating components of the flow field whose statistical properties are dependent on its turbulent kinetic energy.

The Lagrangian autocorrelations $Ru_i(\Delta t)$ are determined from:

$$Ru_i(\Delta t) = \exp(-\Delta t/TLu_i) \quad 6.5$$

where TLu_i are the Lagrangian time scales. These are evaluated as a function of turbulence spectra (e.g., Pasquill, 1962) and take the form:

$$TLu_i = 0.2\, \beta u_i\, \lambda_{mui} / \underline{V} \quad 6.6$$

where λ_{mui} is the peak wavelength in the spectra for each wind component, β_{ui} is the ratio of Lagrangian to Eulerian time scale, and V is the 3-D wind vector. The β term is given by:

$$\beta u_i = 0.6\, V/\sigma'_u \quad 6.7$$

where σ'_{ui} represents the standard deviation of velocity fluctuations. These terms are deduced from planetary boundary layer considerations, and in order to adequately take into account stratification of the atmosphere, the vertical components σ_w and λ_{mw} are stability dependent parameters. The reader is referred to the discussion by Pielke (1984) for more details on the computation of the turbulence terms in this kind of formulation.

The DREAMS-2 model uses the wind field computed previously by DREAMS-1 to advect the Lagrangian particles, while local characteristics of turbulence are given by the dynamic and thermodynamic fields interpolated onto the DREAMS-2 model grid. In order to simulate a plume, or a series of plumes, a large number of individual particles are released from a given source; within the emission region (e.g., the top of a chimney stack) the initial release of particles is assumed to be Gaussian distributed (i.e., bell-shaped in 3-D space). The mean flow and turbulence characteristics in a given area will rapidly modify the initial particle distribution and hence lead to a much more realistic distribution of particles downstream of the flow, than would an analytic Gaussian model or a Eulerian approach to the dispersion problem. In order to estimate immissions of a particular pollutant, it is necessary to count the number of particles in an arbitrary receptor volume. For a given volume $\Delta x \Delta y \Delta z$, the concentration χ is given by:

$$\chi(\overline{x}, t) = \iiint n\, dx\, dy\, dz \,/\, \Delta x\, \Delta y\, \Delta z \quad 6.8$$

with n the number of detected particles. Initial concentration is defined as:

$$\chi_o = \alpha\, Q/V_o \quad 6.9$$

where Q is the initial number of particles released per unit time (equivalent to the emission rate of a gas), V_o is the initial volume into which the particles are released, and α is the number of aerosols, or the volume of gas, that is assumed equivalent to one particle.

For detailed studies, a large number of particles must be released and the particle count needs to be made for small receptor volumes, so that reasonable statistics can be obtained for pollutant concentrations in 3-D space. In general, DREAMS-2 "releases" 11^3 particles (1331) for each point source at each time step. A complete simulation with several continuous emittors in a particular domain will typically involve the computation of several million particles. The receptor volume must, in general be smaller than the computational grid in order to obtain a reasonable distribution of particles within each receptor volume.

It should be noted that chemical dynamics are not taken into account in DREAMS-2, so that the model code is valid only for the study of the physical behavior of non-reactive pollutants. Quite obviously, a model such as the present version of DREAMS-2 cannot be applied to reactive pollutants such as NO, NO_2, or O_3, where chemical reactions occurring in the atmosphere are of primary importance.

6.2.3 Coupling between DREAMS-1 and DREAMS-2

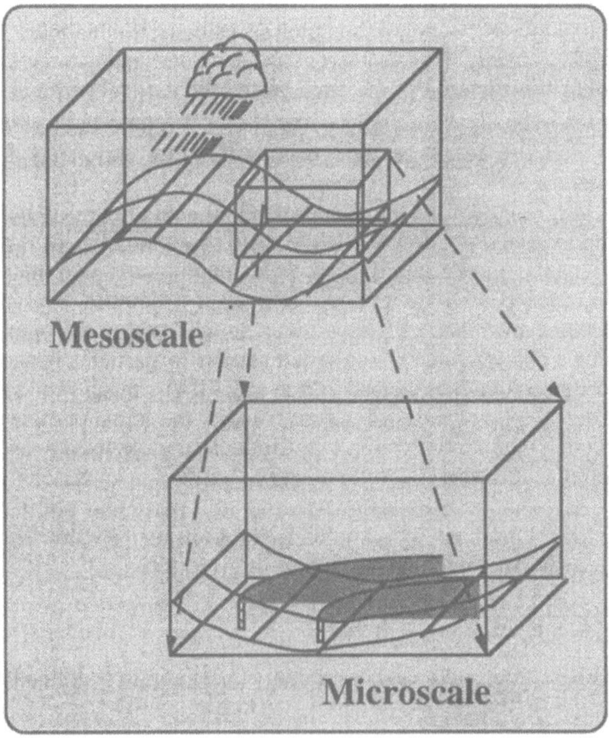

Figure 6.2: Schematic illustration of the coupling between the DREAMS-1 and DREAMS-2 models

As has been previously mentionned, DREAMS-2 uses the dynamic, thermodynamic, and topographic information of DREAMS-1 in order to be able to undertake a simulation. The data from DREAMS-1 can either be transfered directly to an identical domain in DREAMS-2 (i.e., point for point), or can be interpolated onto a finer-scale grid. If a fine-scale grid is selected for DREAMS-2, then the mesoscale model data needs to be interpolated through a 3-D interpolation scheme in which any variable at the nodes of an elementary volume in DREAMS-1 are interpolated according to the following formulation:

$$v_N(\zeta) = \Sigma v_i n_i(\zeta) \qquad \zeta = x, y, z \qquad 6.10$$

v_N is the interpolated value onto the fine-scale mesh; v_i is the nodal value of v and N is the number of nodal points of the element for which the interpolation is being performed (8 for a 3-D volume of the grid system used in DREAMS-1). n_i is the shape or interpolation function which is linear within each element.

Figure 6.2 illustrates schematically the coupling between the two models and their respective range of applications. The microscale model generally has a much finer resolution than the mesoscale model, even if the two domains are coincident.

6.3 Applications of the coupled numerical modeling system

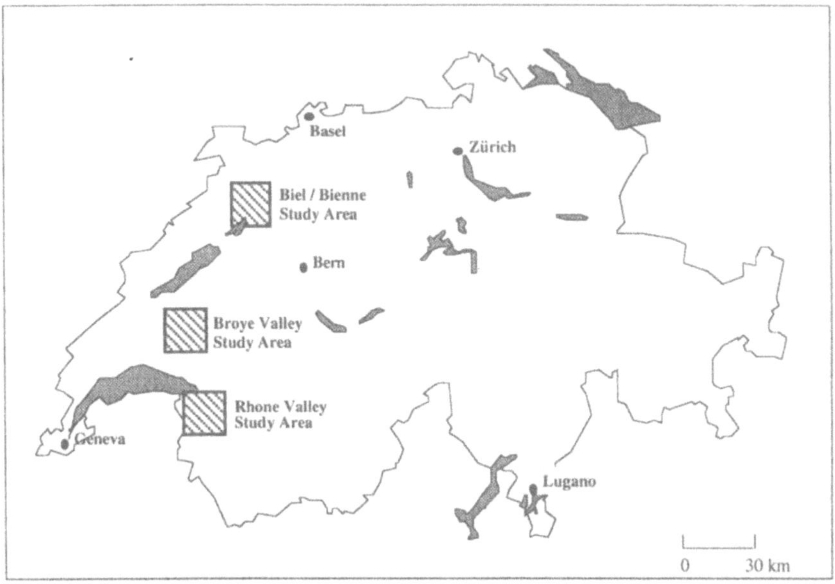

Figure 6.3: Map of Switzerland indicating three of the study areas for which the coupled DREAMS-1 and DREAMS-2 models have been applied (See Sections 6.3.1 - 6.3.3)

When attempting to apply new regulatory measures to abate pollution, local enforcement authorities require two distinct sets of information: those related to the emissions themselves, and those related to air quality in a given region. While in the former case, the compilation of a detailed inventory of emittors is a straightforward though tedious undertaking, the problem of estimating the contribution of each source to regional air quality is much less obvious. As a result, it becomes a difficult matter to impose restrictions on an individual basis unless a method can be devised which can establish source-receptor relations.

The principal aim of the studies described in this Chapter is to try and reproduce numerically certain specific cases of air pollution in order to establish the relative contribution of local sources and unspecified sources outside the study area to ambient air quality in the area under consideration. This is principally achieved through comparisons with field meteorological and air quality data for particular episodes. A generalization of the methodology described here can help lead to the identification of major sources of air pollution in a particular region, and consequently orientate possible abatement strategies and minimum-impact planning for industry and traffic.

Figure 6.3 illustrates the three study areas for which results will be discussed in Sections 6.3.1 - 6.3.3.

6.3.1 The case study of Biel/Bienne

A pollution-related study was carried out for the Swiss city of Bienne (Beniston 1987; 1990; 1991), located in complex terrain at the foot of the Jura mountains. The study consists of an analysis of pollutant transport and dispersion from various emittors located in the city, using a coupled system of mesoscale and microscale atmospheric numerical models. Simulations of atmospheric flow with the mesoscale model over a 20 x 20 km domain (horizontal resolution : 1000 m; vertical resolution : 250 m) are used to initialize a microscale model centered over the city. The domain of this latter model is 4 x 4 km (horizontal resolution : 100 m; vertical resolution : 10 m). Plume trajectories are computed in the micro-scale model, and are a function of the regional-scale flow field previously calculated by the mesoscale model.

Figure 6.4 indicates the 475 m height contour for the Bienne domain - i.e. the 50 m model level - which delineates the areas where flow can occur freely and those where flow is forced over or around the irregular terrain. The Aare river, lake of Bienne, and city of Bienne are superimposed on this diagram.

A number of typical synoptic situations have been chosen. Synoptic wind velocities are specified at 2 m/s. Temperature is characterized by stable conditions within the surface layer (280°K), with neutral stratification between 50 and 1000 m, above which moderate to strong stability prevails.

The specific humidity exhibits a quasi-monotonic decrease with height, from 6 g/kg at the surface to 1 g/kg to the top of the PBL (1600 m). At the ground, roughness length varies from smooth values over the lake (z_0 = 0.002 m) to rough values over the city of Bienne (z_0 = 0.2 m); intermediate roughness lengths (z_0 = 0.02 m) are to be found over the other regions. Surface moisture varies only over the lake of Bienne, where values of specific humidity are saturated at 8 g/kg.

Surface temperature at each model level which intersects the orography is set at the mean value of the atmosphere at the given height. The area representing Bienne has a temperature excess of 2.5 K over the surrounding land. The lake has the same temperature as the land, in order to allow identification of possible heat-island features due to the city.

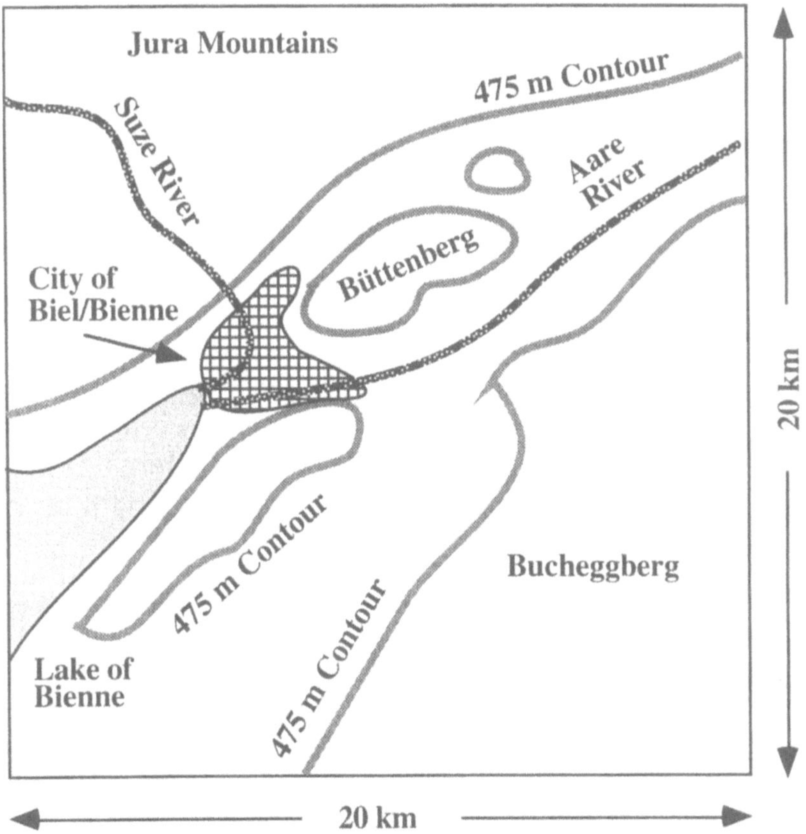

Figure 6.4: Plan view of the mesoscale model domain for the Bienne study

The DREAMS model initialized in this manner reaches quasi steady-state conditions after roughly 40 minutes of model time. The flow adjusts to the perturbations induced by the topography as well as to surface heterogeneities of roughness, temperature, and moisture. All results to be subsequently discussed pertain to steady-state conditions after 90 minutes of model time.

6.3.1.1 South-westerly situation, no clouds

In this simulation, all cloud activity has been suppressed artificially; this enables comparisons to be carried out for a simulation with identical initial conditions, but which include cloud activity.

Figure 6.5 shows the flow field at the 50 m model level; as this is a horizontal cross section, all points where vectors are absent are indicative of terrain levels higher than the computational grid points under consideration. The most obvious characteristic of the flow is the channeling effect by the topography.

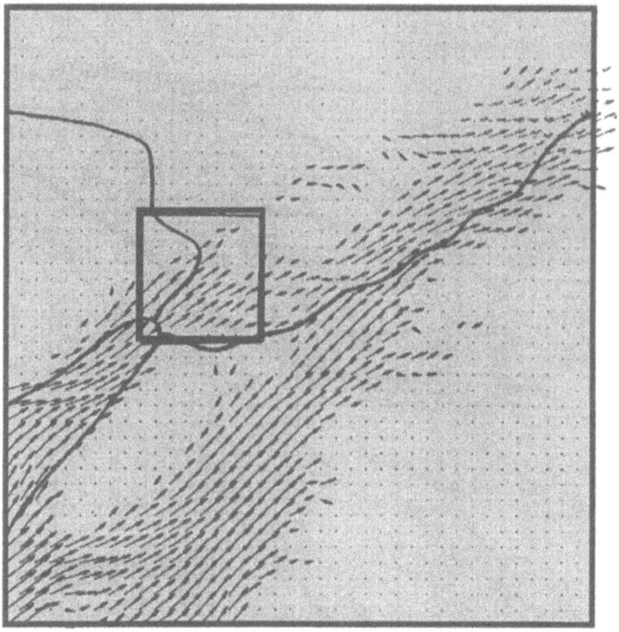

Figure 6.5: Flow field at the 50 m model level; SW situation, no clouds. Square area indicates microscale model domain used for pollution transport studies

The main feature of the flow occurs over the Aare valley, the widest part of the diagonal basin delimited by the Jura mountains to the N-W and the Swiss Middleland to the S-E. A substantial feature is the strong wind field off the lake of Bienne, over Biel itself, and the flow separation around the Büttenberg, a long and low hill located N-E of the city.

Due to differential pressure gradients in the Aare Valley and in the narrow valley encased between Büttenberg and the Jura ridge, most of the air transits around and over this hill to the Aare valley. Nevertheless, there exists a distinct stream of air which flows northeastwards through the narrow valley and converges once again with the main flow within the Aare valley.

Deflection of the air by lateral valleys is observed in this figure, especially in the eastern section of the domain where flow is channelled eastwards through and above valleys oriented in an E-W direction.

The nature of the air flow is characterized by high irregularities in intensity and direction. Regions of confluence and diffluence are a result of the combined effects of the physical presence of hills and mountains, three-dimensional orographically induced waves or cells, and surface roughness and temperature inhomogeneities. In some instances, marked accelerations in the wind field are observed, in particular in the middle zones of the valleys where frictional retardation by the valley sides is at a minimum.

This is particularly well exhibited in Figure 6.6 (left), which is a plot of the u velocity component along an E-W section through the center of the model domain at 50 m height. The maxima of about 2 m/s correspond to the main part of the Bienne and Aare valley flows, whereas the "trough" in between is the slower flow transiting from one valley to another south of the Büttenberg.

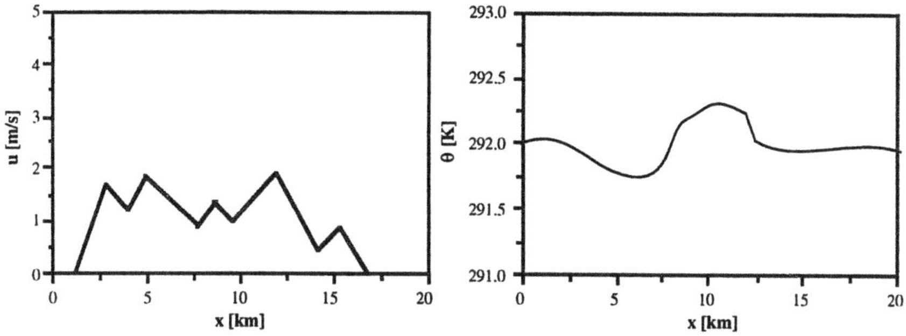

Figure 6.6: u velocity component along E-W section through the DREAMS-1 domain at 50 m (left); θ profile along this section (right)

The variation of temperature along this line, given in figure 6.6 (right) shows small fluctuations which result essentially from surface buoyancy flux contributions. The buoyancy flux is negative at the 50 m level everywhere except above the city of Biel which is warmer than its surroundings. The city effect can be seen in the slight warming of the air at 50 m above and downstream of Biel (due to advection of temperature); elsewhere, the cooling by the surface layer flux exchanges is on the order of 0.1 to 0.2 K. At the 550 m level, the flow exhibits a more general south-westerly tendency than at 50 m, as seen in Figure 6.7.

Some channeling is observed in the Jura mountains especially in the Vallon de St-Imier in the extreme N-W of the domain; this is a depression surrounded by mountains rising to 1300 m or more. Elsewhere, wind vectors are perturbed by flow over underlying hills, and this results in a rotation of 20 - 25° to the right of the flow. Such a deflection reflects an equilibrium between underlying, channelled air, and overlying, free-flowing synoptic winds.

The square area in Figures 6.5 and 6.7 represents the horizontal extent of the microscale model used for local-scale pollution transport and diffusion experiments. The microscale model domain is a 4 x 4 km square (100 m resolution) stretching upwards to 200 m height (10 m vertical resolution).

Data from the DREAMS model is available for each fifth microscale model grid-point; interpolation of the wind field are thus achieved with reasonable resolution. Topography is also interpolated in order to avoid artificial generation by the interpolation scheme of wind velocities at grid-points located beneath the surface.

The flow pattern interpolated in this manner will be discussed later. A number of pollutant sources are located in Bienne, and these can be modeled either individually or as a group. Figure 6.8 illustrates the layout of the principal emittors in the region under consideration; of special interest is emittor 5' which is a producer of SO_2.

The salient features of the geography are also illustrated in this figure, in particular the Aare river and canal, and the extremity of the lake of Bienne. In the present simulation, all emittors have been located at 30 m above ground level, in order to examine multiple plume behavior. The flow field over Bienne interpolated from mesoscale model wind patterns is given in Figure 6.9 (left). It is observed that the flow curves from the S-W to W winds in the eastern part of the domain, which is a feature that was already visible in the cruder grid of the DREAMS model. Channelling to the north and south of the Büttenberg is readily visible in figure 8a, and thereby determines the behavior of plumes located in this part of the domain.

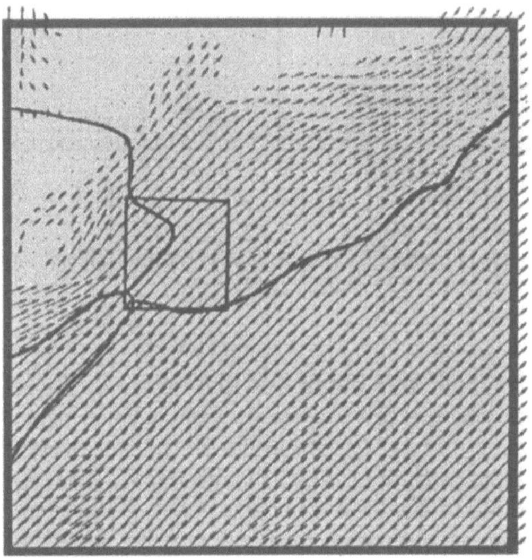

Figure 6.7: Flow field at 550 m model level; SW situation, no clouds

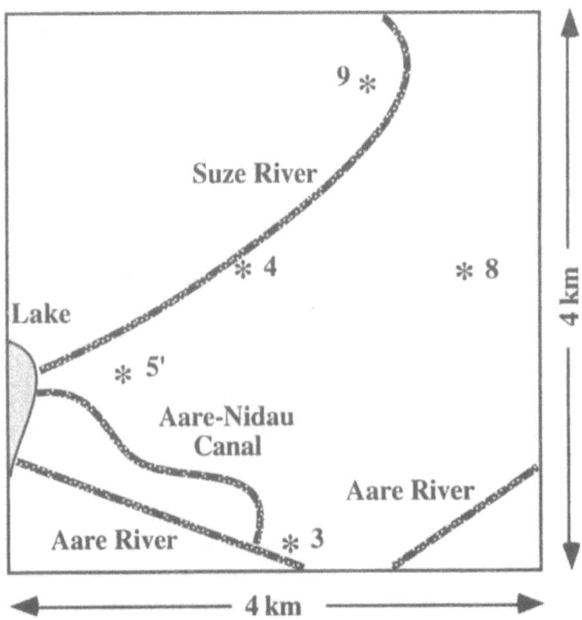

Figure 6.8: Plan view of the microscale model domain; main emittors and geographical features are given

Plumes simulated for the five principal emittors are illustrated in Figure 6.9 (right), for the flow given in Figure 6.9 (left). The continuous emission from each source is identical; isoconcentration lines are relative values. The plume trajectories follow very obviously the curvatures of the mean flow; plumes from sources 5' and 4 join together but are forced around the northern part of the Büttenberg.

For emittors 8 and 9, constriction by valley sides leads to a plume which exhibits little lateral diffusion. Diffusion by turbulent processes is in general low, so that the principal transport mechanism for pollution is advection by the regional scale flow.

6.3.1.2 South-westerly situation, with clouds

Clouds will form in the DREAMS-1 model as soon as a small threshold of convective vertical velocity at the lifting condensation level is exceeded. In this case, cloud formation and growth will occur, resulting in time in a series of complex cloud-induced circulations.

The chief source of moisture required to maintain cloud activity originates from evaporation of the lake of Bienne. Even such a small lake can have significant dynamic effects, as was emphasized in the study by Beniston (1986). A simulation has been carried out with the same initial conditions as in section 6.3.3.1, but here cloud activity is now allowed to take place. The experiment is of interest in determining to what extent a field of cumulus clouds can modify low level flow, and consequently influence plume behavior.

Figures 6.10 (left) and 6.10 (right) show the flow field at 50 m and 550 m for the south-westerly situation in which clouds interact with their environment. In comparison with the case in which no clouds are present (Figures 6.5 and 6.7), it is obvious that the wind-field is significantly perturbed by the clouds. The convergence-divergence fields which were initially responsible for cloud generation are now amplified; vertical recycling of air leads in some cases to reverse flow as in the narrow valley to the north-east of Bienne.

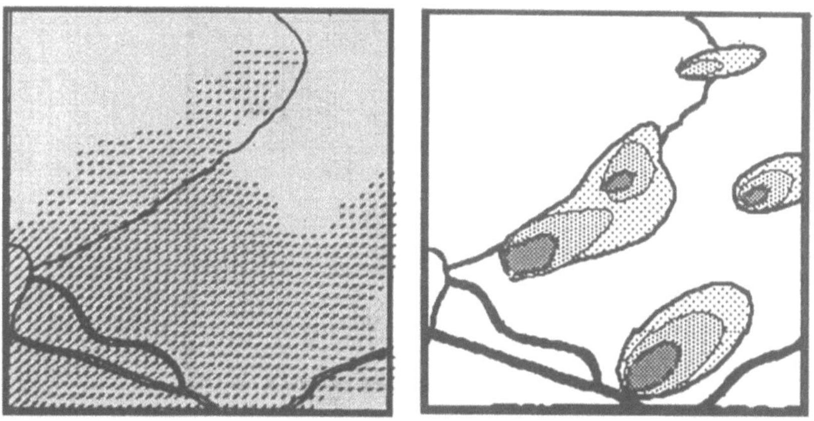

Figure 6.9: Flow field at 30 m in the DREAMS-2 model; SW situation, no clouds (left); concentration distributions for this wind field (right)

Flow across from Bienne to the Aare valley south of the Büttenberg is much less intense because cloud-induced convergence-divergence patterns have led to a significant rotation of the flow in this region. In this instance the clouds have a greater influence on the wind-field than the channeling effect. In the Aare valley itself, flow disruptions are not as marked, though a definite outflow region is to be observed to the north-east.

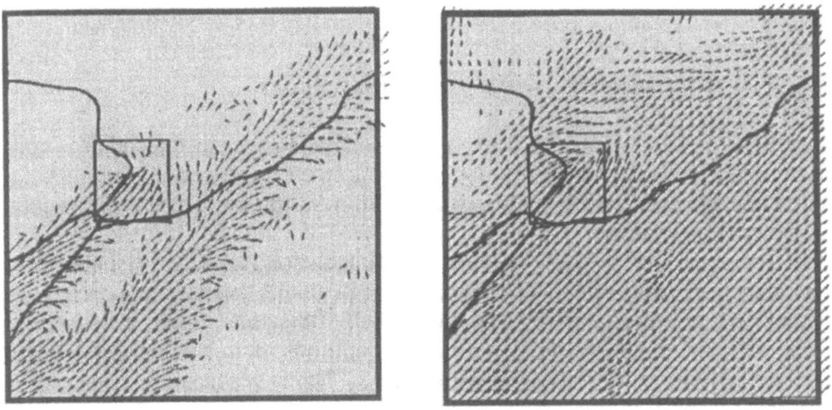

Figure 6.10: Flow field at the 50 m DREAMS-1 model level; SW situation with clouds (left); flow field at the 500 m DREAMS-1 model level (right)

Figure 6.11: Cloud patterns responsible for flow field perturbations seen in Figure 6.10

In Figure 6.10 (right), the 550 m wind-field exhibits regions of stronger confluence/diffluence than in the no-cloud case; however, the general features ressemble those seen in Figure 6.7. The cloud field responsible for airflow perturbations is illustrated in Figure 6.11.

Some of the clouds form above the Jura ridges, but individual cloud cells are seen oriented in the general direction of the main valley; the most important cloud formation is located above and upstream of Bienne. This is a result of the strong moisture contributions from the lake coupled with enhanced buoyancy from the warm city surface. The cloud-mesoscale circulations due to this cloud generates the strong low-level activity of the flow in this region; in particular. the flow between Bienne and the Aare valley is practically decoupled as air is drawn in towards the cloud system.

This feature is reflected in the u-component variations along the same E-W section as in Figure 6.6. Figure 6.12 (left) shows the somewhat greater amplitude of the fluctuations compared to the no-cloud case; valley flow maxima are located once again near the center of the valleys, but in sharp contrast to Figure 6.6 (left), there is practically no flow at this level south of the Büttenberg. It is also interesting to see the behavior of temperature along this section as compared to Figure 6.6 (right) A series of fluctuations are observed in Figure 6.12 (right), with amplitudes of up to 0.6 K; this is a result of vertical redistribution of temperature according to cloud-induced recycling of air.

The principal effect of the cloud activity on the local-scale dynamic field is the convergence of air in the city and the blocking of the flow south of the Büttenberg. The flow at 30 m height is illustrated in Figure 6.13 (left) in which strong convergence is observed in the western part of domain; valley winds transfer air from the northern sectors southwards, while the S-W regional scale flow curves northwards as it approaches the Büttenberg. Figure 6.13 (right) shows the plume characteristics for the same five sources as previously.

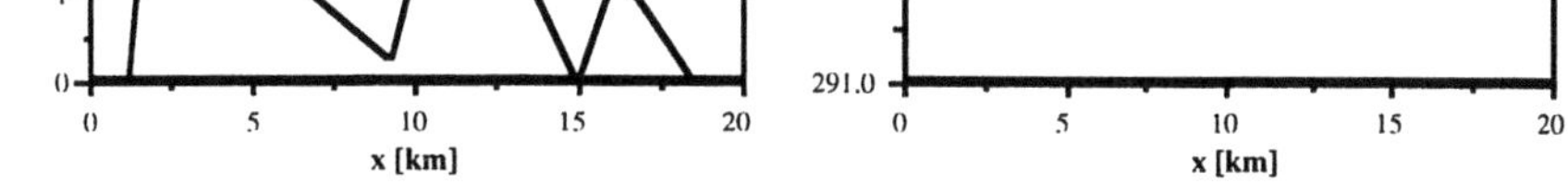

Figure 6.12: As Figure 6.7 (left), except for situation with clouds. As Figure 6.7 (right), except for situation with clouds

In sharp contrast to the behavior observed for the no-cloud situation, plumes exhibits even less dispersion due to the stronger flow. The orientation of plumes from emittors 3, 8, and 9 are directed opposite to one another. The high local concentrations are due to the constriction of the flow. Plumes from emittors 4 and 5' do not join each other as previously since the sources are located on either side of the strong convergence zone, which tends to block the plumes rather than disperse them.

It is clear from this analysis that a change in a particular meteorological process can result in significant modifications to the collective behavior of plumes. Such trajectory changes and concentrations would be very difficult to evaluate with simple models such as Gaussian plume models; however, the implications for pollutant concentration fields at a given point are fundamental.

6.3.1.3 North-east and north-west wind sectors

A common situation, especially during winter months, is the so-called "Bise" which is a north-easterly wind which blows along the Swiss Plateau under the influence of a high-pressure cell located to the north-west of Switzerland. Model results for the bise situation for a no-cloud situation show a reverse-flow with respect to the south-westerly no-cloud situation. Winds are channelled through each of the branches of the Aare and Bienne valleys, with cross-over flow occuring south of the Büttenberg from the Aare valley towards Bienne. Magnitudes of the vector field are comparable to those of the south-westerly flow.

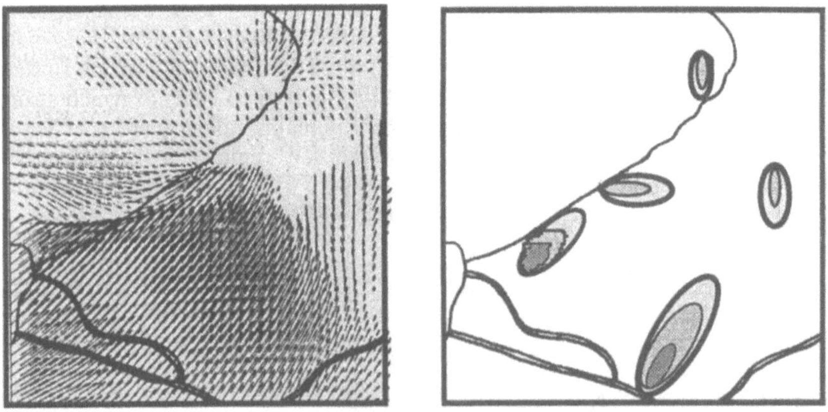

Figure 6.13: As Figure 6.9 (left), except for situation with clouds. As Figure 6.9 (right), except for situation with clouds

Plume trajectories for the steady-state bise situation are essentially in the opposite direction to those of the south-west case, as would be expected, with flow from the city transporting pollutants out over the lake. Flow from the north-west is significantly different from the other cases in that it is normal to the Jura ridges. At the 50 m height, the region of Bienne is in the lee of the mountains so that winds are relatively light (Figure 6.14 (left)).

Large accelerations occur south of the Büttenberg; perhaps the most interesting feature is the flow around the hills south of Bienne, and in particular the strong convergence of the flow in the lee of these hills (i.e. the southernmost part of the Aare valley). Only some small evidence of channeling is visible in certain parts of the valley; most of the air is forced by the synoptic flow across the valleys and over the valley sides downstream of each valley.

The small-scale wind field interpolated by the microscale model (Figure 6.14 (right)) show the influence of the Büttenberg which perturbs and channels the flow originating from valleys within the Jura mountains. Confluence occurs in the central part of the domain, with a northerly wind component clearly visible, forced by the N-W winds to the west of the domain and the N-E component near the Büttenberg. Pollution from the various sources yields a relatively large spread of concentrations as a result of this orographically-influenced flow.

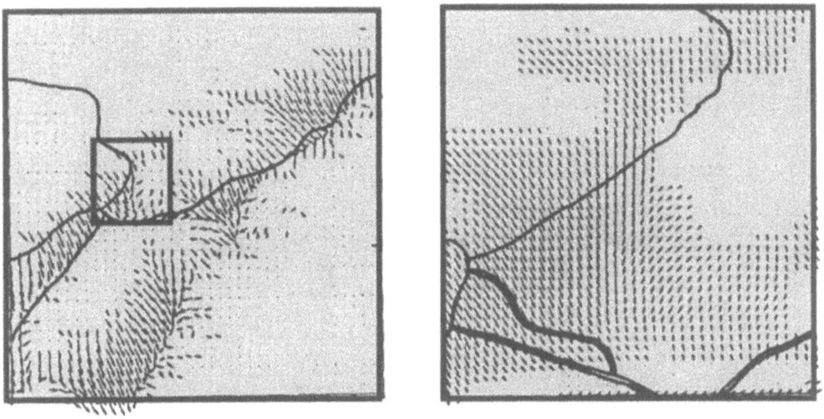

Figure 6.14: DREAMS-1 (left) and DREAMS-2 (right) flows for the NW situation without clouds

6.3.2 A combined numerical and wind-tunnel study of air pollution in the Broye Valley

This case-study focuses on air pollution episodes in the Broye valley, a rural area of Switzerland located to the north-east of the city of Lausanne, in which a number of industrial sources of SO_2 are located. The study comprises numerical simulations of regional-scale meteorology, local plume behavior, as well as a wind tunnel investigation of the same region in order to ascertain the relative advantages and disadvantages of numerical and physical modeling. *In situ* data are available for certain episodes which occured during a very cold period in the winter of 1985, allowing intercomparisons to be made between nature and models. Furthermore, a significant effort undertaken by the authorities of the Swiss Canton of Vaud has led to the setting up of a detailed emissions inventory, which includes industry, individual domestic heating and traffic.

Figure 6.15 provides a 20 x 20 km plan view of the domain included in the DREAMS mesoscale meteorological model used for the numerical parts of the study. Results obtained for several episodes which occurred during the cold winter period of February 1985 will be discussed in this section. The four principal emittors are located in the towns of Moudon (2) and Lucens (2); combined SO_2 emissions from sources in Moudon account for 36.8 kg/h while combined emissions in Lucens are practically identical at 36.2 kg/h. These are characteristic values for daily emissions during the winter of 1985. In February 1985, industrial emissions accounted for 83.8 % of all SO_2 produced in the Broye Valley, while domestic heating sources represented 9.7% and sources due to traffic 6.5%.

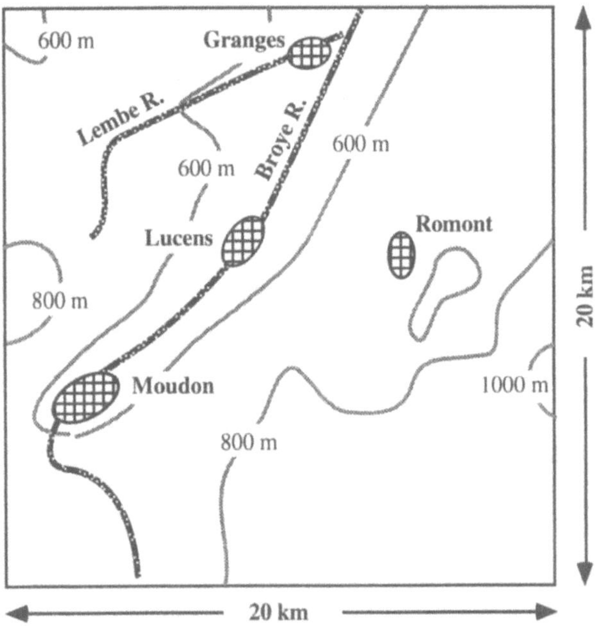

Figure 6.15: DREAMS-1 mesoscale model domain for the Broye Valley air pollution simulations

Of particular interest is the episode on February 21, 1985; at this time, the subsynoptic flow was from the northeast (the cold "Bise" regime typical of a winter high pressure system, flowing along the Swiss Plateau), while within the Broye valley, cold air was observed to drain in the opposite direction. The thermal inversion and dynamic shear layer combined to trap pollutants emitted from local sources in the Broye valley, leading to relatively high SO_2 concentrations. The sketch in Figure 6.16 illustrates this situation, giving the trajectory of a plume which slowly manages to penetrate the inversion layer and is entrained by the upper-level wind-flow.

Another situation considered was that of February 4, 1985. This is a case where the "Bise" (north-easterly) flow extended down to the ground in the presence of quasi-neutral stratification. Light winds were observed in the valley but were directed in the same direction as the stronger winds aloft. In this situation, pollutants were able to spread out vertically as well as horizontally, leading to lower ground-level SO_2 concentrations than in the previous case. Both meteorological and air pollution data were available for this period and emission conditions were known due to the detailed emission inventory established by the regulatory authorities of the Canton of Vaud.

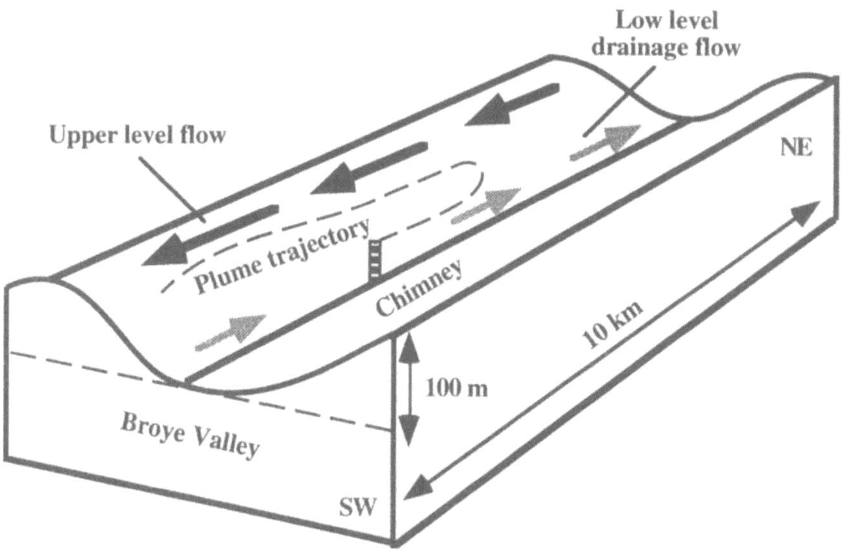

Figure 6.16: Schematic diagram of low-level drainage flow and opposing synoptic flow aloft

6.3.2.1 Model results for the Broye Valley, case study of February 4, 1985

Figures 6.17 (left) and 6.17 (right) represent horizontal cross-sections of wind velocity through the DREAMS-1 model domain at 150 and 250m above the model reference level of 450 m (i.e., at 600 and 700 m ASL) for the simulation of the situation of February 4, 1985. Superimposed on the vector field are some height contours to highlight the main axis of the valley; furthermore, the square area represents the domain of the DREAMS-2 model used in computing pollution transport and dispersion. It can be seen in these figures that the northeasterly "Bise" together with local orographic conditions has generated low-level flow which is clearly channelled by the Broye valley. In the northern segment of the domain, part of the flow is diverted into a side valley, while most of the flow is directed towards the southwest.

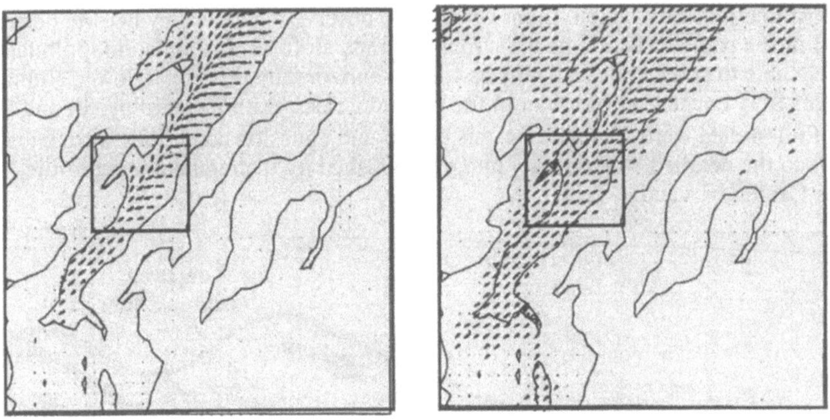

Figure 6.17: Horizontal cross section of wind velocity in DREAMS-1 at 600 m (left) and 700 m (right)

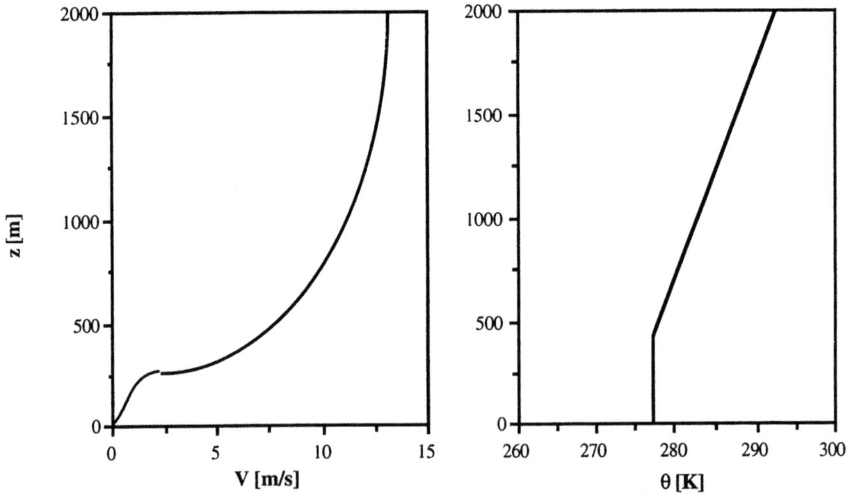

Figure 6.18: Vertical profiles of wind velocity (left) and potential temperature (right)

The vertical profiles of wind and temperature are given in Figures 6.18 (left(and 6.18 (right), respectively, for a grid-point which corresponds to the location of the meteorological observation station used for intercomparison purposes. The wind profile exhibits very low values within the first 400 m (less than 1 m/s), while above this level, wind velocities increase with height to reach some 10m/s at 2000 m ASL. The vertical potential temperature profile in Figure 6.18 (right) is characterized by neutral stratification within the first 400 m of the atmosphere, followed by stable air above. For this case study, the stationary wind-field has been interpolated onto the DREAMS-2 model grid; a horizontal section at a height of 20m is illustrated in Figure 6.19 (left); again, one principal height contour is superimposed to give the outline of the Broye valley in the vicinity of Lucens.

The DREAMS-2 model domain is a 4 x 4 km area with a 100 m grid interval in the horizontal and a 20m interval in the vertical. It is centered on the town of Lucens, where two of the principal SO_2 emittors of the Broye Valley are located. A minor ridge across the valley floor is responsible for the apparent discontinuous nature of the flow in the southern portion of the valley. SO_2 sources imposed at the two emittors in Lucens yield ground-level concentrations given in Figure 6.19 b for this particular wind-flow situation.

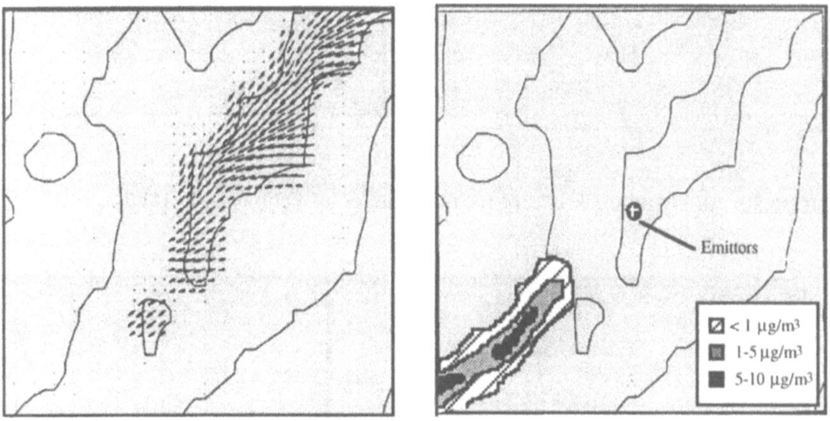

Figure 6.19: Results of DREAMS-2 for wind velocity at 20 m (left) and ground-level SO_2 concentrations (right)

It is seen that the plumes touch the ground some 900 m from the source, and that the SO_2 concentrations are generally low; this is because plume centerline trajectories are between 100 and 200 m above the ground, so that maximum concentrations are not at ground-level.

6.3.2.2 Model results for the Broye Valley, case study of February 21, 1985

Figures 6.20 (left) and 6.20 (right) illustrate the flow situation at 600 and 700 m ASL for the much more complex situation observed on February 21, 1985. Drainage flow occurs at low levels, with contributions coming from side valleys and in some cases from the valley sides themselves, since the ground surface is very cold. Dynamically, these side-valley winds contribute to a general flow acceleration down the main valley, as can be observed in Figure 6.20 (left) towards the exit boundary. 100 m above this drainage flow, the easterly and northeasterly winds characteristic of the "Bise" mesosynoptic flow aloft can be observed (Figure 6.20 (right)).

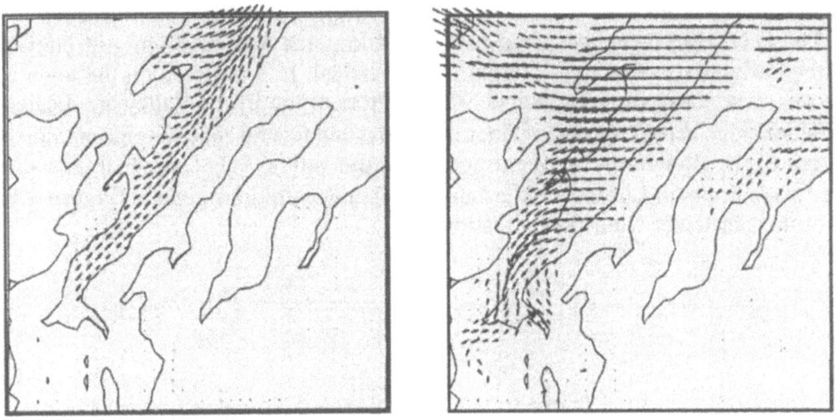

Figure 6.20: As Figure 6.17, except for the situation of February 21, 1985

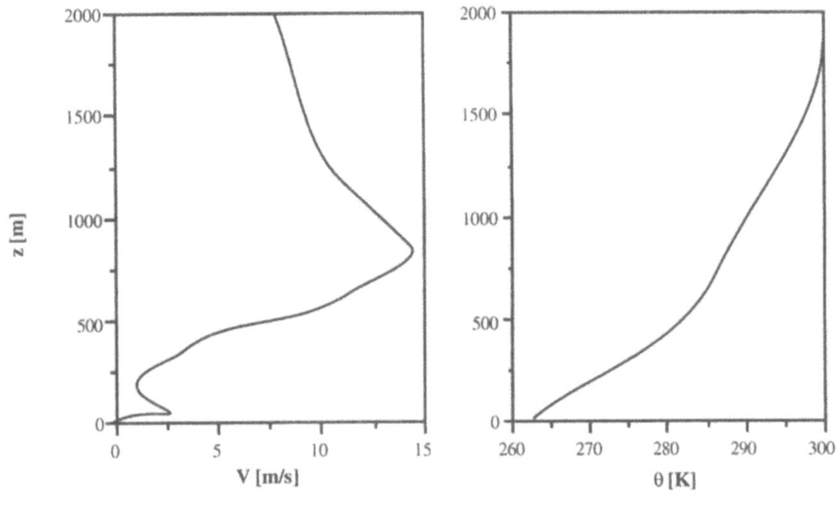

Figure 6.21: As Figure 6.18, except for the situation of February 21, 1985

Figure 6.21 (left) illustrates the vertical wind profile from the surface through to the top of the model at 1800 m above the surface. The drainage flow layer is indentified by the low-level peak of velocity and its upper surface by the minimum at 150 m above the ground. The northeasterly "Bise" strengthens from 200 to 1100 m, reaching nearly 16 m/s, before tailing off to less than 10 m/s at 1800 m. The decoupling of the flow between the lower-level drainage flow and the higher-level northeasterly flow is due to the very strong stability which prevails in the lower atmospheric layers, as shown in the profile of temperature (Figure 6.21 (right)). Such stratification greatly reduces the diffusion terms which govern the parametric equations for turbulence used in the model.

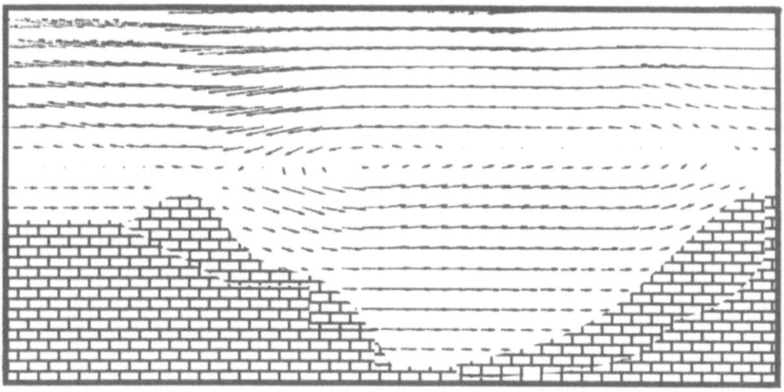

Figure 6.22: E-W vertical section of wind vectors 500 m downwind from the emission sources (DREAMS-2)

These terms, as described by Beniston (1987), are based on first-order closure principles; the diffusion coefficients are stability-dependent. Turbulent exchanges of momentum become so small that no interaction takes place in the vertical, so that the drainage flow "slides" beneath the stronger flow aloft with a minimum amount of dynamic exchange. This is seen particularly clearly in the vertical cross-section of velocity given in Figure 6.22. This is an east-west section through the DREAMS-2 model 500 m downwind of Lucens.

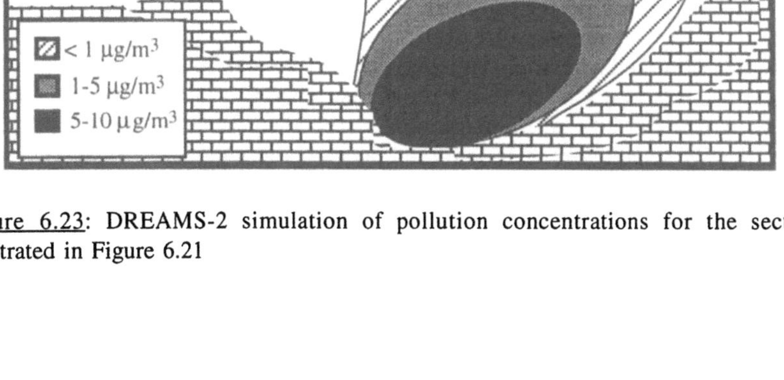

Figure 6.23: DREAMS-2 simulation of pollution concentrations for the section illustrated in Figure 6.21

The low-level drainage flow can be seen as a series of east-directed vectors (flowing into the page when imagined in three dimensions), while the stronger northeasterly flow is observed above (directed towards the west and flowing out of the page). At the interface between these two layers, some slight oscillatory and rotational motion extends over a height of roughly 40 m. This slightly perturbed interface is a result of the strong wind shear which generates a slow roll vortex above the Broye Valley.

Figure 6.23 illustrates the same cross-section as previously but for the simulated plume of pollution emanating from the two sources in Lucens. It can be seen that, due to the high stability and poor dispersion characteristics, the highest concentrations are located close to the ground; the plume is sloping towards the east as a result of the drainage flow. A small proportion of SO_2 penetrates above the interface, and is rapidly transported by the northeasterly flow. Obviously, such a situation renders exceedingly difficult any analysis of the relation between a pollutant source and concentrations at a given point. Studies such as the one described here go some way to resolving such problems even in complex meteorological situations.

6.3.2.3 Comparisons between numerical and wind-tunnel model results

Physical modeling has been used in this study to reproduce local meteorological conditions with a topographic scale-model placed in a thermally-stratified boundary-layer wind-tunnel. The methods employed are based on technological developments aimed at adapting wind-tunnel experimentation to drainage flow situations. The technique is similar to Cermak's (1975) pioneering work, and has been applied to various studies of dispersion in drainage flow situations occurring in complex terrain. A 1:3,000 scale model of the Broye Valley has been utilized, with small smoke generators representing the pollution sources. In very stable stratification, the method is to some extent capable of reproducing observations; the purpose of the next paragraphs is to illustrate the potential for complementarity of numerical and physical modeling.

The complex meteorological conditions of the situation of drainage flow below, northeasterly flow above (February 21, 1985) are here compared to model results.

Figure 6.24 illustrates the vertical profile measured above the meteorological station located in the Broye valley, and the profiles simulated by both types of model. An inflexion point is observed at 160 m above the surface in nature, at 200 m in the numerical model, and at 220 m in the physical model. This inflexion corresponds to the top of the very stable layer in which the valley breeze is flowing in the opposite direction to the "Bise" regime above. Figure 6.25 shows the observed and simulated wind profiles. A peak low level velocity on the order of 2 m/s is observed in all three cases within the first hundred metres. Between 100 and 200 m, the minimum of less than 1 m/s (at the interface between the two wind regimes) is well reproduced by both models. Above 200 m, the sharp increase in wind velocity to the upper-level synoptic value is well taken into account by DREAMS-1; however, a divergence between observed and simulated wind velocity is observed in the physical model above 400 m height. This is a result of the model's rigid upper boundary which causes frictional retardation of the flow at upper levels. This has little incidence, however, on processes taking place at lower levels above the scale model.

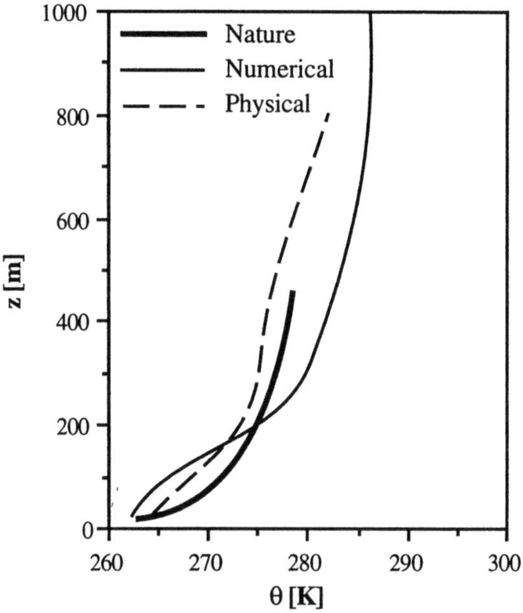

Figure 6.24: Comparisons between model and real vertical profiles of potential temperature for the case of February 21, 1985

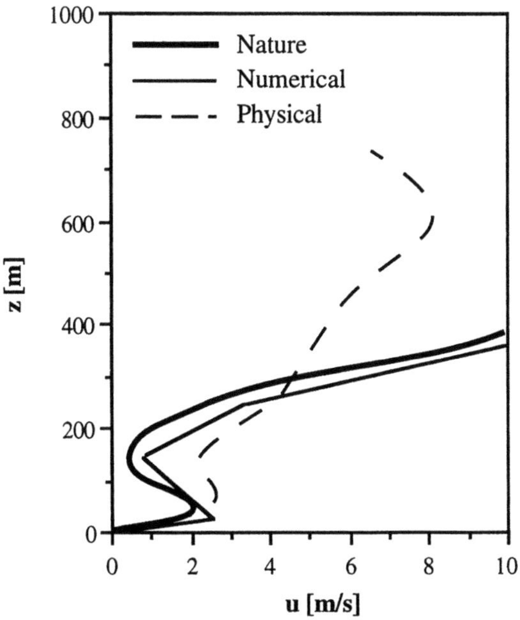

Figure 6.25: As Figure 6.24, except for wind velocity

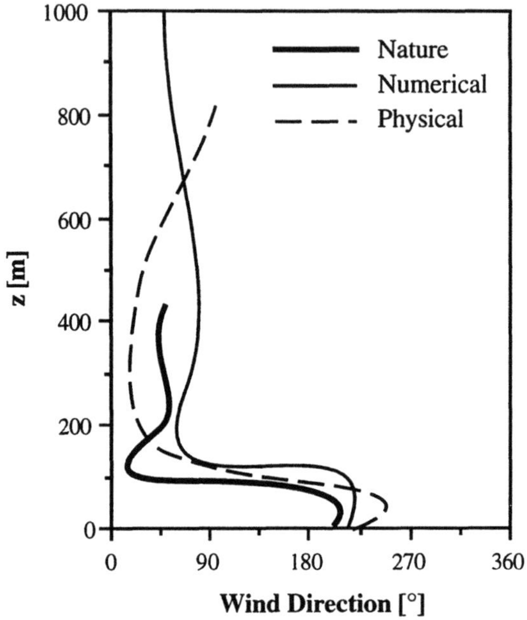

Figure 6.26: As Figure 6.24, except for wind direction

Figure 6.26 gives the vertical distribution of observed and simulated wind directions. Profiles above the scale model are highly irregular up to about 300 m above the surface, while in nature and in the numerical model, the profiles are much smoother. The oscillatory nature of the wind direction profile in the physical model is due to problems of measurement of this parameter at very low velocities and of the influence of local thermally induced turbulence and meandering (in the model, 1 m/s in nature is simulated by a velocity of only 3.5 cm/s according to scaling criteria used for physical models). The discrepancy between numerical model results and observations at the 100-200 m level is a result of the relatively coarse grid resolution of DREAMS-1, which has identified the top of the valley flow layer at 200 m instead of the 160 m seen in nature. Consequently, the change in direction is observed in the numerical model at a point 40 m above the corresponding point in nature.

Vertical concentration profiles for the neutral up-valley wind situation are given in Figure 6.27. Both the physical and numerical models yield profiles which are very similar, with a maximum of about 70 $\mu g/m^3$ at 180 m above the surface. The profile in nature shows a peak of about the same order of magnitude, but located some 50 m above the simulated peak altitude. The measured concentrations comprise 6 sampling points in the atmosphere, 10 for the physical model, and 20 points for the numerical model, so that some margin of error should be allowed for in the *in-situ* measurements.

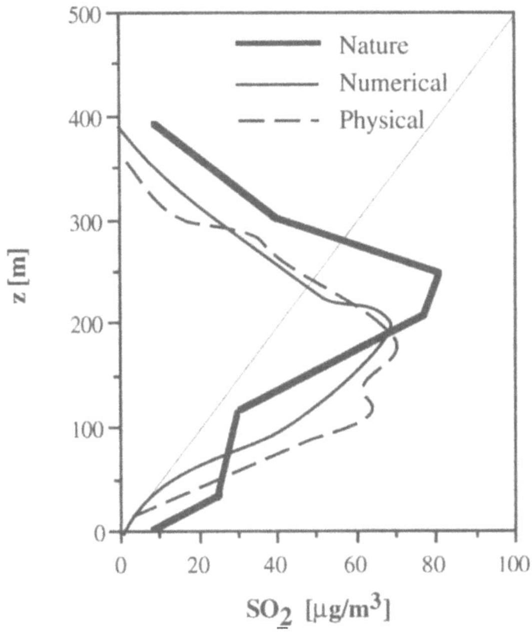

Figure 6.27: As Figure 6.23, except for SO_2 concentrations

It can be seen that if the simulated profiles are displaced upwards so that the peak values are coincident with those observed in nature, then all three profiles would be in almost perfect accord. One reason for the discrepancy in the location of peak concentrations is linked to uncertainties in the exact emission conditions for February 4, 1985. A small uncertainty in emission velocity from a stack can undoubtedly lead to an error of 40-60 m in plume height after a trajectory of more than 1000 m from the source (where the measurements were taken). Furthermore, the non-zero value in the observed profile at low levels is most probably due to the background value of pollution, i.e., that part of the pollution which is not solely a result of local pollution. Given the complexity of the meteorological situations simulated and the various approximations inherent to both physical and numerical models, the results obtained for the Broye valley are in remarkable agreement with reality. This type of modeling can certainly go some way to answering questions related to air quality on a regional or local scale and to the identification of the dominant sources contributing to concentrations in particular locations. In addition, the models allow the possibility of simulations with no background pollution, so that this term can be estimated whenever comparisons with real data can be made.

6.3.3 Intercomparisons between LIDAR measurements and DREAMS simulations: the Rhone Valley case study

6.3.3.1 Introduction

The LIDAR (LIght Detection And Ranging) technique has advanced to a stage such that accurate measurements of air pollution can be performed over a range of different spatial scales and for a number of chemical species in the atmosphere. A DIAL (DIfferential Absorption Lidar) system provides selective measurements of pollutant concentrations over a range of several kilometers, resolved in the same way as a radar and in an interactive manner, i.e., without the usual sampling requirements (Hinkley, 1974; Measures, 1984, 1988; Wolf and Wöste, 1987). It is then possible to obtain 3-D mappings of concentrations, at scales from a few tens of meters up to 15 km, which reflect the propagation and spread of local and regional-scale pollution (Kölsch et al., 1988; Wolf, 1987). Such remotely-sensed measurements represent a valuable complement to numerical simulations of air quality, and provide data of sufficient accuracy to enable intercomparisons to be made between models and observations.

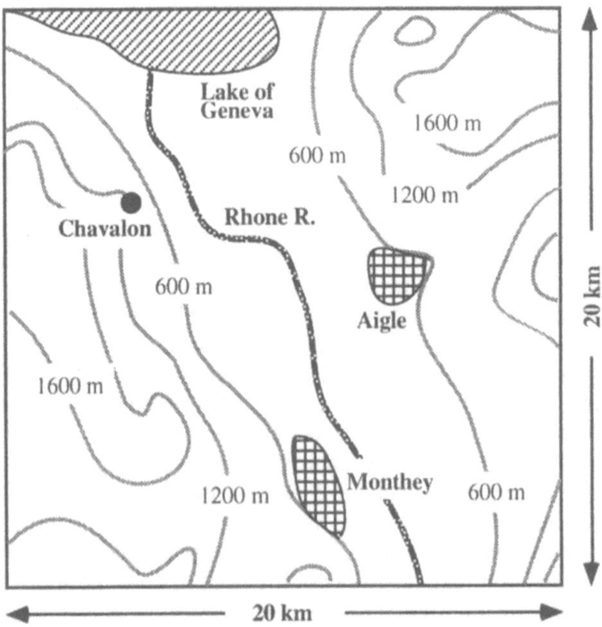

Figure 6.28: Map of the Chablais region (Rhone Valley) studied by Lidar and the DREAMS-1 model

However, despite the fact that the two techniques are obviously complementary, very few studies have been reported on the correlation of lidar measurements and numerical calculations (Benech et al., 1987). The DIAL technique is a powerful tool for verifying model predictions in each spatial dimension, and thus for establishing critical parameters, in particular boundary conditions. Conversely, once these parameters have been determined, theoretical calculations can be used to predict the transport and dispersion at greater distances. These considerations provided an opportunity to perform a comparative study between numerical model calculations and DIAL measurements in the Chablais region; this is the segment of the Rhone Valley which emerges into the Lake of Geneva basin, as illustrated in Figure 6.28. The Rhone Valley is the most important alpine valley in the Swiss Alps and, as such, has significant channeling effects on synoptic or subsynoptic flows.

Two different cases were considered, in particular the spread of a power plant plume (Chavalon), emitting mostly SO_2, at a scale of about 1 km with a resolution as high as 2 m, and the concentration profiles of SO_2 in the Rhone Valley in January and February 1988, at lower resolution but up to distances of 20 km. These two examples, involving widely different scales and concentrations, have been chosen in order to exhibit the versatility and complementarity of both techniques (Beniston et al., 1990).

6.3.3.2 Meteorological conditions and emission sources

Detailed analysis of the meteorology of this region is necessary in order to understand the behavior of pollutants emitted by the numerous sources found in the Chablais region. It is necessary to know whether these pollutants are rapidly dispersed and leave the valley, or whether the ventilation of the region favors stagnation through a diurnal recirculation of pollutants resulting from alternating upvalley and downvalley breezes. Regions up the valley from the Chablais, especially the region of Martigny, are well known for their strong winds. However, in the Chablais region, the situation is substantially different at low levels, because the Rhone Valley broadens out just north of the narrow gorge at St. Maurice. Continuity principles, associated with the fact that downvalley flow is invariably forced above the gorge at St. Maurice, means that only on rare occasions will downvalley winds be strong in the 4-5 km wide Chablais region.

Ventilation in this region is characterized by a diurnal cycle which is typical of thermal currents which generate drainage flows at night and upvalley winds during the day. During summer months, upvalley winds persist until well into the night so that drainage flow can only occur for a short while. In winter, however, downvalley winds are observed more frequently. This cyclic pattern is well observed during periods of fine weather under high-pressure conditions. Other meteorological situations are linked to synoptic situations which can enhance or diminish the strength of thermally-induced flows. In the case of the southerly Föhn, a strengthening of drainage flow is observed, both in duration and magnitude. However, Doppler-Sodar measurements by the Swiss Meteorological Institute (ISM, 1988) have shown that flow reversal could occur, leading to opposing flows at low levels (upvalley) and at higher altitudes (downvalley Föhn).

Air mass circulation in the Chablais region is also a function of the presence of lateral valleys, in particular the Val d'Illiez which merges with the Rhone Valley near Monthey, a region with a significant concentration of pollutant sources, in particular SO_2. The Val d'Illiez is responsible for significant deviations of wind direction in the Chablais region. Measurements made with a Doppler-Sodar at a distance of 1-2 km downstream of the mouth of the Val d'Illiez show that down-valley winds could be deflected by as much as 80°.

Two specific test cases have been chosen for intercomparisons between Lidar measurements and numerical model results, namely small-scale individual plume behavior, and collective plume behavior in the Chablais region. These are the situations of December 15, 1987, and February 14, 1988, respectively. The meteorological conditions of December 15 are characterized by a meandering occluded front stretching from Brittany to the Gulf of Genoa. A weak trough of low pressure located over Italy is generating a southerly flow over the Alps, leading to moderate Föhn conditions in the Rhone Valley. The air temperatures in the warm front ahead of the occluded sector are warm aloft and cooler near the ground, resulting in relatively stable conditions; in the Chablais region, winds are light (4 m/s) from the south, and oriented principally along the main axis of the valley.

The meteorological situation of February 14, 1988, was one of the rare favorable occasions for which lidar measurements and comprehensive meteorological data were available. A high-pressure cell located over central Europe allowed the development of local upvalley and downvalley breezes. Nocturnal drainage flow gave way to upvalley winds by the end of the morning. At the lidar measurement station, located some 5 km from the Lake of Geneva, the flow reversal took place at around 1100 LT, as deduced from observations of a constant-level balloon stabilized between 300 and 400 m above the ground (Fallot and Baeriswyl, 1988). Released in the extreme SE sector of the Chablais in the early morning, the balloon was immobilized and followed circular trajectories for about 90 minutes before being entrained by the upvalley wind in the direction of its point of release. At 1030 LT, at the time of the lidar measurements, winds were still or light, having drained downslope during the previous night. Temperature profiles measured at different points in the Chablais region showed that a temperature inversion formed during the night and persisted till early afternoon; the inversion base was located at about 800 m ASL in the morning, i.e., around 425 m above the valley floor.

For the case of the Chavalon thermal power station (December 15, 1987), a single emission source was taken into consideration; this is a tall stack, which released an average of 350,000 $\mu m^3/h$, of which 550 ppm is SO_2 effluent. Stack exit velocity is 7.5 m/s, and the plume rise is of the order of 140 m. In the Chablais region (February 14, 1988), the major part of the SO_2 emission was associated with industrial activity. As this was a Sunday, a day on which Swiss Federal regulations prohibit heavy commercial vehicles, emissions due to traffic represented a mere 0.5% of total SO_2 sources. On this day, domestic heating accounted for 11% of total SO_2 emissions, while industry represented 88% of the SO_2 sources. The emission register for the Chablais region therefore corresponds essentially to the distribution of the principal large industrial sources. For the case of February 14, 1988, 72% of SO_2 emissions originated in or around the town of Monthey, 12 km south of the lidar measurements station; 20% were emitted near the small town of Roche, downstream of the lidar site, while the remaining 8% were distributed throughout the valley. Table 6.1 summarizes these data.

	Emissions (kg/h)	Emissions (%)
Industry	5 900	88
Domestic Heating	750	11
Traffic	35	0.5
Total	**6 685**	**100**
Monthey region	4 800	72
Roche region	1 350	20
Other	535	8
Total	**6 685**	**100**

Table 6.1: Daily SO_2 emission values for 14 February 1988, according to type of source and region

6.3.3.3 Intercomparisons of the Chavalon power-plant plume

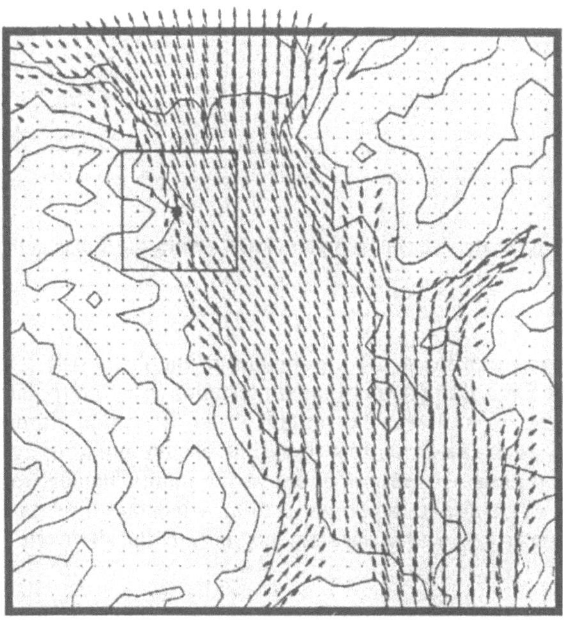

Figure 6.29: Wind velocity vectors at 825 m above sea level

Meteorological conditions observed on December 15, 1987, the date on which lidar data are available, have been used. Winds were blowing at roughly 4 m/s from the direction $180^{o} \pm 30^{o}$. Thermal stratification near the ground was slightly stable, with greater stability above. After initializing the DREAMS-1 model with a southerly wind at all levels, steady state conditions are reached in which stability and especially orography have significantly influenced the flow. Figure 6.29 illustrates the wind field close to the level of the Chavalon power station (825 m ASL). Major height contours are superimposed here in order to identify the principal geographical characteristics. The square area centered upon Chavalon represents the 4 x 4 km horizontal extent of the DREAMS-2 domain used in the subsequent pollution transport and dispersion simulations.

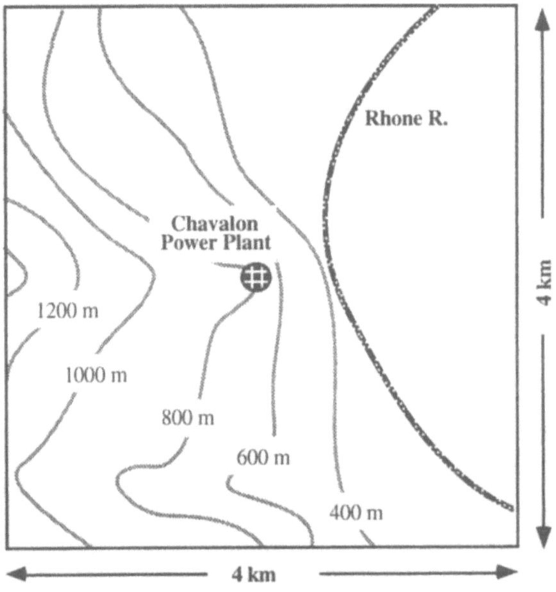

Figure 6.30: Map of the Chavalon region studied with the Lidar and the DREAMS-2 model

Figure 6.30 represents the 4 x 4 km domain of the DREAMS-2 model which is centered on the Chavalon power-generating station. The interpolated wind field at 925 m asl (some 30 m above the summit of the Chavalon chimney) is given in Figure 6.31. Superimposed on this flow field are the main topographic levels of this particular region. It is seen that above the main Chablais region, winds are oriented SSE, while above the power station, winds are directly from the south, which is in good accord with measurements made at the site itself.

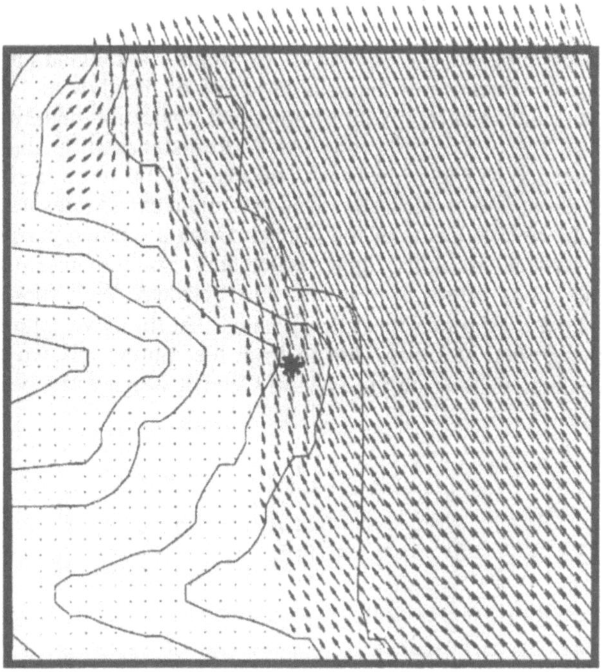

Figure 6.31: Wind velocity vectors at chimney-top height for this domain

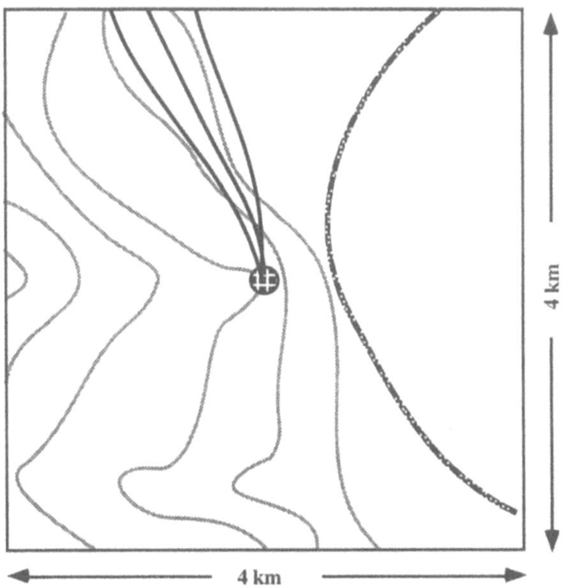

Figure 6.32: Trajectory and spread of the Chavalon plume as simulated by DREAMS-2

When the emission from the chimney stack is included at the appropriate location, the trajectory centerline and spread of the plume are as shown in Figure 6.32. The plume boundaries represent the regions of maximum Lagrangian particle spread from the centerline. Because of the smooth nature of the flow and the isotropic conditions of turbulence, a relatively symmetrical dispersion pattern is observed on either side of the plume centerline. At 2-km distance from the source, the plume has broadened to 600 m in horizontal extent (300 m on each side of the centerline) and about 240 m vertically. This maximum extent does not necessarily correspond to the visible limits of the plume, which are much narrower in both horizontal and vertical extent; the figures stated here simply represent the probable zones where non zero values of SO_2 could be detected. With such a relatively homogeneous flow configuration, it is estimated that the lower plume boundary could touch the ground at about 7300 m from the source; this would give low concentration values of SO_2 on the Lake of Geneva, about 2 km beyond the SE coast of the lake.

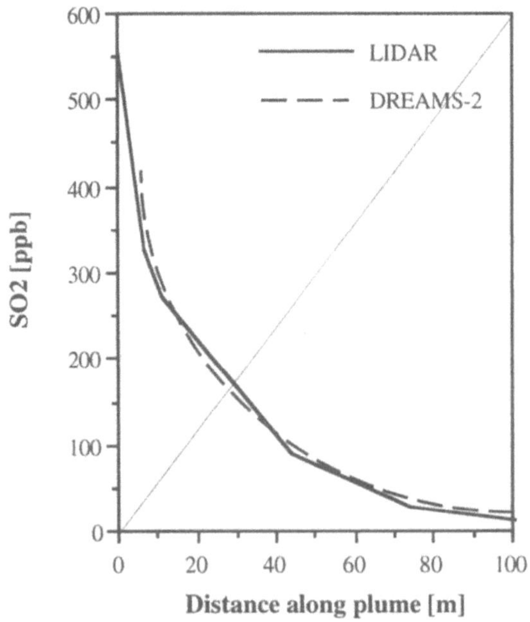

Figure 6.33: Comparison between Lidar and DREAMS-2 for the Chavalon plume

Figure 6.33 illustrates the dilution of SO_2 concentrations along the first hundred meters of the industrial plume emitted by the Chavalon power station. The agreement between lidar data and numerical simulations is excellent, and is made possible only through the careful analysis of a very large number of Lagrangian particles (31^3) emitted per time step, and to the high density of the receptor grid (1 m^3) located along the plume axis in order to estimate dilution. It can be seen, however, that initial concentrations are somewhat overestimated by the numerical model in the vicinity of the chimney exhaust. This difference may result from the presence of numerous aerosols (especially water droplets), over which some SO_2 molecules are adsorbed. The SO_2 concentration, measured by Lidar, is then lowered at the expense of other compounds; this physico-chemical phenomenon is not taken into account by the model. Outside of the 100-m range, it is possible that a difference between observed and simulated SO_2 concentrations may occur, which is one reason why mesoscale pollution intercomparisons are of value.

6.3.3.4 Intercomparisons of regional air pollution in the Rhone Valley

The Lidar technique has often been used for monitoring emissions, but very few experiments have been performed until now to determine immissions. The first large-scale vertical profile of a pollutant concentration performed by lidar is reported here. For this pupose, the lidar unit was located on one side of the Chablais region, at an altitude of about 500 meters, and the opposite side of the valley was used as a topographic target (DARL principle). The length of the optical path (up to 10 km) allowed for a sensitivity as high as 500 ppt of SO_2. The goal of the experiment was to record vertical profiles sequentially, under different meteorological conditions, in order to study the pollution situation in the valley during winter.

The Chablais region is a rather wide channel, in which several smaller side valleys contribute thermally and dynamically to the flow. This often leads to a vertical sequence of multiple stably-stratified layers. The emitted pollution can therefore be trapped in one of these sublayers without any mixing with the others. The pollution concentration is then highly dependant on altitude, and it is of outstanding importance in the context of studies related to vegetation damage to monitor whether maximum concentrations always occur at the same levels.

Of particular interest in this part of the intercomparison study is the case of February 14, 1988, a date on which lidar data are available through a number of cross-valley measurements which yield an integrated cross-sectional value of pollution. On this day, winds were light (3-5 m/s) and blowing downvalley toward the Lake of Geneva. A very strong temperature inversion was located at about 450 m above the valley floor, at 850 m ASL. This inversion naturally plays a fundamental role in the distribution of polllutants within the valley. As previously described, the major contribution to SO_2 pollution in the valley originates from industrial sources located essentially in the southern part of the domain at Monthey. For the present simulations, only seven industrial sources, which represent more than 80% of the emission of SO_2 on February 14, 1988, have been considered. Local emissions due to domestic heating and traffic have not been taken into account due to their relatively low impact on pollution in the valley on this particular day. Low-level domestic emissions tend to stagnate near their sources due to the low ground-level winds, and also because many of the constructed areas are located downstream of the region of interest, i.e., an east-west section across the valley, where lidar measurements were made. The Lidar was located in the vicinity of the Chavalon power station, which corresponds to an east-west line located 14 km north of the model's southern boundary.

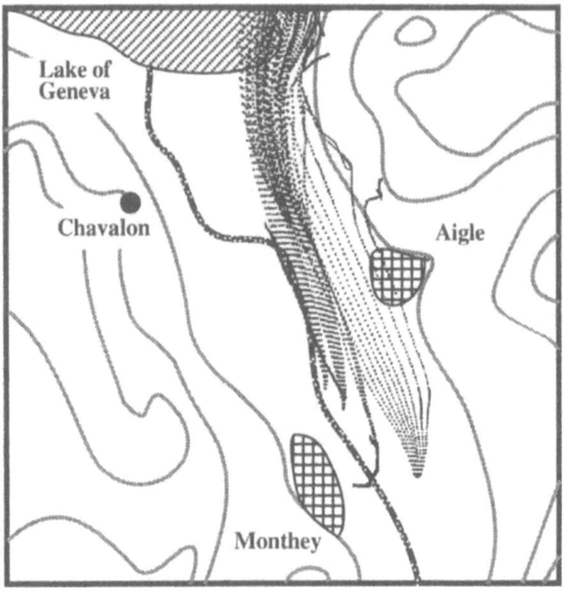

Figure 6.34: Spread of pollutants from the combined emission sources in the Rhone Valley

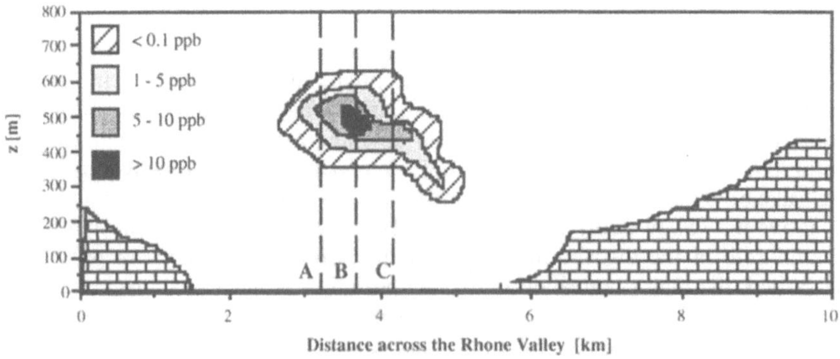

Figure 6.35: Vertical distribution of SO_2 along an E-W section at the level of Chavalon

Figure 6.34 gives an idea of the spread of pollutants from the combined emission sources; only the σ_u moment ($-3 < \sigma_u < +3$ standard deviations) has been illustrated here for simplicity. It is seen in this figure that according to the point of release and to the height reached by the emissions, the plume spread varies considerably from one plume to another. It is of interest to note that the southernmost plume, which rises rapidly into the atmosphere, is initially deflected eastward by the air flowing out of the Val d'Illiez, before being entrained northward in the main stream of the flow. This confirms the earlier remarks concerning the influence of lateral valleys. Furthermore, due to the complexity of the terrain, some segments of plume are deviated into side valleys; such dispersion patterns would be impossible to simulate with analytic Gaussian plume models and with considerable difficulty with Eulerian diffusion models which, for pollution studies of the kind addressed here, are far too dependent on model grid size.

Figure 6.35 illustrates the vertical distribution of SO_2 concentration along an east-west cross section at the level of the Chavalon power station. This field is located between the surface and the summit of the inversion layer. The combined plume axis is tilted toward the west with height as a result of the flow characteristics in this region; columns A, B, and C refer to the locations of the three vertical soundings given in Figure 6.36. It is seen that profile B has a peak concentration of 19.2 ppb at the inversion level at 450 m height. Profile A, located 500 m west of the plume center has a relative maximum of 15.4 ppb also at the inversion level, while profile C, 500 m to the east of profile B, has a peak concentration of 6.9 ppb 250 m above ground level. This lower level of relative peak concentration is due to the tilted attitude of the plume. All three profiles drop to negligible values above 650 m due to the suppression of dispersion by the inversion and the stable layers above.

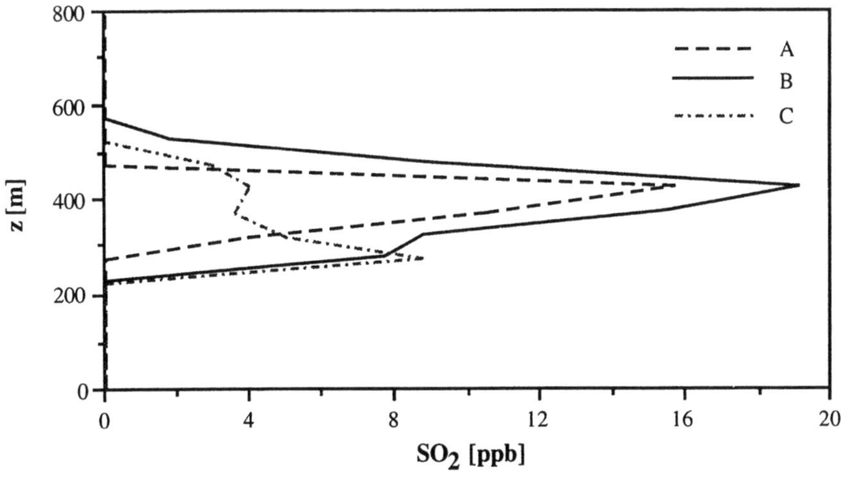

Figure 6.36: Vertical soundings for the three columns A, B, and C identified in Figure 6.34

The calculated and measured vertical SO_2 profiles obtained on February 14, 1988, are compared in Figure 6.37. A path-integrated average value of pollution through the plume has been computed at each level of the model receptor grid mesh, in order to compare these results directly with the integral average values obtained by the Lidar. It is seen in this figure that, even for a receptor located some 12 km from the southernmost sources, the value of peak concentration and its height above the ground are in good agreement with the observations. The Lidar was unfortunately incapable of measuring below a height of 350 m, due to a layer of fog which is impermeable to the laser signal at the wavelengths characteristic of SO_2. Above the region of peak concentrations, which is located at the inversion level through trapping by thermal stability, the Lidar detects non-zero pollution which is not computed by the numerical model. These upper level concentrations are in effect a feature of pollution from the previous day, which was diffused at height but remained at upper levels due to alternating upvalley and downvalley flow, which tends to recirculate the pollutants. The numerical model has computed only the contribution to pollution from emissions on the day chosen for this case study and consequently contains no background pollution, which is a measure of the recent "history" of air quality in the region studied. The absence of background pollution in the model can in such cases lead to results which do not correspond to reality. This problem can be alleviated to some extent by attempting non-steady-state simulations of pollutant transport and diffusion, in order for the model to generate its own background pollution according to the transient motions of the atmosphere in the Chablais region.

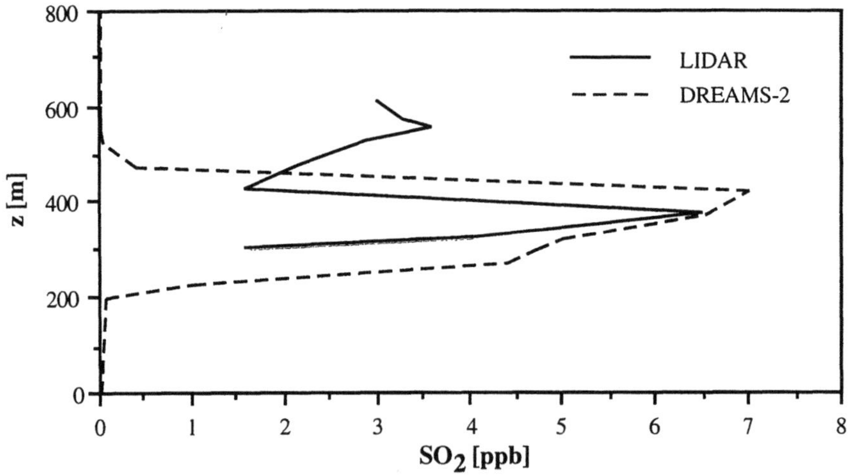

Figure 6.37: Comparison between Lidar and DREAMS-2 for regional air pollution in the Rhone Valley

6.4 Concluding Remarks

The example of air pollution has been chosen to illustrate how atmospheric processes at different scales can be bridged, in particular in the direction of larger-scale to smaller scales. The atmospheric pollution module nested in the mesoscale model simulations has demonstrated the manner in which a small-scale feature will respond to larger-scale forcings. The nested approach proposed here is possible only when the smaller-scales do not feed back to the larger scales, i.e., the system at the small scales is passive with respect to the mesoscale driving forces.

The coupled model simulations have shown that many factors can influence plume behavior and pollutant concentrations. Topography plays a major role in chanelling the flow, but also surface roughness changes, temperature heterogeneities, and cloud activity all play a distinct part in complicating plume transports. In the Biel/Bienne and Broye Valley simulations, model results have proven to be valuable for complex situations and reproduce quite accurately local and regional-scale meteorology, as well as the three-dimensional nature of pollution embedded in mesoscale flows.

Numerical models invariably require verification and calibration with respect to the real world. The combination of advanced remote-sensing techniques with DREAMS-1 and DREAMS-2 simulations has enabled the verification of model simulations under relatively complex atmospheric and geographic condition, and for scales ranging from the local to the regional. The objectives of the study discussed in Section 6.3.3 were not only to illustrate how well numerical model results compare with *in situ* measurements, but also to demonstrate that when models can adequately reproduce phenomena occurring in nature, they consequently have predictive value. This is particularly important when attempting to impose pollution-abatement measures or when undertaking an environmental impact study of a future industrial complex or major highway. In the first case, a correct simulation of overall pollution through such modeling techniques can help in determining the contribution of individual sources to this pollution, by backtracking along the trajectories followed by each plume. In the second case, the predictions of concentrations of a given chemical species will help locate the optimal site of a future chimney stack, industry, or highway as a function of emission conditions and critical meteorological episodes. If predictions show that concentrations of a particular pollutant exceed legal limits, then the modeling technique described here is a valuable tool for finding technical solutions which will reduce immissions to more acceptable levels, such as increasing stack height or imposing traffic limitations. As such, numerical models go a step further than *in situ* measurements by providing a basis for fast and accurate predictions. The LIDAR system used for the intercomparison experiments in Section 6.3.3 is thus of importance in providing the data required for the calibration and validation of model results.

7. Observational Evidence of Regional Climate Change

7.1 Introduction

As mentioned in Chapter 1, simulating climate needs to be closely related to observational evidence in order to verify model performances; in a region as complex as the Alps, where in addition the mountains are poorly (if at all) resolved by a GCM grid, a more complete understanding of regional climatological characteristics is necessary. Before describing the nested GCM + LAM simulations in Chapter 8, it is worthwhile deviating from purely numerical modeling in this chapter, in order to discuss some of our recent investigations related to the of intricacies of regional climate change in regions of complex orography.

A precise understanding of the climatic characteristics of mountain regions is limited by a lack of observations adequately distributed in time and space and insufficient theoretical attention given to the complex interaction of spatial scales in weather and climate phenomena in mountains (Beniston et al., 1994). Meteorological research has tended to focus on the upstream and downstream influences of barriers to flow and on orographic effects on weather systems (Smith, 1979), rather than on micro-climates within the mountain environments themselves. Climatic features relevant to mountain environments include microscale features of the atmosphere which are superimposed on larger scales of motion, and the influence of elements of the surface, such as vegetation and geomorphologic features which can create microclimatic contrasts in surface heating, soil moisture or snow-cover duration (Geiger, 1965). Isolating macro- and microscale processes, in order to determine their relative importance, is complicated by inadequate data bases for most mountain areas of the world (Barry, 1994).

Four principal factors influence mountain climates, namely altitude, continentality, latitude and topography. These climatic factors in turn influence vegetation type and cover, hydrology, and sometimes also geomorphic features. In particular, vegetation distribution in mountain regions is so closely linked to climatic parameters that vegetation belt typology has been widely used to delineate climate influences (Troll, 1968; Lauer, 1979; Monasterio, 1980; Quezel and Schevok, 1981; Klötzli, 1984, 1991, 1994; Ozenda, 1985; Quezel and Barbero, 1990; Rameau et al., 1993). Although it would appear simple to discuss how these typologies might be modified as a result of changing global climate, the climate system and its interaction with such landscape components as vegetation, geology, topography and soil are highly nonlinear (Beniston, 1993; 1994).

Most mountain regions except Antarctica are inhabited, and account for about 20% of the terrestrial surface of the globe. Ives (1992) has pointed out that mountains provide the direct life-support base for about a tenth of humankind and indirectly affect the lives of more than half, in particular because mountains are the principal source regions for the world's rivers; climate change would have influences perceptible well beyond the boundaries of the mountain ranges themselves. Mountains are characterized by sensitive ecosystems, enhanced occurences of extreme weather events and natural catastrophes; they are also regions of conflicting interests between economic development and environmental conservation. Once regarded as hostile and economically non-viable regions, mountains have in the latter part of the century attracted major economic investments for tourism, hydro-power, and communication routes. In the context of climate change, significant perturbations can be expected to natural systems and these will inevitably have an influence on the economy of mountainous regions. Global change will probably exarcerbate the present conflictual situation between environmental concerns and economical considerations.

Despite their relatively small surface area, mountains are an integral part of the climate system. As a physical barrier to atmospheric flows, they perturb synoptic patterns and are considered to be one of the trigger mechanisms of cyclogenesis in mid latitudes. Because of significant altitudinal differences, mountains such as the Himalayas, the Rockies, the Andes, and the Alps exhibit within short horizontal distances climatic regimes which are similar to those of widely separated latitudinal belts; they also feature increased diversity of species, communities and habitats over lowland environments. Mountains are also a key element of the hydrological cycle, being the source of many of the world's major river systems, as already mentioned.

Abrupt changes in existing temperature and precipitation patterns which have led to the present distribution of vegetation, ice, snow and permafrost zones, would impact heavily on the unique features of mountain environments. This in turn would lead to significant perturbations to the existing socio-economic structures for populations living within the mountains themselves, and indirectly to populations living outside of these zones but dependent on them. Mountains around the world are also significant to local and global economies because of the mineral and timber resources they contain, as well as their hydro-power potential. Regional climate change could directly affect the viability and health of commercial timber production and, in an indirect manner, add cost to mining and other mineral extraction and processing activities in mountainous areas.

Beniston (1994) and Beniston et al. (1995) have reported on the potential impacts of climate change in mountain regions, and examples of systems in mountain areas which could be significantly disrupted or perturbed by putative climate change consecutive to increases in greenhouse gases of anthropogenic origin, are given below.

7.1.1 Impacts on water resources

Climate change may be characterized by changes in seasonal or annual precipitation, proportions of solid to liquid precipitation, or frequencies of extreme events, and these changes will inevitably have an impact on the timing and intensity of peak river discharge. Whatever the directions and magnitudes of change, mountain communities and those downstream need to be prepared to implement flexible water management strategies which do not assume that recent patterns will continue.

Events in recent history may provide useful guidelines for developing such strategies (Glantz, 1988).

Simulating precipitation patterns in mountain areas is today exceedingly difficult, because many precipitation events are local or mesoscale in nature and are poorly resolved by forecast models. Predicting the future trends of precipitation over areas as small as the Alps, for example, is practically impossible with present-day GCMs, so that little can be inferred for the impact of global warming on precipitation patterns. However, this is one key parameter which urgently requires quantification, as any significant change in rainfall or snowfall will have wide-ranging effects on natural and economic environments, both within the mountains themselves and far downstream in the river basins which they control. This is one reason why downscaling approaches, including the physically-based nested modeling technique, are the focus of much scientific attention.

7.1.2 Impacts on snow, glaciers, and permafrost

The IPCC has attempted to assess some of the potential effects of a warmer atmosphere on the mountain cryosphere. According to a report by the National Research Council (1985), a 1° C increase in mean temperature implies that the present snowline could rise by about 200 m above present levels in the European Alps. Kuhn (1989) has estimated that, for a 3° C warming by the middle of the next century, ice would no longer occur below the 2500 m level in the Austrian Alps, with the consequence that 50% of the remaining ice cover would disappear. It is, however, difficult to estimate the exact response of glaciers to global warming, because glacier dynamics are influenced by numerous factors other than climate, even though temperature, precipitation, and cloudiness may be the dominant controling factors. According to the size, exposure, and altitude of glaciers, different response times can be expected for the same climatic forcing. For example in Switzerland, a number of glaciers are still advancing or are stationary despite a decade of mild winters, warm summers, and lower than average precipitation.

A rise in the snowline would be accompanied by ablation of present permafrost regions. Zimmermann and Haeberli (1989) show that permafrost melting as well as the projected decrease in glacial coverage would result in higher slope instability, which in turn would lead to a greater number of debris and mud slides, and increased sediment loads in rivers.

7.1.3 Impacts on forests and natural ecosystems

Species will react in various manners to global climate change; increases can be expected for some species while others will undergo marked decreases. Under conditions of rapid warming, it is quite likely that certain ecosystems may not adapt quickly enough to respond to abrupt climate changes; this could be the case of forests which may not be able to migrate fast enough, or of species with very local ranges which have nowhere to migrate to. The possibilities of migration of species to higher altitudes, in order to find similar conditions to those of today, will probably be limited by other factors such as soil types, water availability, species competition and the human barriers to migration such as settlements and highways.

Analogies with migrations of the past are hazardous at best because of human perturbations to the environment which were not present millenia ago, and because the expected climatic change will be at a rate unprecedented in the last 10,000 years. The disappearence of certain types of protective vegetation cover, particularly forests, from mountain slopes would render these more prone to erosion, mud and rock slides, as well as avalanches.

7.1.4 Impacts on mountain economies

The environmental impacts of a warmer atmosphere would have a number of consequences for the regional economy of a mountain region. Increases in debris flows and avalanches would be responsible for greater incidence of damage to buildings and traffic routes, vegetation, and the obstruction of rivers with corresponding flooding. In the severe flooding episode of August 1987 in Central Switzerland, road and rail traffic on one of the densest north-south communication routes across the Alps was severely disrupted.

Increases in such episodes in future would have severe economic consequences not only for the region itself, but also for all users of such vital communication links. The additional risks due to climate change would sharply raise insurance premiums to cover the higher frequency of damage to property, as well as the basic costs of civil engineering works necessary for the protection of settlements and communication routes. Such costs would weigh heavily upon regional mountain economies.

Tourism may also suffer from climate change in mountain regions; in particular, uncertain snow cover during peak winter sports seasons may discourage tourists to come to the mountains. However, in recent years, numerous winter sports resorts have been facing financial difficulties even during favorable winters. Many are now re-organizing their sports and cultural infrastructure in order to attract vacationers whatever the climatic conditions.

Hydro-electric power is a key element of mountain-based economy in the Alpine regions, as is the case in other mountains of the world. In New Zealand, over 70% of electricity is generated by hydro-power (Fitzharris, 1989), particularly in the Southern Alps. Accelerated glacier melting would result in an increase in electricity production and a reduction in water storage requirements, at least until the glaciers find a new equilibrium level. However, if the glacier melting is intense, then river flooding and damage to dam structures may occur, offsetting the temporary 'benefits' of increased storage and production capacities.

In many developing countries, mountains provide food and fuel needed for human survival. Disruption of mountain resources needed for subsistence would have major consequences, but the response of mountain populations to such perturbations are difficult to estimate.

7.2 Regional climate change in Switzerland: observational evidence

The climate of the Alpine region is one of considerable interest and compexity, not only because of the physical influence of the Alps on synoptically-induced weather features (gravity waves, wave breaking, blocking situations, etc.), but also because a number of different climatological regimes converge on the region. The complexity of Alpine climates in terms of macro-scale features is brought about by the competing influence of Mediterranean, continental, Atlantic, and Polar regimes. Any response to global climate change will probably result in the altered frequencies of these principal regimes, possibly leading to an amplification of the regional response of climate variables beyond the average changes estimated by the IPCC.

7.2.1 Secular climate change in Switzerland

The climatological stations of Zurich (569 m above mean sea level), Säntis (2500 m), Davos (1590 m) and Lugano (276 m) have been selected on the basis of the length of their time series (uninterrupted daily records from January 1, 1901 - December 31, 1992) and their geographical location. The rationale is to identify, at least to a first order, whether there is a consistency in the data at the four locations, or whether they exhibit altitudinal or geographical particularities (Beniston et al., 1994).

Figure 7.1 illustrates the locations of these four stations in Switzerland. The principal climatological variables studied include mean, daily maximum and daily minimum temperature, 24-hour precipitation sums (7 am to 7 am totals), relative sunshine duration (the proportion of sunshine duration with respect to its maximum possible site-specific duration), total snow depth, and pressure.

Pressure is essentially determined by the synoptic scale and is far less sensitive to local and regional effects than temperature or precipitation; of the climatological statistics available for this study, pressure is the only variable capable of providing a link between the large and the regional scales.

When working with long time-series, the question invariably arises with regard to the homogeneity of a particular data set. Over a period as long as the 92 years investigated here, certain instruments have been changed or upgraded, the location of the instruments may have been shifted, the environment of the measurement site will have certainly have experienced some modifications, and biases imposed by the different observers over the period of the record may enter into the data sets.

Despite these constraints, it is generally recognized that the SCDB is one of high quality and reliability; for example, when instruments such as sunshine recorders were changed, the old and the new instruments were placed side by side for a period of several months for calibration purposes. Site changes have also occured, in particular in Zurich in the 1950s; however, the temperature and pressure data sets contain corrections taking into account this change, in particular for the altitudinal differences between the old site and the present site. Although it could be argued that the time series at Zurich and Lugano will inevitably be contaminated by urban heat island effects, this does not appear to be a dominant feature of the series, since as will be seen the increase of temperature at these locations is not significantly higher than at the mountain stations where the urban signal is absent.

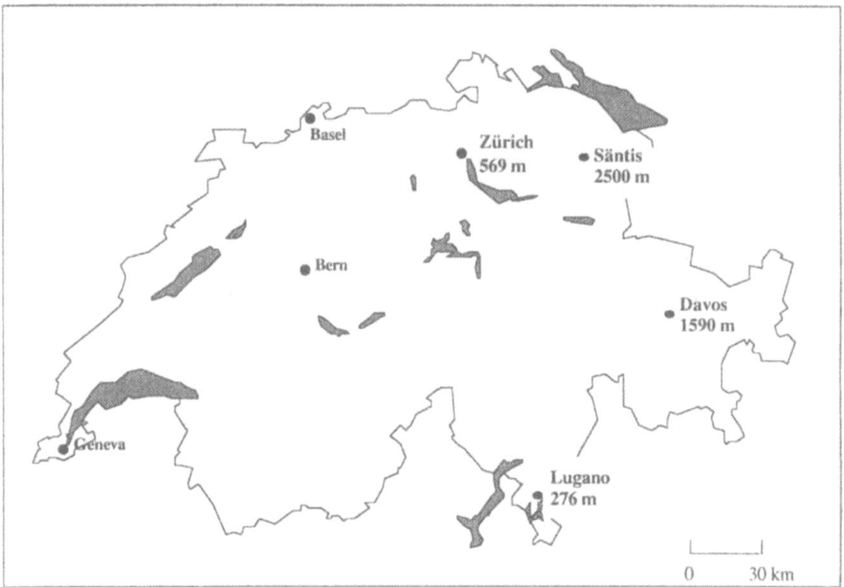

Figure 7.1: Locations of the four stations analyzed in the present study: Zurich, Davos, Lugano and Säntis

It should be furthermore noted that while the climatological stations at Zurich and Lugano are indeed located in an urban environment, both Lugano and, to a lesser extent Zurich, are small cities which do not exhibit the large urban sprawl characteristic of major industrialized cities and where the urban heat island effect is readily identifiable. Furthermore, as will be seen in the discussion, the data sets at the four very different study locations (in terms of their geography) exhibit a very large degree of consistency, indicating that for daily observations over a period of close to a century, contamination of the data resulting from one or more of the abovementioned factors are relatively minor. The information provided by the SCDB can be considered to be at least as reliable as the data sets used to investigate global warming by other authors (for example, Jones et al., 1986; Boden et al., 1990, to cite but two examples)

7.2.1.1 Temperature changes

Temperature is one of the most widely used climatological variables in climate change and associated impacts studies, not only because it is one of the easiest to measure directly and to reproduce in numerical models but also because it is, with precipitation, one of the principal controling factors on a wide range of climate-dependent environmental and ecological systems.

Figure 7.2 provides a graphical representation of the evolution of daily minimum temperatures at Davos from the beginning of the century to the end of 1992. Based on the daily temperature values, mean annual statistics have been established; a five-year running mean has been applied to the data in order to filter out some of the high-frequency modes inherent to the interannual variability. At all four stations, minimum temperatures have increased on average by about 2 K between 1901 and 1992, though the manner in which this warming is achieved is not homogeneous; the low-level stations (Zurich and Lugano) exhibit a continuous increase in temperature, whereas at Davos a considerable fraction of the increase in minimum temperature has been attained by 1930 and has since then tended to oscillate around a quasi-constant value; at the high-elevation Säntis, warming has taken place in an accelerated manner in the decade of the 1980s. It should be noted that in all four records, an event lasting about 15 years occured from the early 1940s to the late 1950s, charaterized by warming in the initial phase of the event and cooling of similar magnitude towards the end of the event. Certain stations exhibit other long-term cycles of between 8 and 12 years' duration, but none are as clearly marked as the 1940 - 1960 departure.

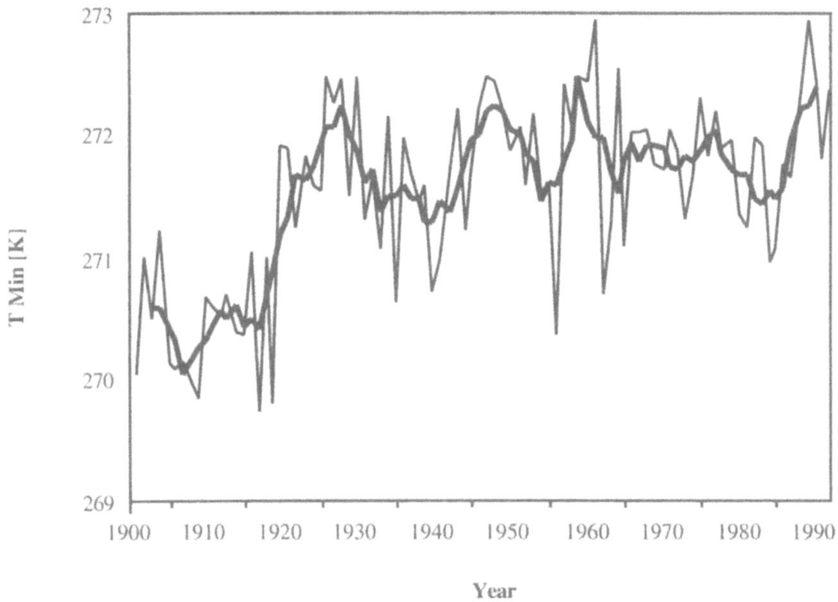

Figure 7.2: Minimum temperature trends at Davos

Figure 7.3 illustrates the evolution of maximum temperatures at Davos. At the low-altitude locations, maximum temperatures exhibit a decreasing trend especially between 1940 and 1980 of between 1 and 2 K. In Davos, maximum temperatures remain essentially unchanged up till 1940 then undergo a series of major decadal-scale fluctuations. Säntis maxima exhibit very similar behavior to the minima, with a strong increase of maximum temperature in the 1980s. The major departure beginning in the early 1940s, present in the minimum temperature records, is visible here also. Because minimum temperature trends are changing at a rate different from that of the maximum temperatures, the diurnal range of temperature has tended to decrease at all stations; in Zurich, for example, the daily temperature range on an annual average basis exceeded 9 K at the beginning of the century and has dropped to about 7.5 K today. The fact that the diurnal range and annual variance of temperature is decreasing has been reported in other data analysis studies for Northern Hemisphere locations (e.g., Karl et al., 1984), and therefore provides some degree of consitency between the Swiss data and other regional or supra-regional data. It is also consistent with model studies of greenhouse-gas induced warming, where the major effect would indeed be on nocturnal temperatures, even though these model simulations generally tend to underestimate the trapping of terrestrial radiation by greenhouse gases (Rind et al., 1989).

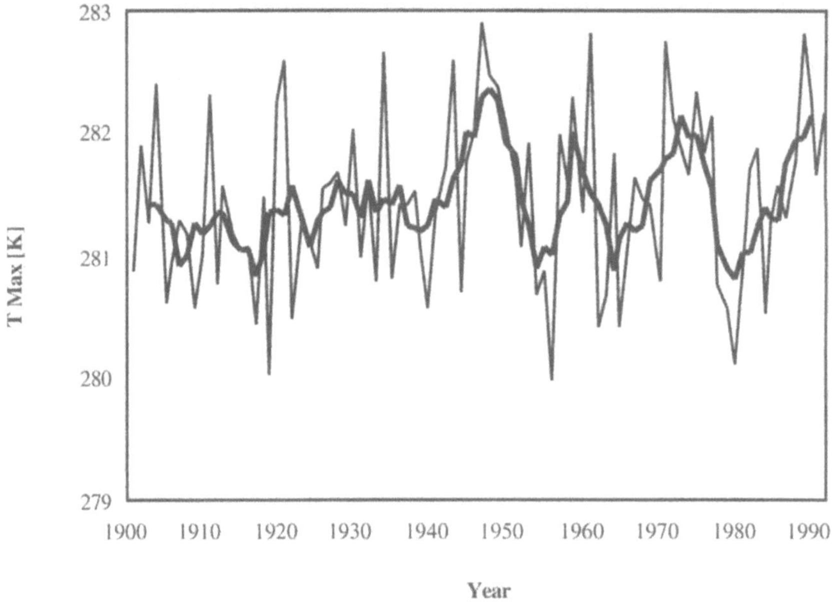

Figure 7.3: Maximum temperature trends at Davos

7.2.1.2 Analysis of precipitation records

The analysis of precipitation is made difficult by the fact that the daily precipitation data (24-hour precipitation sums) available on the SCDB is raw, uncorrected data. It is a well-known fact that for technical reasons, precipitation at high mountain elevations can be underestimated by as much as a factor of 2 (e.g., recent papers by Kirchhofer and Sevruk, 1992, and Sevruk et al., 1993). The instrumental errors are often linked to microscale dynamics in the highly complex orography in the vicinity of a high-elevation station, where strong eddies and accelerating airflow over mountain crests will determine the manner in which precipitation will be intercepted by a rain-gauge. Additional problems related to discontinuities in precipitation measurements resulting from even minor site changes and instrument changes contribute to the uncertainties in precipitation measurements.

The measurement site in Zurich changed in the 1950s and the precipitation records on a daily basis have certainly been affected as a consequence of this change. However, use of yearly precipitation totals to some extent alleviates this problem, as these values will in general be representative of more than just their immediate measurement environment (the exception being if much of the precipitation is in the form of convective precipitation whose intensity can vary considerably within very short distances).

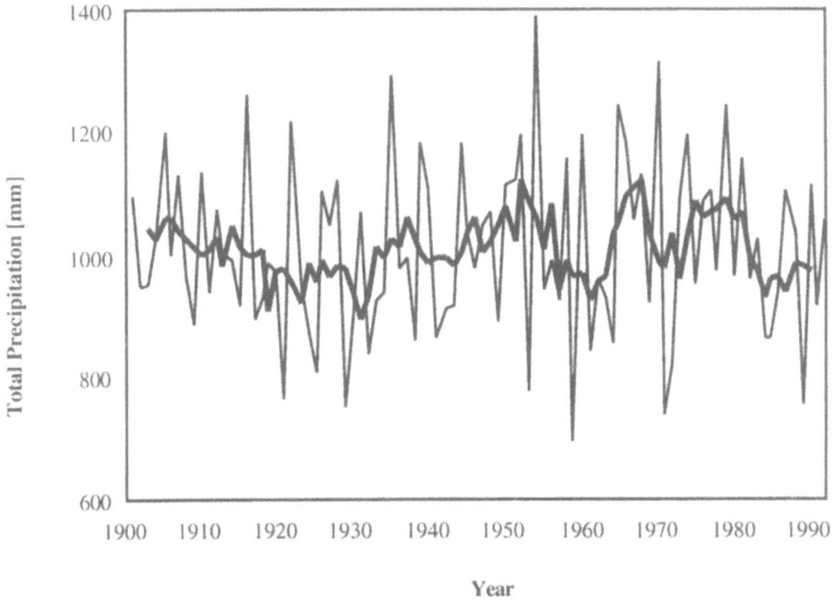

Figure 7.4: Precipitation trends at Davos

Civil engineering work at he summit of Säntis in the last few years will also have inevitably contaminated the precipitation data, so that the Säntis records are given here without any guarantee as to their statistical validity in the trend analysis. Corrections are at present possible for monthly and yearly precipitation values but not for daily data, nor consequently for the daily extremes which are probably the most important control parameter on a wide range of systems. For the purposes of this paper, however, the precipitation data still has a measure of usefulness in bringing to light certain aspects of secular changes in regional climates.

Figure 7.4 illustrates the annual precipitation totals based on the daily 24-hour sums in Davos. At all four stations, there is no significant trend over this century, and the low-frequency curves exhibit cycles on a 10 - 20 year time scale around a relatively constant mean value. Precipitation records at Säntis exhibit a decreasing trend from the early 1920s to the late 1980s with an increase in the last decade; due to problems of measurement inherent to high-elevation stations, the tendencies shown here may be as much the result of instrumental error as a physically-based climatological signal.

All four stations exhibit very noisy precipitation time-series, reflecting the strong interannual variability in the Alpine region. Precipitation may be triggered by a number of mechanisms, which occur on different time and space scales, from the large synoptic scale associated with the passage of a frontal system to the very local influence of orography or convection (which themselves may be amplified by synoptic disturbances). As a result, it is very difficult to decouple one effect from another solely on the basis of annual values; it is consequently equally difficult to determine any particular trend associated with global warming, for example.

An analysis of extreme precipitation amounts, i.e, the highest 24-hour precipitation total, also exhibit a very strong interannual variability and no significant trends. Extreme record for Lugano, where episodes of precipitation in excess of 100 mm/day, generally associated with strong downpours in unstable convective situations are, if anything, less frequent than in the early part of the record.

7.2.1.3 Analysis of sunshine duration hours

Relative sunshine duration hours may be defined as the percentage of sunshine duration intercepted by a sunshine recorder relative to the total possible sunshine duration. This parameter is dependent not only on the season but also on site-specific features, such as the location of the climate station in the vicinity of mountains or other obstacles which would mask the sun (i.e., the sky-view factor).

With the exception of Säntis, a general decrease of annual mean sunshine hours is observed at all locations. A logical conclusion for the trends of maximum temperature is that the decrease in maxima observed elsewhere than Säntis could be explained by this decrease in sunshine hours; furthermore, in the absence of nocturnal cloud amount, the values of relative sunshine hours point to an increase in cloudiness over Switzerland.

It could be hypothesized that cloudiness would persist during the night hours, thereby trapping outgoing infrared radiation and resulting in the observed increase of minimum temperatures. At all stations, sunshine increased in the 1980s, reaching average values not experienced since the early 1970s (i.e., the dip in relative sunshine amounts was one of the most marked and persistent periods in the record between 1970 and 1990).

7.2.1.4 Analysis of snow records

Snow is a particularly important feature in the Alps because it is a controling factor on a wide range of systems, ranging from hydrology to ecosystems. Its presence or absence may in many cases determine the evolution of plant species during their annual growing period, or the runoff and flow in many of Europe's major river systems. The amount and duration of snow in the Alps also has economic significance in terms of tourism and hydro-power in particular.

Figure 7.5 gives an overview of the occurence of snow cover in Davos during the period from 1931 - 1992, for three thresholds of total snow depth (25, 50, and 75 cm respectively). Snow records in digital form unfortunately do not start before 1931; in this figure, the year 1931 should be understood as the winter season beginning in the fall of 1931 and stretching into the spring of 1932. The extreme right-hand record is for the winter of 1991 - 1992.

The principal comment which can be made with respect to snow depth is that, even in a high-altitude resort such as Davos, there is a significant interannual variability. All winters in the period have experienced at least the minimum threshold of snow cover (25 cm), but snow has been more abundant in certain years, in particular in the early 1950s, the mid 1960s and mid 1970s, and the early to mid-1980s. The lack of snow in the latter part of the 1980s, culminating in the winter of 1989 - 1990, which raised considerable media interest and led to financial difficulties for low-altitude resorts, was not in itself an exception. There have been periods (1931 - 1934; 1964; 1971 - 1972; 1978 - 1979) in which snow has been lacking in terms of snow duration and abundance.

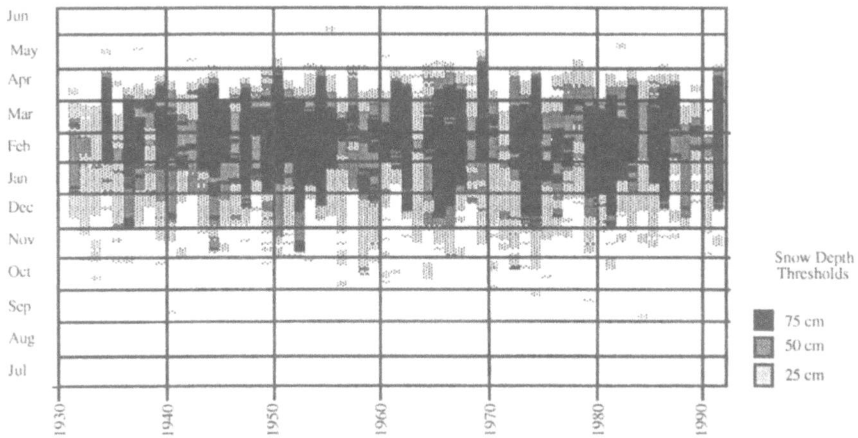

Figure 7.5: Occurrence of snow cover at Davos (1931 - 1992 records)

One explanation for the increasing concerns in mountain communities regarding periodic lack of snow may be found in the human perception of climate. During the 1960s, snow was generally present and, in terms of snow depths was perhaps more abundant than in recent years. As this was a boom period for mountain resorts and a time when investments in tourist infrastructure were high, that particular decade was regarded as the norm as far as winters are concerned. The data analysis shown here suggests that, in the present sampling period, the decade of the 1960s was rather the exception in terms of snow abundance, and certainly the 1940s were marked by less snow duration and less snow depth than the 1980s, widely regarded as anomalous in the sense of lack of snow. The cultural and perception factors have been widely surveyed by Rebetez (1994) and shed light on climate issues in Switzerland.

7.2.1.5 Analysis of pressure records

The analysis of surface pressure allows a link to be made between a climatological variable which is essentially representative of synoptic-scale atmospheric processes, and those other variables just described, which are generally more representative of the regional and local scales.The discussion will focus on pressure in Zurich only, since the geographical variability of pressure in an annual average context is small.

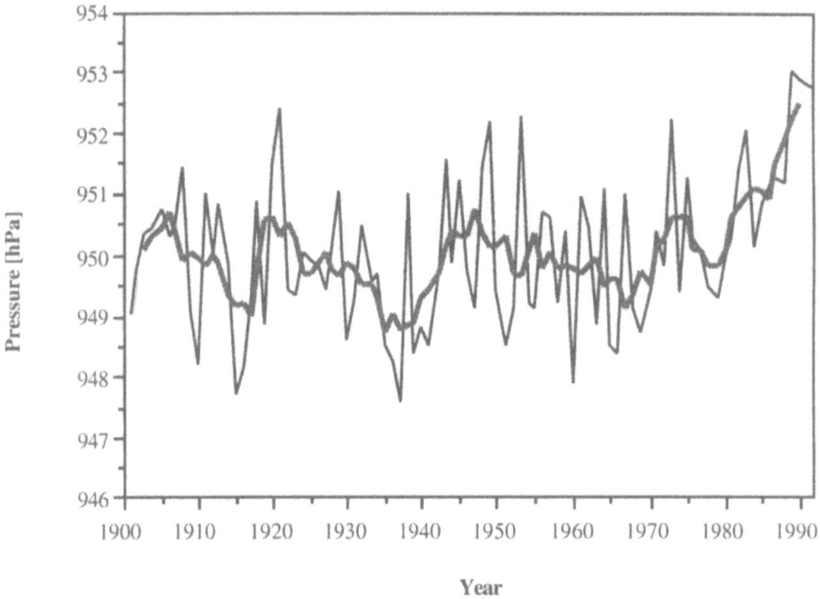

Figure 7.6: Mean pressure trends in Zürich

Because pressure is an integral value which describes the mass of the atmosphere above any particular point, it is only in short-lived circumstances that a strong pressure gradient will occur between adjacent locations.

High-frequency pressure differences characteristic of different weather situations (passage of frontal systems, foehn situations, etc.) where the pressure gradient across the Alps can be high, are filtered out when analyzing mean annual data. The pressure data measured at the Zurich station can therefore be considered representative of the large-scale pressure field over Switzerland.

Figure 7.6 illustrates the evolution of average surface pressure measured at Zurich, i.e., not reduced to its sea-level value. It is seen that for the most part of the century, the low-frequency component of pressure (5-year running mean) has fluctuated within the range of 949 - 951 hPa. Since the beginning of the 1980s, however, the pressure field has exhibited a totally anomalous behavior, in that it has increased continuously to reach values beyond 952 hPa. The 952 hPa annual average level was reached only four times previously this century, but since 1987 has exceeded this value and has continued to increase into the 1990s. This remarkable increase of pressure in the latter part of the record is related to a greater frequency of blocking high episodes in the Alpine region. At no other time in the record has the rise in surface pressure been as long and uninterrupted in the 5-year average curve, nor has the amplitude of the increase been as large. The change in the latter part of the century may be indicative of a shift in the relative occurences of typical weather regimes over the Alpine region in recent years, i.e., a manifestation of climate change over the Alps.

7.2.2 Links between regional climate change and global climate change

As a means of feeding into the next sub-section concerning possible causal relationships, it is of interest to see the extent to which the temperature trends observed in the Swiss data are consistent with global trends, despite the many uncertainties inherent to both the instrumental record and to model simulations.

7.2.2.1 Comparisons with the global record

For comparison purposes, the Swiss data are presented in the same manner as the Jones et al. (1986) analysis, i.e., in the form of anomalies from the 1951 - 1980 climatological average. The results are illustrated in Figure 7.7. What stands out clearly in this figure is that the amplitudes on the regional scale are much larger than on the global scale. The Swiss data is characterized by a number of periods of significant temperature fluctuations, lasting between 8 and 15 years approximately. In some cases, the Swiss data is out of phase with the global record (in the 1940s, 1960s and 1970s). Since the mid-1980s, the increasing temperature trend for the Swiss records is synchronous with the global warming trend, and the rate of warming far exceeds that of the global tendencies. The Swiss data indicates that the 1940s were the warmest decade of this century, more so than the 1980s.

The Swiss data is characterized by a number of periods of significant temperature fluctuations, lasting between 8 and 15 years approximately. In some cases, the Swiss data is out of phase with the global record (in the 1940s, 1960s and 1970s). Since the mid-1980s, the increasing temperature trend for the Swiss records is synchronous with the global warming trend, and the rate of warming far exceeds that of the global tendencies. The Swiss data indicates that the 1940s were the warmest decade of this century, more so than the 1980s.

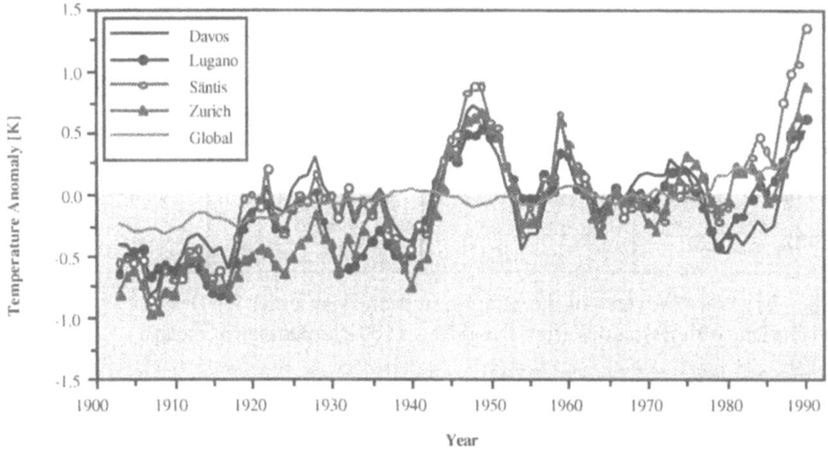

Figure 7.7: Comparisons between global temperature change and changes observed in Switzerland

In terms of record temperatures, Table 7.1 shows that among the five warmest years of the century, only one lies in the 1983 - 1992 period at Davos, Lugano, or Zurich, unlike the global record, where six of the warmest years of the century occur between 1983 and 1992 (Houghton et al., 1992). The exception here is Säntis, where indeed three of the warmest years are recorded in 1983, 1989, and 1992, with very significant departures from the 1951 - 1980 climatological mean. The five coldest years, in terms of their annual average, are generally confined to the first twenty years of the century, with the exception of 1956 which experienced a very cold month of February, with temperatures below freezing (even for maximum temperatures) throughout the month; in Zurich, the monthly mean maximum temperature was 9.4 K below the 1951 - 1980 norm for February, and the mean monthly minimum was 11.2 K below the norm. This cold spell was confined to the part of Switzerland located to the north of the Alpine chain, as can be deduced from the Lugano records which do not include 1956 among the statistics for the 10 coldest years. In terms of record-breaking temperatures for an individual day, the years of 1985 and 1987 are amongst the coldest at all locations except for Lugano. With these notable exceptions, however, there is no doubt that the beginning of the century was on average colder than the latter part of the record.

There have been a number of studies of trends on global, hemispheric, and continental scales, many of them linked to the IPCC process. These studies have been instrumental in highlighting some of the features of observed climate variability through exhaustive data analysis (Boden et al., 1990; Jones et al., 1986; Jones, 1988; Jones and Kelly, 1983; Jones and Wigley, 1990; Jones et al., 1990; Wigley and Jones, 1981; Wigley and Raper, 1990). A first-order analysis of the global or hemispheric data emphasizes the rise in average temperature since the beginning of the industrial era, with a slight cooling between the 1940s and the 1970s.

Davos		**Lugano**		**Säntis**		**Zurich**	
Year	Anom.	Year	Anom.	Year	Anom.	Year	Anom.
1961	1.31	1943	.91	1989	2.04	1990	1.36
1947	1.21	1949	.90	1992	1.74	1961	1.16
1989	1.11	1990	.89	1920	1.34	1949	1.05
1934	.81	1961	.79	1983	1.32	1947	.97
1948	.80	1945	.70	1959	1.24	1959	.95

Table 7.1: Warmest years of the century, in decreasing order for Davos, Lugano, Säntis and Zurich, with deviations from the 1951 - 1980 climatological mean

It is of interest to analyse whether the global signals of climate change this century are also discernible in the Swiss data. Figure 7.8 provides a decade-by-decade illustration of the warming rates of average annual temperature for this period at all the locations studied. It is seen that until the 1940s, the rate of change of temperature per decade was on the order of 0.5 K or less, without any particular consistency of the sign of the temperature change according to the location, i.e., two stations can be seen to be in phase one decade and out of phase the next. In the 1940s, however, there is a very significant change occuring, with all four stations exhibiting the strongest warming of the century, exceeding 1.5 K/decade at Säntis and Zurich (i.e., more than 3 times the previous warming or cooling rates). This sudden jump in the warming trends has resulted in some of the warmest years in terms of their annual average, as seen in Table 7.1. The warming is much less pronounced in the 1950s and reverts to a cooling in the 1960s and 1970s, reaching or exceeding 1 K/decade. The 1980s exhibit a return to strong warming, around or above 1 K/decade. When combined with the strong warming of the 1940s, the net effect over the period of record is a warming, since the cooling trend in the 1960s and 1970s does not compensate for strong warming of the 1940s and the 1980s. The analysis undertaken by Balling (1992) to the effect that most of the 20th Century warming occured in the first half of this century in the Northern Hemisphere does not appear to be the case for the Swiss records; strong warming has occured in two distinct decades, namely the 1940s and the 1980s. The beginning of the century is characterized by small warming/cooling rates while the latter part exhibits larger rates of temperature change.

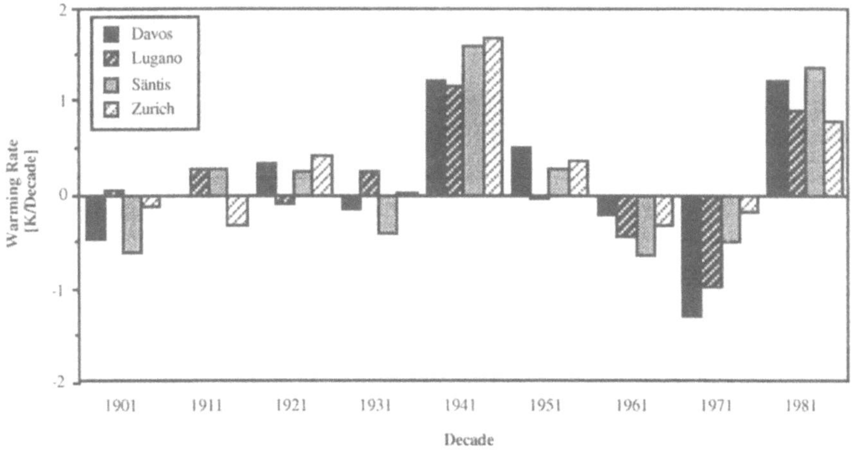

Figure 7.8: Decadal warming rates observed in Switzerland

7.2.2.2 Possible causal mechanisms of observed climate change in Switzerland

There are numerous studies in the literature which attempt to link observed climate variability at certain locations in space with parameters representative of the atmospheric general circulation and variations of those parameters. The work done on the ENSO phenomenon and the definition of the Southern Oscillation Index (SOI) has brought to light the dependence of precipitation régimes in the Pacific region (Indonesia, Oceania, South and Central America), for example, to changes in the SOI (Bjerknes, 1969; Barnett, 1978; Diaz and Markgraf, 1991 among many others). Research into climate variability in the Northern Hemisphere has been conducted by a number of researchers on the basis of the Pacific/North American (PNA) teleconnection patterns (*inter alia*, Trenberth, 1990; Hansen et al., 1993; Wallace and Zhang, 1993). Hansen et al. (1993) note that blocking events over Europe have contributed disproportionately to large departures from normal regional surface pressure, temperature, and precipitation. As will be demonstrated here, the fact that most of the anomalous mid-troposphere circulation events occur during the winter season (from November through March) means that the climate variables on the regional scale should exhibit more anomalous behavior in this period than during the summer months. Indeed, studies of hemispheric climate trends show that much of the average surface air temperature increase is associated with wintertime warming, especially over the high-latitude continents (Jones and Briffa, 1992).

This interplay between global or hemispheric forcings and regional climate response will be returned to in the next section, but we shall first present here the conclusions regarding the link between the time series of pressure as measured in Zurich which, for reasons justified in the previous section, can be considered to be representative of the large-scale pressure field over the region. The index chosen as indicative of the synoptic-scale is the North Atlantic Oscillation Index (NAO). The NAO Index is the difference in sea-level pressures between Punta Delgada in the Azores and Akureyri in Iceland; it is a measure of the strength of the Westerlies over the North Atlantic. If the NAO Index anomaly with respect to the 30-year climatological mean period (1951 - 1980) is positive, then this is indicative of a higher-than normal occurence of Westerly winds across the North Atlantic.

Period	R	Significance
1911 - 1930	.241	Low
1931 - 1950	.656	High
1951 - 1970	.176	Low
1971 - 1990	.847	Very High

Table 7.2: Correlation between the North Atlantic Oscillation Index (DJF values) and the observed DJF surface pressure at Zurich for different periods of the record; significance test is at the 1% level

Figure 7.9 shows the relationship between the wintertime (DJF) anomaly of the NAO Index and that of the Zurich DJF pressure anomaly. In both cases, the strong decadal variability can be observed with fluctuations in both anomaly sets lasting about 20 years through to the end of the 1960s. The apparent regularity which has prevailed until this time changes considerably after the early 1970s, after which the fluctuations are of shorter duration; more significantly, there is a marked upward trend towards strong positive anomalies, particularly for the Zurich data set. While the two DJF curves bear a remarkable similarity, a more detailed analysis of the two data sets highlights an even more interesting phenomenon: the low correlation in the first part of the record and also between 1951 and 1970, the higher correlation in the period 1931 - 1950, and the highly significant correlation in the last 20 years of the record. The correlation coefficient between the two curves for four twenty-year periods is given in Table 7.2, along with the significance of the t-test at the 1% significance level. The present analysis emphasizes the fact that there is a switch occuring in the general circulation patterns over the Alpine area (more probably over Western Europe, which then impacts on flow patterns over and around the Alps), with alternation between pressure fields embedded in a zonal flow described by the NAO index and those in which the pressure field is controlled by other large-scale circulation patterns. The periodicity is on a time-scale of about 20 years according to Table 7.2, and confirmed by Figure 7.9.

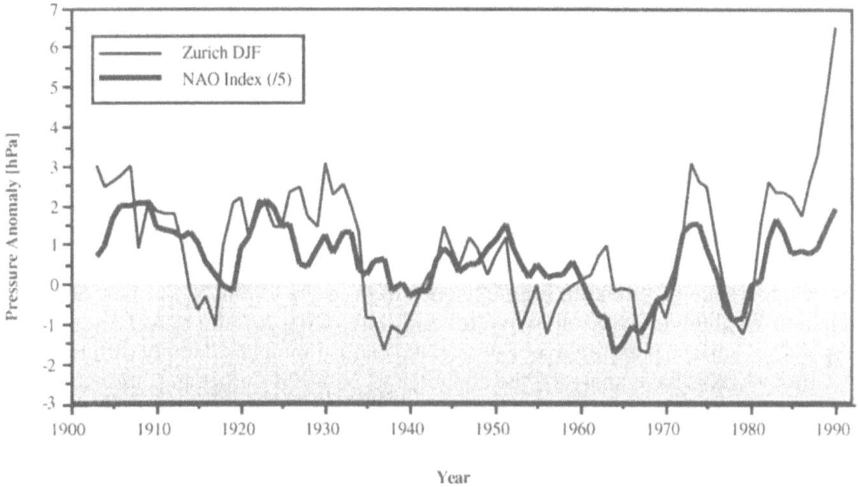

Figure 7.9: Links between surface pressure in Zürich and the NAO Index (DJF anomalies)

Decade	Persistence
1901 - 1910	6.3
1911 - 1920	5.6
1921 - 1930	5.3
1931 - 1940	5.9
1941 - 1950	5.8
1951 - 1960	3.2
1961 - 1970	4.5
1971 - 1980	4.9
1981 - 1990	11.6

Table 7.3: Average persistence of high pressure situations in Zurich (days/year for the 965 hPa level)

The table implies to some extent that periods in which the NAO Index anomaly is positive are accompanied by above-normally persistent periods of high pressures. Furthermore, in the 1980s, where the increase in surface pressure over Switzerland is quite remarkable, the blocking highs are far more frequent than at any other time of the century, with a decadal frequency 2 - 3 times greater than during the previous decades; the 1980s account for almost 25% of the total observed blocking highs this century. Indeed, 1989 is a record-breaking year with 30 days of persistent high pressure beyond the 965 hPa threshold, followed by 1990, 1991, and 1992. This sequence of four successive years accounts for over 16% of the blocking episodes this century.

It should be stressed that a strong positive NAO Index can result from either a deepening of the Icelandic Low or a strengthening of the Azores High; synoptic chart evidence for the 1980s suggests that the positive NAO Index in the 1980s are linked to an extension of the Azores High into southern and central Europe, thereby explaining the persistence of periods of high pressure in the winter months over the Alps; baroclinic activity over the North Atlantic is deflected over more northerly latitudes of Europe. A deepening of the Icelandic Low, on the other hand, which can lead to similar positive values of the North Atlantic Index, appeared to be a more common occurence in the years between 1930 and 1950. This led to more westerly flow régimes over the Alpine region and thereby in a lower occurence of blocking high episodes than in the 1980s. This short analysis helps to explain the reason why, for similar values of the NAO Index, the response of surface pressure over Switzerland can exhibit considerably different.

Other workers have analyzed the connections between European stations and the NAO index, such as Glowienka-Hense (1990), Rogers (1984), Rogers and van Loon (1979) and, more recently, Hurrell (1995; 1996). These studies indicate that Southern European locations have a relatively low teleconnection signal. The signal is, on the contrary, very much in evidence in the Scandinavian region, where above-normal precipitation events since the early 1980s are clearly associated with the persistence of a strongly positive NAO index during this period (Hurrell, 1995). The Alpine region nevertheless seems to be located within the boundaries of the region which exhibits significant correlation with the NAO index, as has been confirmed by Weigel (1994), who emphasizes the key role of the Iceland Low in this context. Gribi (1995) has also confirmed that, particularly in the winter months (December to March), the teleconnection signal is high and statistically significant for the climatological stations of Altdorf, Lugano and Säntis.

It is true that correlations between oscillation indices and weather statistics tend to diminish rapidly from one location to anther; this is for example the case for the ENSO signal, which is well correlated to precipitation events in northern Spain (e.g., Fraedrich and Müller, 1992), but is insignificant in the Alpine region located less than 800 km away.

Indeed, in this study, some attempt to quantify the possible effects of ENSO on regional data have not been successful, although the fact that in the 1980s three tropical Pacific Warm Events occured with no Cold Events in between (Trenberth, 1990) raises questions as to the possible influence of events far removed in space on regional climatological processes in Switzerland; the signal-to-noise ratio in the data which has been analysed is too low for any meaningful conclusions to be reached.

7.2.3 Response of climate on the regional scale to larger scale regimes

This sub-section will investigate in some more detail the statistical features of this data to provide an illustration of how these variables change under different synoptic conditions discussed in the previous section, above and beyond their interannual and decadal-scale fluctuations. The periods of lowest and highest frequency of blocking episodes (which coincide with the lowest and highest correlations between the NAO Index and the Zurich time series of pressure), respectively 1951 - 1960 and 1981 - 1990, will be analyzed in this context. Figure 7.10 provides an illustration of the changes in the daily pressure field which have occured between these two decades; the curve is obtained by subtracting the decadal-average daily pressure values for the 1950s from those of the 1980s. It enables one to identify the shifts in pressure anomalies throughout the year. This, and subsequent figures for other climatological variables have a 5-point filter applied to the data in order to remove high-frequency noise.

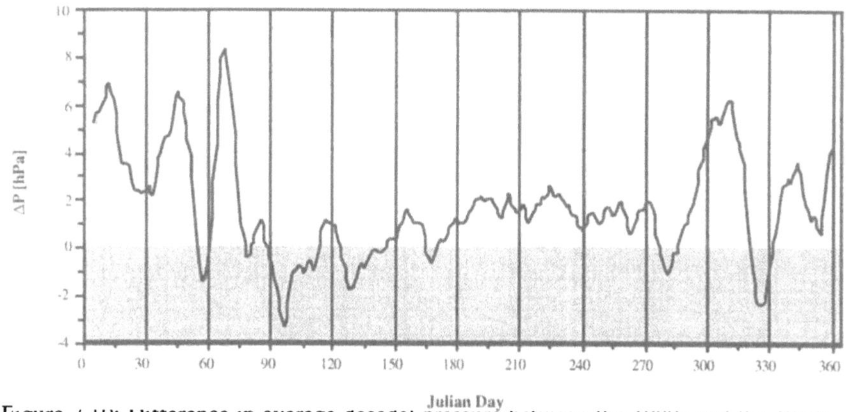

Figure 7.10: Difference in average decadal pressure between the 1980s and the 1950s

It is not surprising from the discussion on mean pressure tendencies to see that at almost all times, pressure is systematically higher in the 1980s than in the 1950s; this figure highlights the fact that the pressure increase in the 1980s over the 1950s occurs particularly in the fall and winter months (early November to late March), with anomalies reaching up to 8 hPa; in the spring and summer months, the pressure excess in the 1980s is much more modest and there are even periods of negative pressure anomalies from April to June.

It can be expected that the sensitivity of the different climate variables be felt more strongly during the winter season than during the rest of the year. However, the relationship between pressure and other variables is not quite as straightforward and it is worthwhile giving in this context a brief qualitative description of a rather unique feature of regional climate in winter under high-pressure conditions in Switzerland. Very often, extended periods of high pressure are characterized by a cold air pool trapped above the Swiss Plateau, i.e., that part of the country located between the Jura Mountains and the Alps (Figure 7.1), stretching from the Lake of Geneva in the SW to the Lake of Constance in the NE. The average elevation of this zone ranges from 400 - 700 m, and the depth of the cold air pool averages between 100 and 600 m; the upper boundary is characterized by a sharp temperature inversion, and this discontinuity is generally marked by the top of stratus formed by condensation of moist air trapped within the cold air pool. In the calm dynamic situation of a high-pressure system, the cold air has difficulty in evacuating out of the Swiss Plateau because of the bordering orography; the Jura and the Alps converge in the region of Geneva and act as a physical barrier to the cold air outflow. It is only with the advent of a perturbed synoptic regime that the cold air pool may be broken up. Above the inversion and the fog or stratus layer, conditions are usually very clear in the mountains, and can be relatively mild during the day because of the subsidence of air within the high-pressure system. Because of reflected solar radiation at the top of the stratus layer, conditions beneath the stratus may be colder during the day in Zurich, Bern, or Geneva, than they are at high-elevations such as Davos. At night, temperatures in the higher elevations will usually be cold as a result of strong radiative loss to space which more than offsets the subsidence warming; below the stratus layer, temperatures will not change very much from their daytime values because of the opaqueness of the stratus clouds to outgoing infrared radiation.

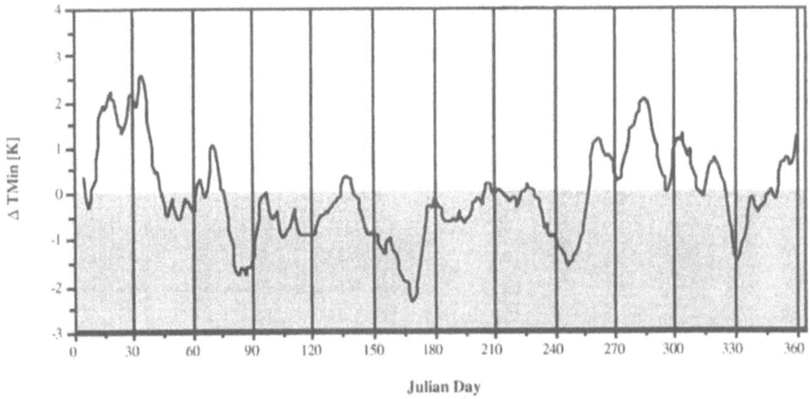

Figure 7.11: As Figure 7.10, except for minimum temperatures (Davos)

7.2.3.1 *Response of temperature*

Correlation between:	R
Zürich - Davos	0.769
Zürich - Lugano	0.593
Zürich - Säntis	0.661
Davos - Säntis	0.747

Table 7.4: Correlation analysis for the minimum temperature differences between the 1980s and the 1950s for the four climatological stations

While these figures allow somewhat more insight into the manner in which changes have occured by analysing not only a mean change in annual values but also the change across the range of temperatures, it is also useful to investigate the manner in which climate change between the 1950s and the 1980s has occured in time over the year. Figure 7.11 gives a description of minimum temperatures in Davos analogous to the pressure anomaly given in Figure 7.10, i.e., a difference of decadal averages of daily minimum temperature between the 1980s and the 1950s. From previous analyses, it has been shown that minima have risen substantially between the 1950s and the 1980s, and an integral value over time for the four curves would simply confirm this fact. Figure 7.11 shows that much of the warming occurs in the early fall and in the winter months. Analysis of the other stations shows that, except for Lugano, some periods exhibit an average cooling, such as April, June, or December. The correlation between these curves (Table 7.4) is high between Zurich and Säntis and Davos, and somewhat lower between Zurich and Lugano. The correlation analysis serves to illustrate that, at least to the north of the Alps, the same basic processes are taking place; the weaker correlation with the Lugano data, which will be seen to be systematic, is indicative that Lugano is more clearly subjected to other climatic régimes, in particular Mediterranean influences, than the other stations.

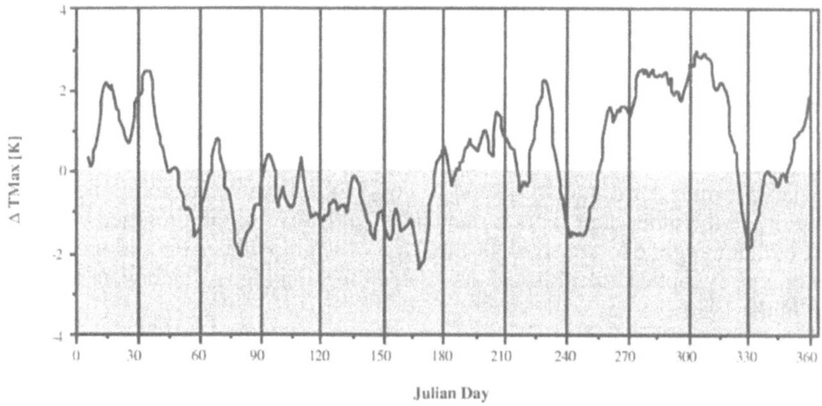

Figure 7.12: As Figure 7.10, except for maximum temperatures (Davos)

Correlation between:	R
Zürich - Davos	0.808
Zürich - Lugano	0.683
Zürich - Säntis	0.721
Davos - Säntis	0.944

Table 7.5: As Table 7.4, except for maximum temperatures

Figure 7.12 is analogous to Figure 7.11, but represents the anomalies of maximum temperatures for all four stations. Negative anomalies occur particularly from late winter to mid-summer, but are present at other periods of the year also. The correlation analysis is given in Table 7.5 and reflects an essentially similar picture to that of Table 7.4. The correlation between the two high-elevation stations is remarkable and serves to support the hypothesis that the same climatic change is leading to the same kind of response within the Alps.

7.2.3.2 Response of precipitation

An analysis of the anomalies of precipitation in the 1980s and the 1950s exhibit the same kind of periodicity as the temperature data, although the series tends to fluctuate around zero mm for most of the year in Zurich and Davos. The summer months exhibit somewhat negative anomalies which can be explained by the reduction of convective precipitation events resulting from the higher average pressures in the 1980s. There is a sharp transition from negative to positive anomaly at the end of August; a secondary peak of higher precipitation in the 1980s compared to the 1950s occurs towards the end of November and is linked to the reversal of positive pressure anomalies which take place during this period, i.e., by assuming that on average during the 1980s, the dip in pressures represents the passage of frontal systems and associated precipitation events.

However, other sharp pressure transitions, such as at the end of February and from March to April do not lead to the same precipitation response. As seen in the correlation analysis given in Table 7.6, Lugano does not exhibit the same behavior in precipitation anomalies. It has by far the greatest amplitude in the series and is often out of phase with the other stations. Late fall is clearly drier in the 1980s while April and May tend to be wetter, a feature which is not as outstanding in the other data. This is once more indicative of the presence of a different climatic régime to the south of the Alps which modulates the influence of the zonal-type synoptic situations and associated blocking highs which were experienced in the 1980s.

Correlation between:	R
Zürich - Davos	0.670
Zürich - Lugano	0.419
Zürich - Säntis	0.672
Davos - Säntis	0.714

Table 7.6: As Table 7.4, except for 24-hour precipitation totals

It is worth commenting here on the quite remarkable results obtained in Table 7.6; in a region as complex as the Alps, where the orography may not only amplify or suppress precipitation of frontal or convective origin, but also exhibit strong regional differences according to exposure, slope orientation, and altitude, it is surprising to see that to the north of the Alps, a large fraction of the observed precipitation changes which have occured between the 1950s and the 1980s can be identified at Zurich, Davos and Säntis. This similarity of precipitation observed at sites which are very different geographically implies that synoptic systems, rather than locally-forced orographic or regional convection, stand out more clearly in the 1980s than in the 1950s as the dominant precipitation mechanism. This is perhaps not surprising for a mid-latitude region, but it is nonetheless interesting to be able to identify the synoptic forcing in records which one could expect to be strongly contaminated by regional and local factors.

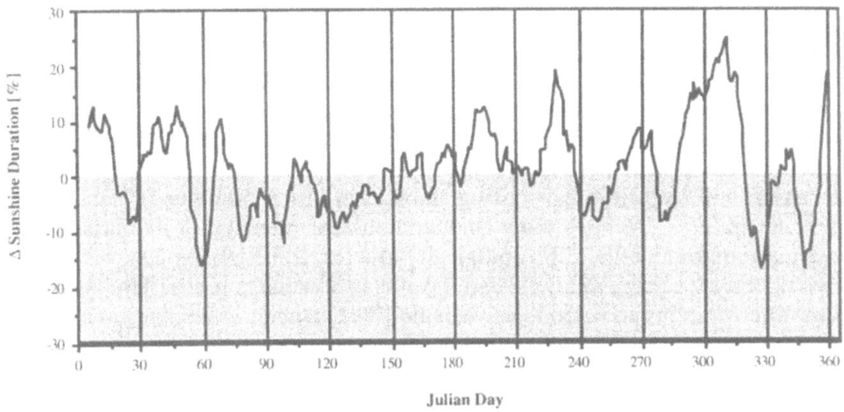

Figure 7.13: As Figure 7.10, except for sunshine duration (Davos)

7.2.3.3 Response of sunshine duration hours

The anomaly of relative sunshine duration between the 1980s and the 1950s at all four stations shows that spring is characterized by lower sunshine hours in the 1980s, whereas early fall, winter, and mid-summer (July and August) exhibit higher values (Figure 7.13). As with the precipitation data, Lugano tends not to follow these patterns; in particular late March and early December show strong positive departures of sunshine duration. This is due to the fact that these are periods in which the likelihood of synoptic disturbances embedded in a westerly airflow crossing Switzerland is high; in such situations, the Alps act as a barrier to the flow and mitigate the effects of such disturbances to the south of the Alpine chain.

Correlation between:	R
Zürich - Davos	0.622
Zürich - Lugano	0.367
Zürich - Säntis	0.608
Davos - Säntis	0.865

Table 7.7: As Table 7.4, except for sunshine duration

Table 7.7 confirms the fact that the Lugano statistics are different from the other stations, while the two high-altitude stations of Davos and Säntis show remarkable agreement due to the fact that if sunshine is present at one site, it is likely to be present at the other site during periods of extended high-pressures.

7.2.3.4 Response of snow cover

Although it is tempting to use snow as an indicator of climate change, it is a very sensitive variable which can be influenced by a wide range of factors other than simply temperature; for example the snow statistics will be influenced by the amount of snowfall at the onset of the winter season. A small initial quantity of snow is likely to melt quickly in the presence of sunshine, so that until there is more than a critical amount of snow on the ground, there will be little chance of significant accumulation. On the other hand, heavy snowfall at the beginning of the season will lead to longer-lasting snow even in the presence of sunshine or mild temperatures. Because snow amount holds the 'memory' of its initial conditions more than the other climatological variables, it is perhaps less valuable in terms of climate change analysis, and it is not as obvious to repeat here the analysis of differences in snow depth between the 1980s and the 1950s; the entire record for Davos (1931 - 1992) will therefore be discussed here.

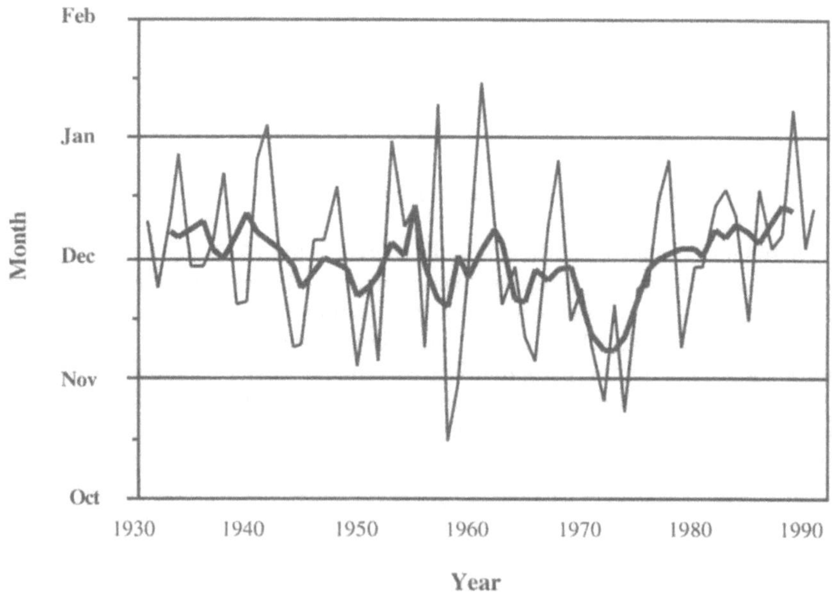

Figure 7.14: Period of snow cover in Davos for the 25 cm depth threshold (1931-1992 statistics)

In terms of the response of snow depth to the observed climate change, a certain number of features nevertheless stand out; periods of persistent high pressures, especially towards the end of the 1980s, are generally accompanied by low snow depth. This reflects the fact that snow precipitation during episodes of blocking high pressures was insufficient to allow snow to accumulate to the depths generally attained in other years of greater snow abundancy.

Figure 7.14 shows that the beginning of the snow season in Davos (for the 25 cm threshold) underwent a change from the 1930s to the early 1970s when the snow appeared as much as one month earlier in the early 1970s compared to the early 1930s; in the last 20 years, however, the beginning of the snow season has reverted to its values at the origin of the time series. In Figure 7.15, the duration of snow cover in excess of 25 cm has an inverse relationship, i.e., there was a marked increase in the duration of the snow season from 1930 - 1970 from less than 130 days with *continuous* snow cover exceeding the selected threshold value to over 160 days in the mid 1960s and early 1970s. The duration of continuous snow cover has reverted to its 1930s values in the 1980s, although the data is considerably biased by the winter of 1989/90 which was the period of least snow duration of the entire record (only 98 days of continuous snow cover). The aforementionned interannual variability is remarkable in this period, since the winter following 1989/90 was characterized by over 150 days of continuous snow cover in excess of 25 cm. Similar conclusions can be reached for the beginning of the snow season and the duration of snow for the 50 cm and 75 cm thresholds. The variability is amplified for the higher threshold values, and for the 75 cm limit, duration varies from 0 to 148 days (average amplitude between 30 and 90 days). The duration of 'deep snow' underwent four major cycles (amplitude approximately 40 days and duration between 8 and 15 years) between 1940 and 1980; on average, duration of 'deep snow' is longer in the 1980s than in the 1930s.

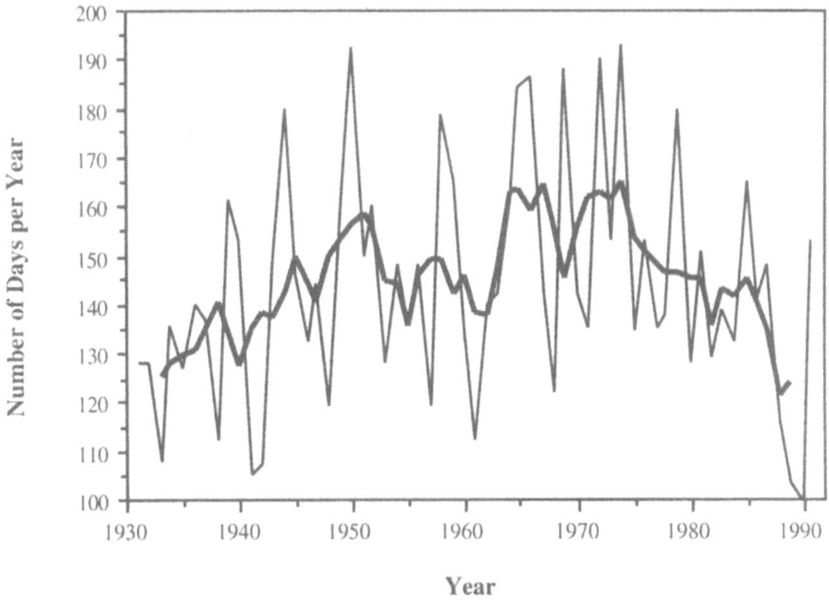

Figure 7.15: Duration of snow cover in Davos (1931-1992 statistics)

Snow is probably the least reliable climatological parameter, not only because of its strong year to year variability in the onset, duration, and total depth of snow cover, but also because it is a very regional feature. At equivalent altitudes in other parts of the country, or even at neighboring locations with different geographical aspects (slope orientation and exposure), the situation can be markedly different. An analysis we have made of the snow data for Montana (Canton of Valais) shows that the duration of snow cover has progressively increased from 1930 to the present time. However, at an altitude barely 100 m lower than Davos, the south-facing resort of Montana experiences between 30 and 50 days less snow cover (at the 25 cm threshold) than Davos.

7.2.3.5 Links to synoptic-scale influences

To this point, it has been established that there has been a regional response to changes in two periods of differing climatic regimes, namely the 1980s (characterized by generally high pressures and in phase with the NAO Index) and the 1950s (out of phase with the NAO Index and exhibiting low persistence of high pressures). In this final section, it is shown statistically that the observed anomalies between these two periods are indeed a function of the large-scale forcing associated with changes in the NAO Index.

It was seen in the discussion that the periods of anomalously high pressures occured essentially from late fall to late winter, and spring and summer months exhibited pressure anomalies of much lower amplitude. The months of January and July have therefore been selected to represent these periods of very high and low pressure anomalies, respectively. Correlations have been established between the pressure anomaly (i.e., the synoptic forcing term) and the variables representative of the regional response (i.e.,daily maximum and minimum temperatures, precipitation, and relative sunshine duration), and are illustrated in Figures 7.16 and 7.17 for January and July, respectively.

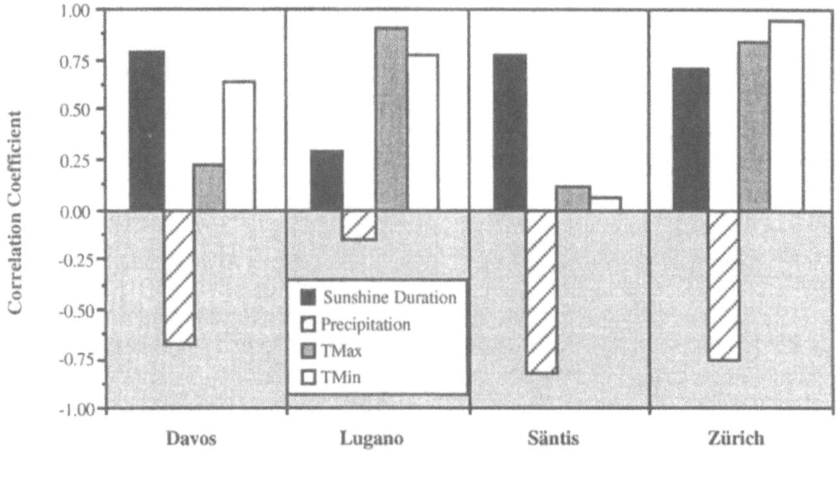

Figure 7.16: Correlation between pressure anomalies and temperature, precipitation and relative sunshine duration anomalies in January

Relative sunshine duration and precipitation anomalies are particularly well correlated with the pressure anomalies in Davos, Säntis, and Zurich, and much less so at Lugano. This indicates that the Lugano records are less influenced by westerly regimes than the other stations, and are more subjected to Mediterranean-type climates; high pressure over the northern part of Switzerland does not necessarily lead to the formation of stratus or fog as is the case in Zurich, so that insolation and precipitation in Lugano are essentially decorrelated from the pressure anomalies.

Correlations between pressure and temperature anomalies are high at Lugano and Zurich, and low at the two high-elevation stations. Temperatures during winter in the mountains can be affected by a number of local factors such as slope orientation and exposure. According to the postition of a high-pressure cell, temperatures can either be relatively mild, as in the case of an extension of the Azores High over Europe which would advect mild air at height, or cold as in the case of a high-pressure cell located over Scandinavia, Eastern Europe or Russia, where cold-air advection would be the dominant feature. While the presence of a persistent high-pressure zone leads to a significant increase in sunshine duration, the presence of sunshine is not in itself the unique controling factor on temperature; as a result, high-pressure systems can lead to a wide range of temperature responses at high elevations. A winter high-pressure system influencing Zurich, whether it be of a "warm" Azores-High type of a "cold" Russian or Scandinavian type of anticyclone, will generally be accompanied by persistent stratus formations or, at best, misty conditions which strongly attenuate the incoming solar radiation. Such a situation will essentially modulate the amplitude of diurnal temperature fluctuations; since there is a direct link between the overcast skies and pressure, it is not surprising, therefore, that the relation between pressure and temperature is also high in Zurich.

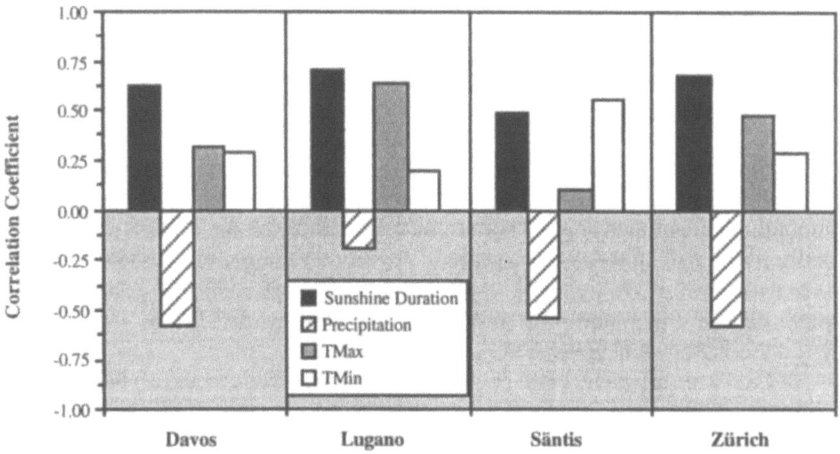

Figure 7.17: As Figure 7.16, except for July

The high correlation between pressure and temperature anomalies in Lugano can at first sight seem unusual, since Lugano is only partially influenced by Atlantic regimes. A persistent high-pressure system will generally result in relatively mild conditions south of the Alps. A "warm" high pressure system will lead to an influx of mild Mediterranean air to the southern foot of the Alps, while a "cold" system will generate a weak spill-over of air over the Alpine barrier which then warms by adiabatic compression as it descends towards the plains of Lombardy.

In July (Figure 7.27), when the pressure anomalies are lower, correlations between this quantity and the anomalies of the other variables are substantially less pronounced than in January. This is principally because the atmosphere is much less stable in the summer months, so that a number of regional factors contaminate the data and reduce the overall correlation between the synoptic and the local regional parameters. In particular, July and August are periods in which the probability of convective instability and precipitation is at a maximum. While such instabilities can be triggered by synoptic systems, they remain an inherently regional feature capable of determining the intensity of precipitation and associated temperature changes. From this type of analysis and the available data, it is difficult to dissociate the relative magnitudes of regional features such as convective instability or orographic forcing from the synoptic forcing. As a result, the broad correlation between pressure and the other variables is not as significant as for the January data.

As a final comment on the response of regional climate variables to the pressure anomalies, a close examination of the different figures presented in this section emphasizes the fact that the amplitude of fluctuations is greatest in the fall and winter months, with the possible exception of precipitation. This is also when the pressure fluctuations are at a maximum, and if one were to remove or damp out the anomalies for this period in the year, the amplitude of the response of the regional variables would also be reduced. This would be reflected in the annual average trends discussed in the first part of the paper where the rise of temperature in the 1980s, in particular, would have been of a much more modest nature.

7.2.4 Altitudinal behavior of minimum temperature anomalies in Switzerland

As a follow on to the detailed analysis of regional climate change this century in Switzerland described in the previous sub-section, Section 7.2.4 reports on the climatological anomalies observed in the Swiss Alps for the 15-year period from 1979 - 1993. This most recent period of the secular temperature record has been chosen because it exhibits two very distinct series of cold and warm winters, which may be illustrative of a shift in climate patterns during the 15-year sampling period (Beniston and Rebetez, 1995).

The focus in this paper will be on the minimum temperature record, because minimum temperature trends for Switzerland are far more significant than the corresponding trends for the maxima. This is a conclusion which has been reported by a number of investigators, such as Balling (1992) and Karl et al. (1993). In the latter publication, the authors stress the fact that, over a significant part of the continental land masses in both hemispheres, minimum temperatures have risen over the last 40 years at a rate three times faster than the maxima. The figures cited are 0.84 K and 0.28 K for the period from 1951 - 1990, respectively.

Because of the asymmetry between the relative trends of the maxima and minima, it is hypothesized here that minimum temperature trends may provide clearer evidence for an early detection of anthropogenically-induced climate change than the maximum temperature trends. However, the analysis of mimimum temperature records for climate change studies is not trivial, particularly in complex mountain regions such as the Alps, as there are a number of factors (cold air pools, stratiform clouds and fog, valley breezes, etc.) which tend to contaminate the observations and thereby obscure any climate signals of global origin. This section will therefore attempt to disentangle some of these complicating factors in such a manner that significant shifts in climate may be more readily identified.

Barry (1992) has stated that in addition to global- or regional-scale forcings, climate in mountain regions is controled by four principal factors, namely altitude, latitude, continentality, and topography. The geographical characteristics of minimum temperatures will be illustrated in terms of Barry's criteria, namely location (latitude, continentality) and site (altitude, topography), based on a set of 88 Swiss climatological stations. An overview will be given of the trends which have been observed during the 15-year sampling period, and then an analysis will be made of the regional characteritics of both the "warm" and "cold" anomalies of this period.

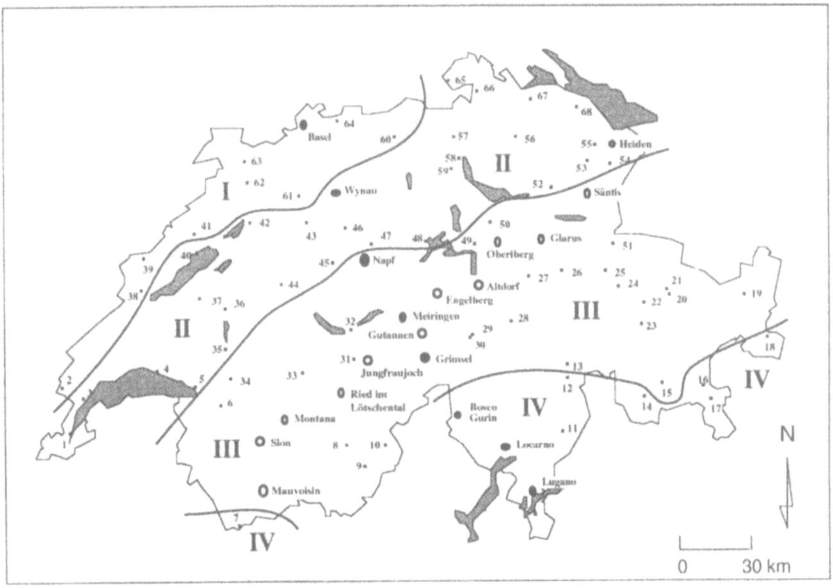

Figure 7.18: Map of Switzerland, illustrating the 88 climatological stations used in this study and the four principal climatological regions of the country. The stations are listed in Table 7.8 in the text. Closed circles represent the "N-S transect" stations discussed in the text, and the open circles the "E-W transect" stations. I: Jura Mountains; II: Swiss Plateau; III: Alps; IV; Southern Alps

One of the motivations behind such a study, in addition to investigating some of the fundamental processes involved, is to link observational data to climate model data. High-resolution climate model experiments conducted over the Alpine area (Beniston et al., 1993; Marinucci et al., 1995) have shown that there exist significant discrepancies between model results for current climate and climatological observations in the Alps and surrounding lowlands. If we are to have greater confidence in model projections for enhanced greenhouse gas scenarios in a complex region such as the Alps, regional climate models should be in a position to simulate not only the means of climate variables in a particular region, but also the variability around those means. This variability can be altitudinally-dependant, as will be shown in this paper, and it is essential for models to reproduce this dependency if their results are to be meaningful. The altitudinal gradients of not only the means but also the anomalies have a number of consequences for the response of different environmental systems in the mountains, particularly snow, ice, and hydrology, but also vegetation and forests (Beniston, 1994). Any studies of the possible impacts of climate change on these systems requires climatological data from high-resolution models which are capable of reproducing these features in complex terrain.

7.2.4.1 Climatological stations and distribution

Up to 88 climatological stations in Switzerland have been used in this analysis. Their distribution is illustrated on the map in Figure 7.18, which depicts the four principal climatological zones defined by Baeriswyl and Rebetez (1995):

- The Jura Mountains: this mountain range runs along the Western boundary of Switzerland. It consists of a karstic-type landscape with a series of parallel ridges separated by wide, flat valleys. The average elevation is in the vicinity of 1,000 m, with crests often exceeding 1,500 m
- The Swiss Plateau: the Plateau lies between the Jura and Alps, and is oriented SW - NE from the Lake of Geneva region to the Lake of Constance. The elevation range is between 400 and 800 m; the topography is rarely flat and consists of rolling hills and valley systems
- The Alps: these mountains cover over 60% of the surface area of Switzerland and consist of several ranges, the principal ones being the Valais or Pennine Alps, the Bernese Oberland, the Central Alps, and the Grisons. Altitudinal differences are extreme, with many mountain summits exceeding 4,000 m
- The Southern Alps: this region encompasses not only the southern part of Switzerland (Tessin and the Italian-speaking valleys of the Grisons) but also a number of mountain passes located on the divide between the Mediterranean basin and the northern Alps, such as the Grand Saint Bernard or the Bernina, whose climatological charateristics are more commonly associated with those of the Southern Alps

While a more detailed partitioning of climatological regions is possible, the minimum four regions used here have been deemed sufficient for the purposes of this investigation.

Table 7.8 provides an overview of the station elevations for the four climatological regions just described. The data has been compiled on the basis of the Swiss Climatological Data Base, a computerized data set managed by the Swiss Meteorological Institute (Bantle, 1984).

The Jura stations comprise 10 data points, the Plateau 29, the Alps 39, and the Southern Alps an additional 10 points. For these 88 stations, uninterrupted climatological data exists for at least the 15-year sampling period, allowing exhaustive statistical analyses to be made on unique spatial and temporal scales.

Jura		Plateau		Alps		Southern Alps	
Station	Alt	Station	Alt	Station	Alt	Station	Alt
Balmberg (61)	1075	Altstätten SG (54)	474	Adelboden (33)	1355	Bosco Gurin (NS)	1505
Basel (NS)	317	Bern-Liebefeld (44)	570	Altdorf (EW)	451	Gd St Bernard (7)	2479
Chaumont (41)	1141	Biel (42)	434	Alvaneu (23)	1175	Grono (11)	357
Delémont (63)	416	Broc (35)	680	Andermatt (30)	1442	Löbbia (14)	1420
La Brévine (39)	1042	Changins (3)	435	Arosa (22)	1847	Locarno (NS)	379
Dôle (2)	1672	Ebnat Kappel (52)	629	Bad Ragaz (51)	496	Lugano (NS)	276
La Fretaz (38)	1202	Fribourg (36)	634	Château d'Oex (34)	980	Bernina (16)	2256
Les Rangiers (62)	856	Genève (1)	430	Chur (24)	586	Poschiavo (17)	1078
Rheinfelden (64)	271	Güttingen (68)	438	Davos (20)	1590	S Maria/M (28)	1390
Unterbözberg (60)	514	Haidenhaus (67)	694	Disentis (28)	1180	S. Bernardino (12)	1628
		Hallau (65)	450	Einsiedeln (50)	910		
		Heiden (EW)	811	Elm (26)	962		
		Huttwil (46)	639	Engelberg (EW)	1018		
		Langnau (45)	695	Glarus (EW)	470		
		Luzern (48)	456	Grächen (8)	1617		
		Menzberg (47)	1035	Grimsel (NS)	1950		
		Montreux (5)	408	Gütsch (29)	2288		
		Neuchâtel (40)	487	Guttannen (EW)	1055		
		Oeschberg (43)	482	Hinterrhein (13)	1619		
		Payerne (37)	491	Interlaken (32)	574		
		Pully (4)	461	Jungfraujoch (EW)	3572		
		Schaffhausen (66)	457	Mauvoisin (EW)	1841		
		St. Gallen (55)	664	Meiringen (NS)	832		
		Stein AR (53)	786	Montana (EW)	1495		
		Tänikon (56)	536	Mürren (31)	1639		
		Uetliberg (59)	810	Napf (NS)	1408		
		Wynau (NS)	422	Oberiberg (EW)	1090		
		Zürich Airport (57)	431	Ried/Lötsch. (EW)	1480		
		Zürich SMA (58)	569	Saas Almagell (9)	1680		
				Säntis (EW)	2500		
				Schwyz (49)	448		
				Scuol (19)	1295		
				Sepey (6)	1267		
				Sils Maria (15)	1802		
				Simplon Dorf (10)	1495		
				Sion Airport (EW)	483		
				Tierfehd (27)	810		
				Vättis (25)	948		
				Weissfluhoch (21)	2540		

Table 7.8: List of climatological stations and altitudes. Numbers refer to the locations of these stations on the map in Figure 7.18; NS refers to stations on the North-South transect, identified by station name and closed circles in Figure 7.18; EW refers to the East-West transect, identified by station name and open circles in Figure 7.18

7.2.4.2 Temporal distribution of temperature

The minimum temperatures exhibit in general an abrupt increase in the winter and spring anomalies for the years between 1988 and 1990, then again in 1992 and 1993 at the high elevation sites, whereas prior to 1988 anomalies were, in broad terms, lower than average if one excepts the particularly warm summer of 1983. Zürich exhibits also a very warm summer period in 1991 which is not reflected quite as much at the other locations. The minimum temperature anomalies are seen to be particularly strong for the winter period (DJF), followed by the spring (MAM). Temperature anomalies exceed 2 °C at low elevation sites and 4 °C at higher elevations. This is consistent with other analyses of climate data in Switzerland which have pointed to the fact that temperature variance generally increases with altitude (Rebetez, 1994).

The strong departures from the 15-year 1979 - 1993 mean period in three successive years at the end of the 1980s and again in the early 1990s is indicative of a significant shift in climate features over this period; these are linked to changes in the westerly flow entering Western Europe. The positive anomalies correspond to periods of persistent high pressure over the Alpine region, related to surface pressure fields over the western Mediterranean and the Iberian Peninsula which were systematically higher than their climatological averages during these months. The persistence of high-pressures over the Alps and the associated subsidence fields resulted in large temperature departures from the mean period with severe consequences for winter snow and hydrology. These were shown in BRGM to be closely linked to the North Atlantic Oscillation Index (NAO), i.e., the pressure difference between the Azores and Iceland which provides a measure of the strength of flow over the Atlantic. The pressure statistics emphasized the appearance of a new and anomalous behavior in the 1980s, where pressure reached annual average values far higher than at other times this century; winter (DJF) pressure anomalies are particularly high in the 1980s and are well correlated with the NAO Index at this time. This points to shifts in North Atlantic westerly flow fields to more northerly latitudes as being the principal mechanism influencing the Alpine region over the sampling period.

A careful analysis of the data shows that the individual months which exhibit the strongest departures from the mean are the January minima; this is also the month which generally experiences the lowest annual temperatures in Switzerland. Figure 7.19 shows the evolution of January minimum temperatures for each year of the sampling period at the four stations. It is seen that a significant shift occurs from 1987 onwards, with a remarkable increase of winter temperatures until 1990, after which time they remain at an anomously high level; it would appear that the system has not recovered from whatever climate mechanisms triggered the warming trend in the latter part of the 1980s. A similar analysis of July trends (not illustrated here) indicates that apart from a very warm summer in 1983, no major anomalies in the rest of the period are apparent; in fact, the maxima begin to decrease from 1991 to 1993. In terms of the annual mean temperatures which were discussed in Section 7.2.3, it is clear that the strong positive temperature anomalies of the winter period are generating the bias in the annual record and are the principal contributor to overall warming in the 1980s.

It is apparent from this discussion that there is an altitudinal dependency of warming with altitude, i.e., the higher the elevation, the greater the positive winter minimum temperature anomaly. This is indicative of one or several large-scale forcing mechanisms which become more perceptible in the observations the higher the elevation; at high altitudes, stations are closer to the free atmosphere and are less likely to be contaminated by local site features, particularly in stable winter situations. Because it is difficult to support this hypothesis on the basis of only four sets of observations, it will now be investigated on the basis of the 88 observational stations introduced in Section 2, representative of the four broad Swiss climatological zones.

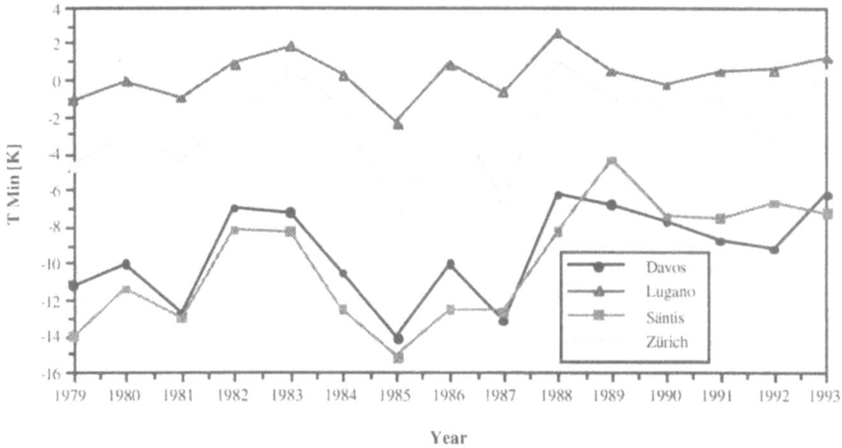

Figure 7.19: Average January minimum temperature trends during the period 1979 - 1993 for Lugano, Zürich, Davos, and Säntis

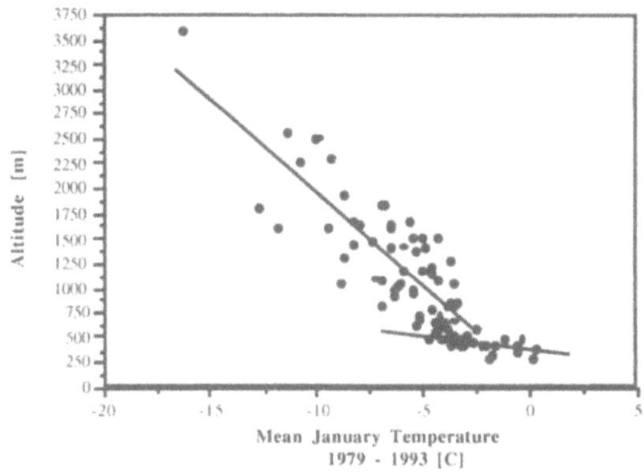

Figure 7.20: Altitudinal distribution of mean minimum temperatures at the 88 climatological stations, based on the 15-year averages of the period 1979 - 1993

7.2.4.3 Spatial distribution of temperature anomalies

Based on Figure 7.19, January minimum temperature series can be broadly separated into two five-year periods, one with "cool anomalies" from 1983 - 1987, and one with "warm anomalies" from 1988 - 1992. Figure 7.20 illustrates the distribution of the mean 15-year average January temperatures with height; at low elevations, the average lapse rates are very high and are close to 5 K/100 m. The range of temperatures here reflects different geographical conditions: some stations are located south of the Alps, such as Lugano, where milder temperatures can be expected, while others are located in the cold air pool which frequently forms over the Swiss Plateau in winter, accompanied by fog or stratus cloud conditions. At higher elevations, the average lapse rate becomes steeper and attains a characteristic value of about 0.5 K/100 m; the spread of data around this mean vertical temperature gradient is mainly indicative of specific site conditions which determine long-term means; some stations may be located on north or south-facing slopes, on mountain summits, or at the bottom of valleys where temperature inversions may often occur.

Figure 7.21 illustrates the altitudinal distribution of minimum temperature anomalies for the "warm anomalous" period, with individual symbols representing stations in each of the four climatological zones; Figure 7.21 (left) shows the distribution for the entire set of stations, and Figure 7.21 (right) for the Alpine stations alone. The correlation coefficients between temperature anomalies and altitude for the entire data set and for the four climatological regions are given in Table 7.9.

Data Set	R	t-test (99% level)
All Stations	0.63	7.52 (2.44)
Jura (without La Brévine)	0.88	4.90 (3.50)
Plateau	0.05	0.26 (2.77)
Alps	0.80	8.11 (2.71)
Southern Alps	0.84	4.38 (3.36)

Table 7.9: Correlation coefficient for relationship between minimum temperature anomalies and altitude for the average 1988-1993 Januaries. t-test of R at the 99% significance level is provided; in brackets, limit of t-distribution above which R is significant

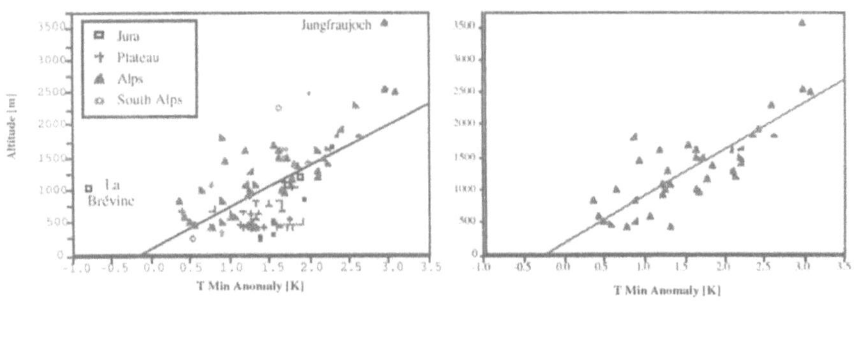

Figure 7.21: Left Height-temperature anomaly relationship for the "warm" Januaries of the 5-year period from 1988 - 1992 at all 88 climatological stations. Right As left panel, except for the 39 stations in the Alpine climatological region

Figure 7.21 confirms that the anomalous behavior of the minima is altitudinally-dependent, i.e., the higher the elevation, the greater the positive anomaly. The scatter around the trend line is related to local site characteristics. For example, the Jura station of La Brévine (cf. Figure 7.21 (left)) is located at the bottom of a closed valley where strong temperature inversions frequently occur in winter; cold air pools tend to persist over long periods and will only disappear if a significant external perturbation occurs. The persistent cold air pools in winter explains the strong departure from the other stations; La Brévine holds the cold record for Switzerland, with temperatures slightly below -40 °C recorded in 1985 and 1987.

Many of the locations to the left of the linear regression line in both Figures 7.21 (left) and 7.21 (right) are stations which are located in valleys subject to strong and persistent temperature inversions; those to the right, which exhibit warmer anomalies than those predicted by the regression are located in regions more commonly exposed to wintertime stratus, which traps outgoing infrared radiation at night and maintains nocturnal temperatures warmer than they would be in the absence of the cloud cover; some locations in this section of the diagram are also subject to occasional warming föhn effects. The Plateau station data lie almost all to the right of the regression line; the Plateau and some of the side valleys feeding into it are the most subject to persistent stratus cover, which tends to have an average thickness of 200 - 500 meters and an upper limit of between 800 - 1400 m. The stations lying beneath the stratus layer are essentially decouled from processes occuring at higher elevations, as exemplified by the significant change in mean temperature lapse rates illustrated in Figure 7.20.

In terms of sunshine duration, values for the November - February period were particularly low in both Neuchâtel and Zürich (less than 20 % in January) for the 5-year period, and relatively high in both the Jura station of La Dôle and the Alpine station of Davos (70% of maximum possible sunshine). The very low values over the Swiss Plateau are symptomatic of the presence of persistent stratus; the cloud cover in such situations is generally more dense during the nocturnal hours than during the day, so that for the winter period, the use of sunshine duration as a measure of persistance of fog or stratus beyond just the daylight hours is justified. Conversely in the mountains and in the absence of cloud-bearing synoptic perturbations, clear skies during the day will invariably be followed by clear skies at night.

The representativity of the Jungfraujoch station (indicated in Figure 7.21), far removed from the predictor line, is open to question; the observation station is located on the northern side of the Sphinx Observatory, itself situated on a pass subject to a bimodal flow distribution, i.e., either northerly or southerly flow. It is suspected that the lack of sufficient exposure of the instruments to many of the characteristic flow régimes is responsible for the bias in the height/temperature anomaly relationship.

When analyzing a single January, namely the highly anomalous winter of 1989, the relationship between temperature anomaly and altitude is even more remarkable, as illustrated in Figure 7.22. Correlation coefficients are provided in Table 7.10.

The relationship between minimum temperature anomalies for this period and the altitude of the station is remarkable, even when taking into account the bias generated by the Plateau stations under the influence of the persistent winter stratus cover. This seems to be the only local site condition of importance in this study.

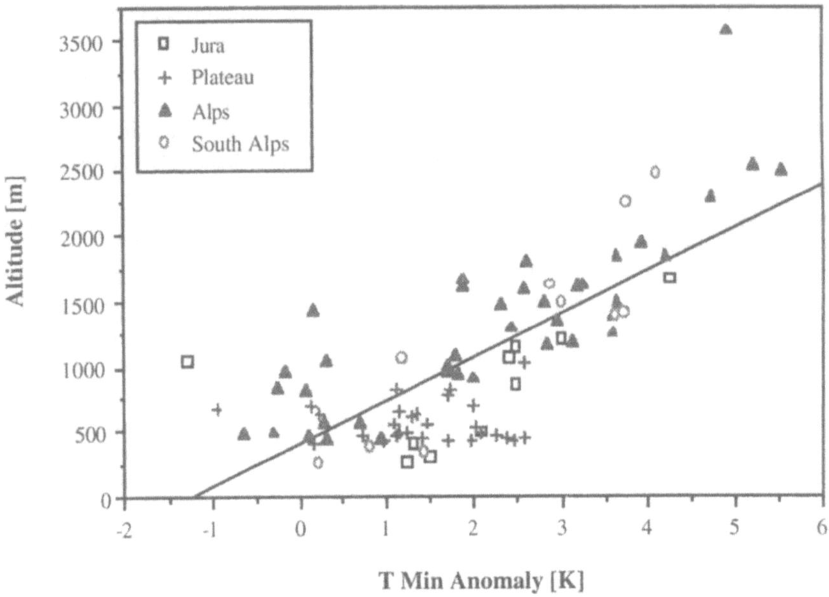

Figure 7.22: Height-temperature anomaly relationship for the very warm January of 1989 for all 88 climatological stations

Data Set	R	Significance (99%)
All Stations	0.74	10.20 (2.44)
Jura (without La Brévine)	0.95	48.05 (3.50)
Plateau	0.00	0.00 (2.77)
Alps	0.83	9.05 (2.71)
Southern Alps	0.89	5.52 (3.36)

Table 7.10: As Table 7.9, except for January 1989

A similar analysis can be made of the "anomalously cool" period from 1983 - 1987; the results for minimum temperature anomaly vs height are illustrated in Figure 7.23; the corresponding correlation analysis table is given in Table 7.11.

Data Set	R	Significance (99%)
All Stations	-0.57	6.43 (2.44)
Jura (without La Brévine)	-0.89	5.16 (3.50)
Plateau	-0.22	1.17 (2.77)
Alps	-0.75	6.90 (2.71)
Southern Alps	-0.87	4.99 (3.36)

Table 7.11: As Table 7.9, except for 1983-1987 Januaries

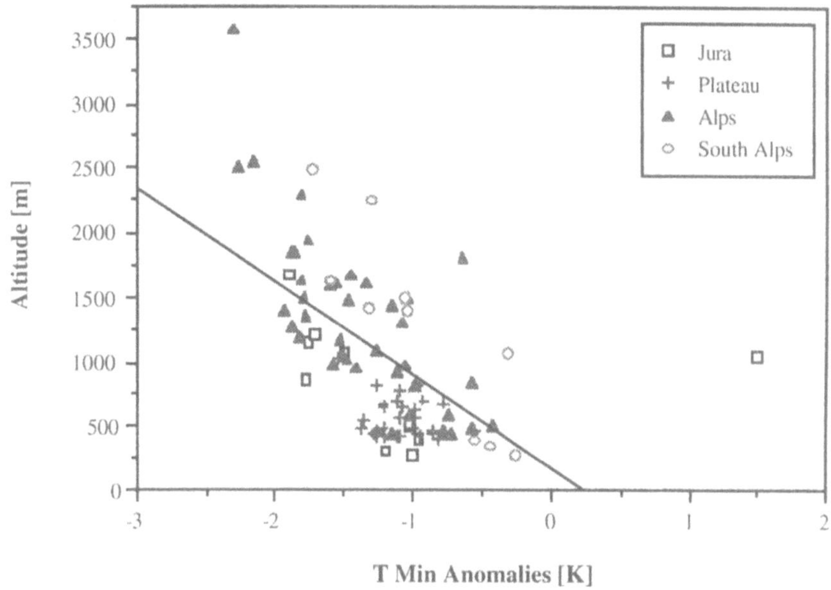

Figure 7.23: Height-temperature anomaly relationship for the "cold" Januaries of the 5-year period from 1983 - 1987

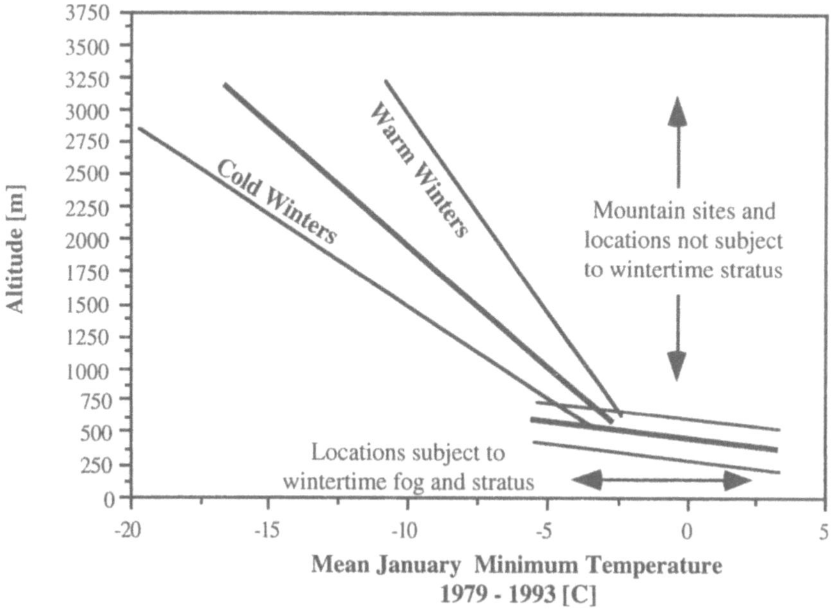

Figure 7.24: Schematic diagram to illustrate the switch of temperature anomalies about the mean between warm and cold winters. The bold line represents the mean 15-year minimum temperature lapse rate for the 1979 - 1993 sampling period

The figure shows that during cold winters, the anomalies are reversed with respect to warm winters, i.e., the higher the elevation, the more negative the anomaly. It is also noteworthy that the distribution of data around the regression line is also completely reversed with respect to Figure 7.21; stations which were warmer than expected from the predictor line are, during cold winters, colder than expected from the predictor. This is mainly due to the fact that even if the winters are characterized by stratus cover, they are also accompanied by very cold low-level temperatures and often snow on the ground which influences the net radiation balance and contributes to low nocturnal temperatures. Similar conclusions can be reached when interpreting data for individual "cold" Januaries during this period, such as the very cold winter of 1987 (not shown). Relationships between anomalies and height are statistically more significant for individual Januaries than for their average 1983-1987 values, a conclusion already reached when examining the warm anomalies (cf. Tables 7.9 and 7.10). The switch mechanism between anomalous warm and cold winters is sketched in Figure 7.24.

This leads to the conclusion that the vertical temperature lapse rate with height tends towards lower stability during warm winters and higher stability during colder winters; the lower the elevation, the more the anomaly becomes damped out by surface, topographic and local climatic effects, as one would expect. The density of the Swiss climatological network allows this damping to be quantified, i.e., the trend line provides an attenuation gradient of the large-scale signal. It is also remarkable that the warm winter anomalies are characterized by a switch in the sign of this gradient, from negative to positive. Indeed, analyses of the mean January anomaly gradients over the 15-year period show these to be on average largely negative, thus indicating that the warm winters of the second half of the 1980s and the beginning of the 1990s are very much out of the ordinary. While it has been speculated that high elevations may indeed be more sensitive to climatic fluctuations than lower sites where climate variables become contaminated by local features, this study allows a quantitative evaluation of the relationship between a large-scale signal and the location at which it is measured.

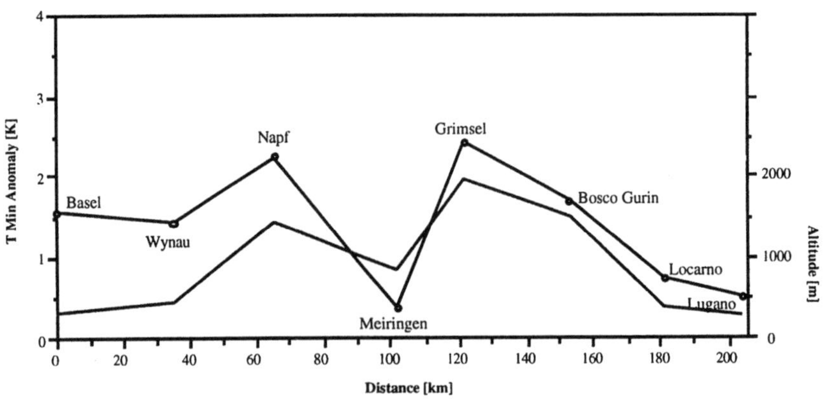

Figure 7.25: Distribution of minimum temperature anomalies for the "warm" winters of 1988 - 1992 along the N-S transect illustrated in Figure 7.8. The altitudinal profile (thin line) is also shown to provide an illustration of the height-anomaly relationship

7.2.4.4 *Influence of latitude and continentality on temperature anomalies*

In order to investigate whether the influence of latitude (north or south of the Alps) or longitude/continentality may also influence the nature of temperature anomalies, the "transect stations" illustrated in Figure 7.18 will be used; these comprise 8 stations running NW - SE from Basel to Lugano, and 12 stations distributed SW - NE from Mauvoisin (Valais Alps) to Heiden (close to the Rhine River and Liechtenstein). The NS line cuts across all four climatolgical regions, whereas the EW section is confined to the Alpine zone only.

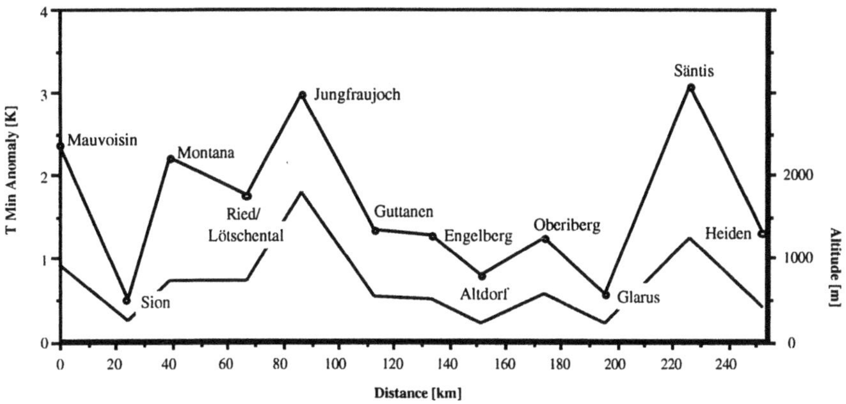

Figure 7.26: As Figure 7.25, except for the E-W transect

Figures 7.25 and 7.26 provide an overview of average 1988 - 1992 January anomalies along the NS and EW transects, respectively. The altitudinal dependency of the anomaly is clearly visible in both figures, and it is apparent that there are also regional differences. These can be quantified to some extent by plotting the ratio of anomaly to height against NS or EW distance, as shown in Figure 7.27. If the anomalies were everywhere perfectly correlated to altitude, then the resulting line linking this anomaly/height ratio and distance would be horizontal. The deviations from the ideally horizontal line are due primarily to site influences, as at Basel and Wynau for the NS transect, related to the presence of stratus at this time of the year, and to some extent latitude (in the case of the NS transect). The deviations for the anomalously warm winters are not large along the EW transect and reflect the relative homogeneity of the Alps as a region. There is little evidence of a significant continentality influence along these transects, as differences of the T/z ratio over this transect are relatively minor between one station and the next. For the NS transect, on the other hand, the differences between Plateau locations (Basel and Wynau), which are frequently beneath fog and stratus clouds, and locations south of the Alps (Locarno and Lugano) not generally subject to persistant stratus, stand out clearly in this picture.

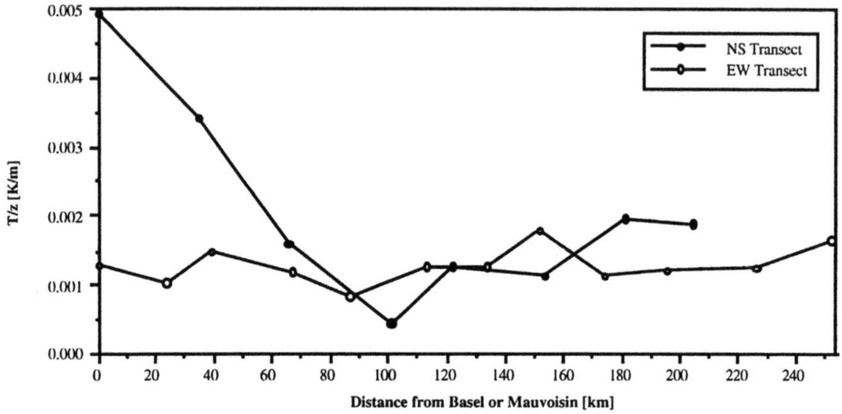

Figure 7.27: Distribution of minumum temperature anomaly:altitude ratio along both the N-S and E-W transects as a means of highlighting regional differences in the data

7.2.5 Shifts in extremes in relation to changes in means in the Alps

One of the numerous problems associated with projections of climate change in the next century under conditions of enhanced atmospheric greenhouse gas concentrations is the manner in which extreme values of temperature and precipitation may change (Giorgi and Mearns, 1991; Beniston, 1994; IPCC, 1996). In terms of the response of a number of environmental and socio-economic systems, extreme climatological events tend to have a greater impact than changes in mean climate, which is commonly the basic "quantity" for discussing climatic change.

It is therefore of interest to determine whether there is a link between means and extremes, because if such relationships exist for a particular region, it would be possible to determine statistically how extremes may shift in a changed climate. Such links may perhaps also ultimately help improve the simulation of extreme events in high-resolution climate models, which are not currently capable of reproducing certain types of extreme events, even at high resolution (IPCC, 1996).

There is speculation that a warmer climate will induce more extremes of storms, floods, droughts, and heat waves, though this is by no means certain (IPCC, 1996); for example, recent high-resolution climate model simulations have shown that the frequency of tropical cyclones in a 2 x CO_2 atmosphere (i.e., an atmosphere whose concentration of equivalent CO_2 is double that of the pre-industrial era) may in fact diminish in the Northern Hemisphere compared to today (Bengtsson et al., 1995). In a changed climate, meteorological extremes may also occur in regions which are today less prone to such events, and vice versa.

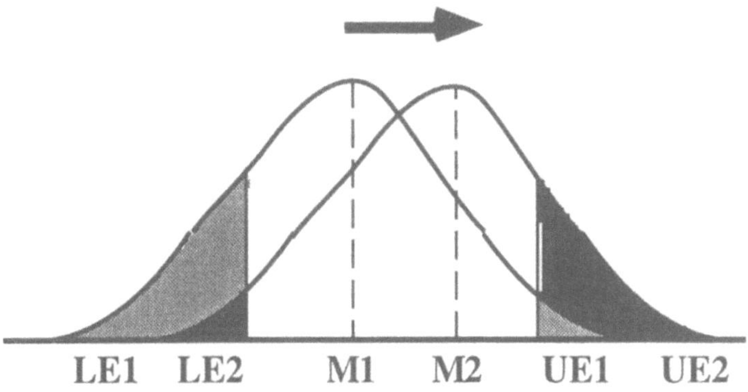

Figure 7.28: Illustrative example of possible shifts in temperature distributions and their extreme limits, for a change in mean temperature

It has been suggested (Mitchell et al., 1990; Warrick and Barrow, 1991; Wigley et al., 1992) that, in the case of temperature, a shift in the mean may be accompanied by a shift in the extremes (indeed, it may be the modification of the extremes which is responsible for a change in the mean). A simple illustrative example is given in Figure 7.28, which shows a hypothetical temperature distribution with mean value M1 whose lower and upper extremes are LE1 and UE1, respectively. LE1 may represent, for example, conditions under which frost damage to crops may be significant, and UE1 may be the region in which heat stress on crops is excessive; the white area in between these extremes would represent adequate to optimal growth conditions. If climate were to warm to a mean value M2, it is possible that the entire distribution may follow the shift in mean, resulting in extremes LE2 and UE2. Again taking the example of growth conditions for crops, the lower extreme LE2 may be considered as an improvement in terms of frost damage; UE2, on the other hand, would represent an increase in the frequency of heat waves, with an enhancement of heat stress on crops. Of course, as has been pointed out by Mitchell et al. (1990), the extremes can also change wihout a shift in the mean, through a redistribution of the variance of the data. It is also equally possible that a shift in means may lead to a skewed distribution, in which one extreme may change disproportionally with respect to another, with contrasts in the potential impacts induced by the lower and upper extremes. Mearns et al. (1984), on the basis of studies carried out for the US Midwest concluded that the relationships between changes in means and those of the probabilities of extreme events are strongly non-linear, with small changes in average temperatures resulting in large changes in event probabilities.

The notion of changes in extremes as a function of changes in means is not new; however, the behavior of extremes is likely to be linked to the geographical region considered. In the Swiss Alps, there exists a large number of climatological data sets for the 20th century, which allow a verification of hypotheses relating means and extremes. The Alps are characterized by a high complexity of climate, not only related to topographical effects, but also to the competing influences of a number of climatological regimes which converge on this sensitive mid-latitude region. In recent years, the Alps have experienced unusual weather conditions, such as a run of mild, snow-free winters (e.g., Beniston, 1996), extended periods of drought (particularly south of the Alps), episodes of extreme precipitation (e.g., Gotthard Region of Central Switzerland, 1987; Pfister, 1995; Brig, Canton of Valais, 1993; Kunz and Rey, 1995; Locarno, Lake Maggiore, 1994), and high-intensity storms (Vivian, February 1990; Widmann and Schaer, 1996). There has been some speculation that these extreme events are linked to the increases in global mean temperatures and could be attributed to a human-induced "greenhouse signal", although this is by no means confirmed. It is quite possible, however, that such extremes may occur more frequently in a future, warmer climate.

It is therefore of interest to investigate, on the basis of climatological data from several Swiss observational sites (where digitized climatological data is available daily since 1901), the manner in which the frequency distributions of temperature depend "statically" on mean conditions at different altitudes, and also "dynamically" in terms of the shifts in mean temperatures which have been observed this century (Beniston and Rebetez, 1996). The analysis will further discuss how these changes in temperature extremes are linked to precipitation, and the implications for future climate change as projected by numerical modeling techniques (Beniston et al., 1995; Marinucci et al., 1995; Rotach et al., 1996) in the Alpine area will be addressed.

7.2.5.1 Links between extremes and means

Beniston et al. (1994) undertook an exhaustive analysis of secular climate change in Switzerland at four stations with daily climatological records spanning the period 1901 - 1994 (cf. Section 7.2.1 in this chapter). These same four stations will be used for the analysis of extremes; their location can be seen in Figure 7.1. The altitude of the observational sites ranges from 276 m at Lugano to 2500 m at Säntis; the four sites are representative of different regional climatic regimes in Switzerland, namely the Swiss Plateau (Zürich), the Mediterranean side of the Alps (Lugano), a high Alpine valley (Davos), and an elevated mountain top (Säntis).

Extreme temperatures are here defined as those whose values are less than the lower 1 percentile of the frequency distribution or greater than the upper 1 percentile. The 1 % limit can be considered in this context to be an extreme value with a significant impacts potential on a range of environmental systems. In terms of precipitation, for example, such an extreme limit is often closely linked to the triggering of natural hazard events, such as mud slides and debris flows in mountain regions (Rebetez et al., 1996).

Station	Altitude [m]	T Min [C]	T Max [C]
Davos	1590	-1.6	8.29
Lugano	276	7.4	16.98
Säntis	2500	-4.4	0.84
Zürich	569	4.9	13.45

Table 7.12: Average values of minimum and maximum temperatures for the period 1901 - 1994

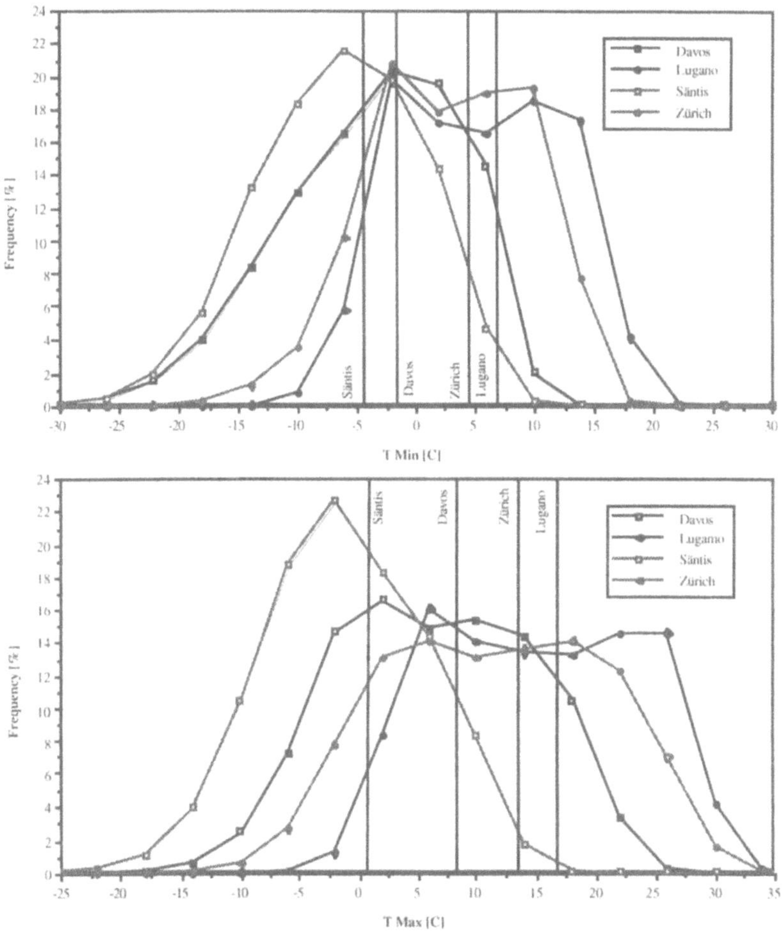

Figure 7.29: Upper) Minimum temperature distributions and their mean values at the four stations; the curves represent average values for the period 1901-1994; Lower) Same as upper, except for maximum temperatures

Table 7.12 lists the average minimum and maximum temperatures for all four stations for the period 1901 - 1994. Figures 7.29a and 7.29b illustrate the mean frequency distributions of mimimum and maximum temperatures at these four observational sites, respectively, where the mean values listed in Table 7.12 have been highlighted. These distributions are generally far removed from ideal normal distributions at low elevations (Lugano and Zürich), because of a number of contaminating factors associated with urbanization, low-level fog or stratus formations in winter which bias the distributions, and other perturbation features characteristic of the lower boundary layer. The mountain sites, particularly Säntis, are more symmetrical in their distribution; there is a tendency towards negative skewness in the minima and positive skewness in the maxima.

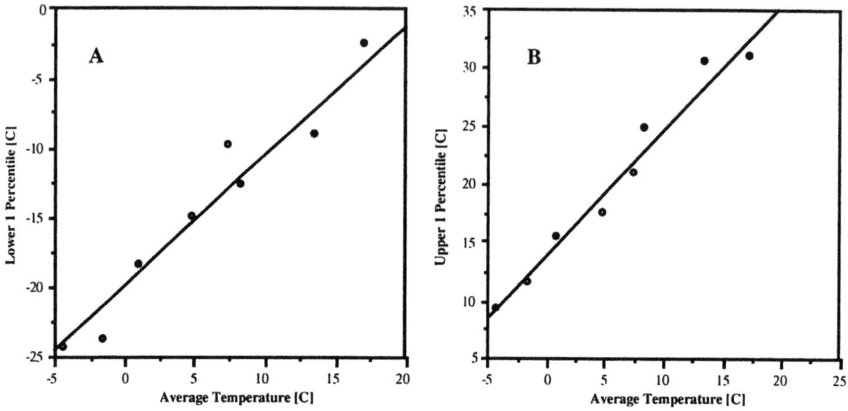

Figure 7.30: A) "Static" relationship between the lower 1 percentile of minimum temperature, B) mean minimum temperature at the four stations

If plotted against average temperature, the lower and upper extremes of the minimum and maximum temperatures exhibit a quasi-linear profile, as illustrated in Figures 7.30a and 7.30b. For average climatic conditions this century, therefore, a shift in the mean temperature value of 12° C, representative of an altitude difference of about 2000 m, is accompanied by a shift in the lower and upper extreme range of both the minima and the maxima of roughly the same amount. In spatial ("static", i.e., not time-dependent) terms, therefore, there is indeed a justification for supposing that a shift in the mean value will be accompanied by a corresponding shift in the extremes. The question of interest here, however, is to determine the manner in which the extremes respond to a "dynamic" (i.e., time-dependent) shift in the mean values, since this will shed light on the possible range of extremes which may be expected in a changed global climate in the 21st century.

Figure 7.31 shows the change in mean minimum temperatures which have occured this century at the four observational sites, averaged over 5-year periods beginning in 1901. 5-year averages allow smoother frequency distributions for temperatures to be generated than individual years, where a particular distribution may be biased by unusual events. A general increase in the minima is observed, with a temperature excess in the warmest period (1991-1994 average) of more than 2 C over the coldest period (1906-1910 average).

This regional-scale secular warming is far greater than the observed global or hemispheric average warming, as has been pointed out by Beniston et al. (1994). The shifts in the frequency distribution of minimum temperatures between the coolest and warmest period sampled here are illustrated in Figure 7.32.

In all cases, the translation of the extremes towards warmer temperatures is clearly observed. At the higher elevation sites, this is accompanied by a peak frequency which regresses towards lower temperatures (but not to the extent that it counteracts the overall strong warming), whereas at the lower elevations, the distribution tends to become smoothed out in the warmer period compared to the "double maximum" profiles observed in the 1906-1910 period.

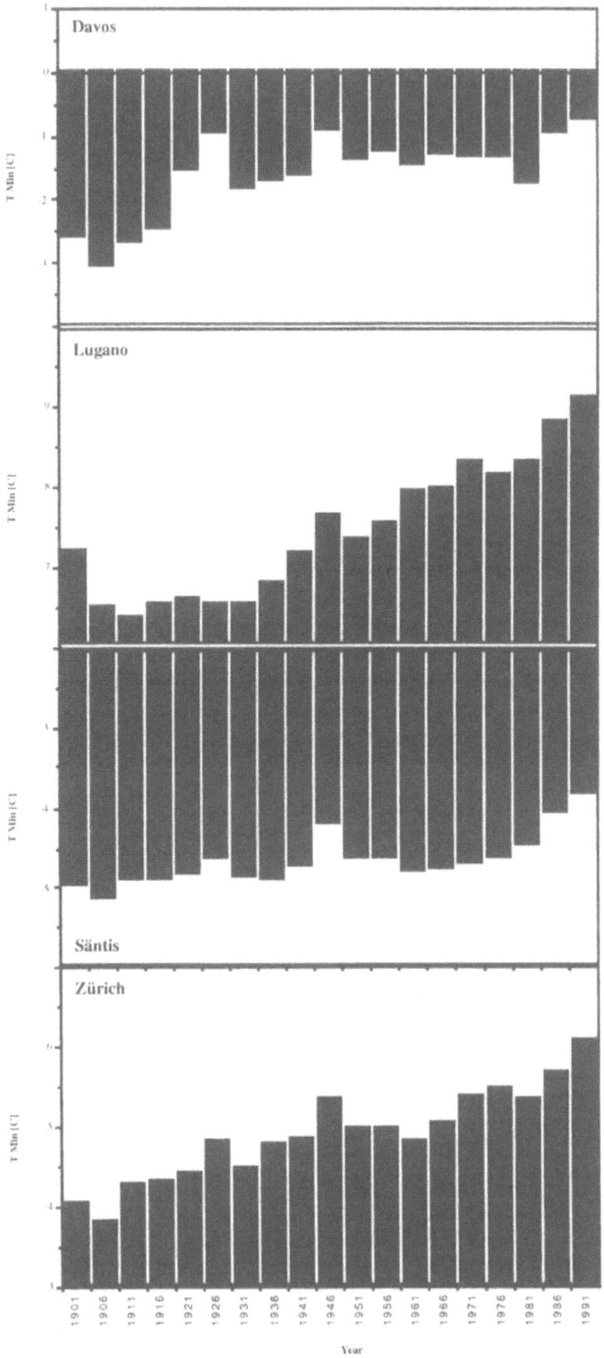

Figure 7.31: 5-year average mean minimum temperatures from 1901 - 1994 at the four stations

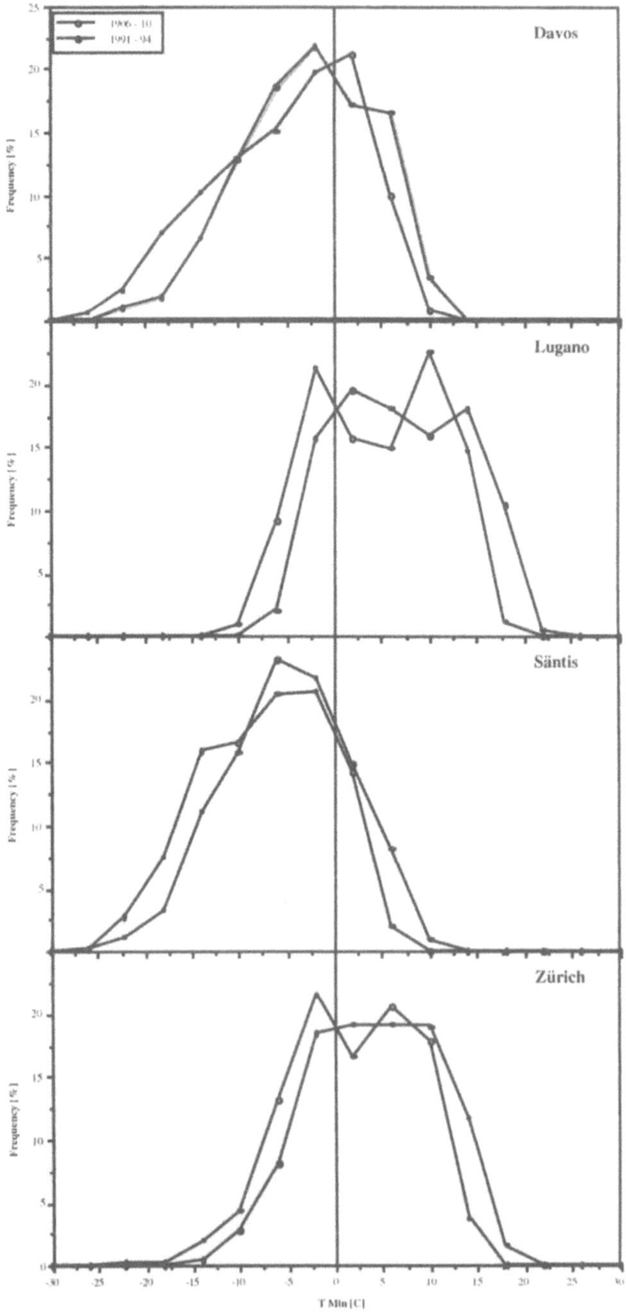

Figure 7.32: Shifts in frequency distribution profiles of minimum temperature at the four stations, between the coldest 5-year period (1906-1910) and the warmest period of this century (1991-1994). "Cold" and "warm" periods here refer to minimum temperature records

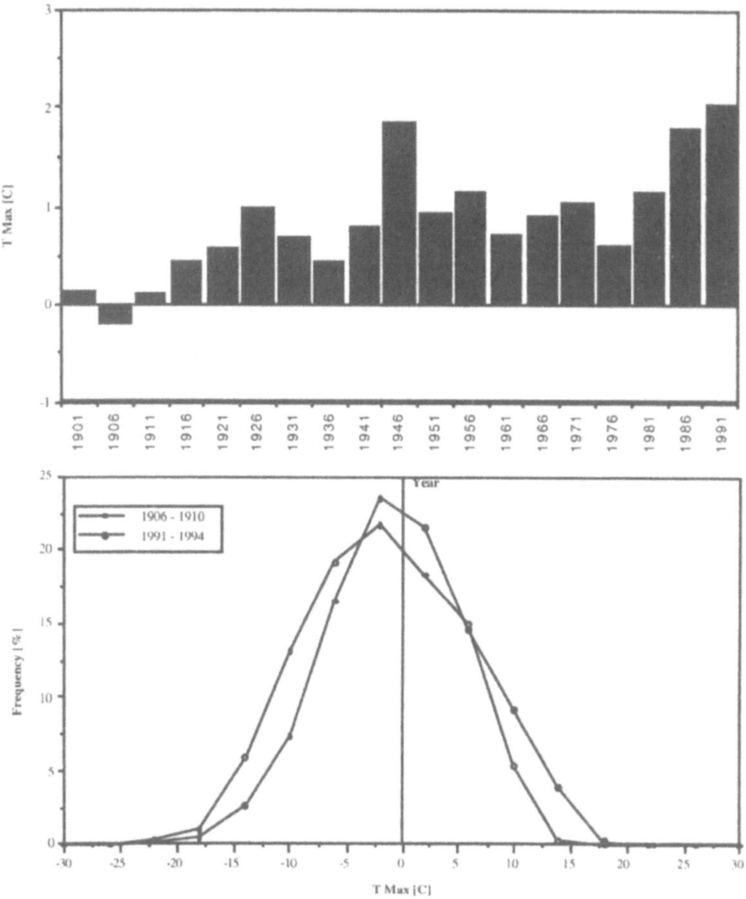

Figure 7.33: Upper) 5-year average mean maximum temperatures from 1901 - 1994 at Säntis. Lower) Shift in the frequency distribution of maximum temperature at Säntis between the coldest and warmest periods of the century for maximum temperatures

Maximum temperatures do not follow the clear warming trend of the minima in the 20th century (Beniston et al., 1994; Beniston and Rebetez, 1995), except at Säntis. The other sites exhibit little or no significant trend in maxima between the beginning and the end of the century, even though there are fluctuations of considerable amplitude. For sites other than Säntis, the warmest period in terms of maximum temperatures occurs in 1946-1950, and the coldest period in 1976-1980. Taking the example of Säntis, Figure 7.33a shows that the 1906-1910 period and the 1991-1994 period are, respectively, the coldest and warmest periods of the 20th century, as was the case for the minima. The translation of the extremes, and indeed of the entire distribution, is observed in Figure 7.33b, in an essentially identical manner to the shifts in minimum temperature distributions.

The behavior of the extremes in relation to changes in the means allows us to draw up a statistical link between these two sets of variables. This is illustrated in

Figure 7.34, where the lower 1 percentile of minimum temperature and the upper 1 percentile of maximum temperature are plotted against the mean annual minimum and maximum temperature, respectively, for the available data this century. These sets of extremes are those which may have a significant impact on environmental systems, as already discussed for Figure 7.28. Figure 7.34 shows that, while the scatter may appear large in some instances, the relationship is clearly linear and the correlation coefficients between means and extremes, given in Table 7.13, while highly variable, are nonetheless significant in terms of the Student t-test at the 99% significance level.

Station	TMin	TMax	t-test T Min	t-test T Max
Davos	0.57	0.56	7.97	7.83
Lugano	0.78	0.88	19.10	37.14
Säntis	0.54	0.81	6.61	22.59
Zürich	0.62	0.79	9.66	20.16

Table 7.13: Correlation coefficients between mean minimum temperatures and the lower 1 percentile of the minimum temperature distribution, and mean maximum temperatures and the upper 1 percentile of the maximum temperature distribution. For the t-test, any term exceeding 2.64 is statistically significant at the 99% level

Unlike the "static", altitude-dependent shift in extremes discussed in the previous sub-section, where there was an almost 1:1 relationship between extremes and means, the relation between extremes to time changes in means is on the order of 1.2:1 - 1.7:1, i.e., the shift in extremes at the 1 percentile level is amplified relative to the shift in means, particularly for the maximum temperatures. This suggests that the changes in extremes are contributing more to the mean changes than other segments of the frequency distribution. Similar conclusions were drawn by other studies, in particular that of Mearns et al. (1984) who analyzed climatological data in the US corn-belt.

A similar attempt to correlate mean annual average temperatures to the extremes leads to much less convincing results, principally because the extremes of each set of temperatures are related not to the mean but to the respective minima or maxima. If the behavior of the two sets of temperature had exhibited similar secular trends, then of course the link between mean temperature and extremes would have been more coherent. This implies that, if these relationships were to persist in a warmer global climate, the extreme maxima would increase even more rapidly than the extreme minima, leading to a skewed overall distribution and a number of consequences for systems sensitive to elevated temperatures.

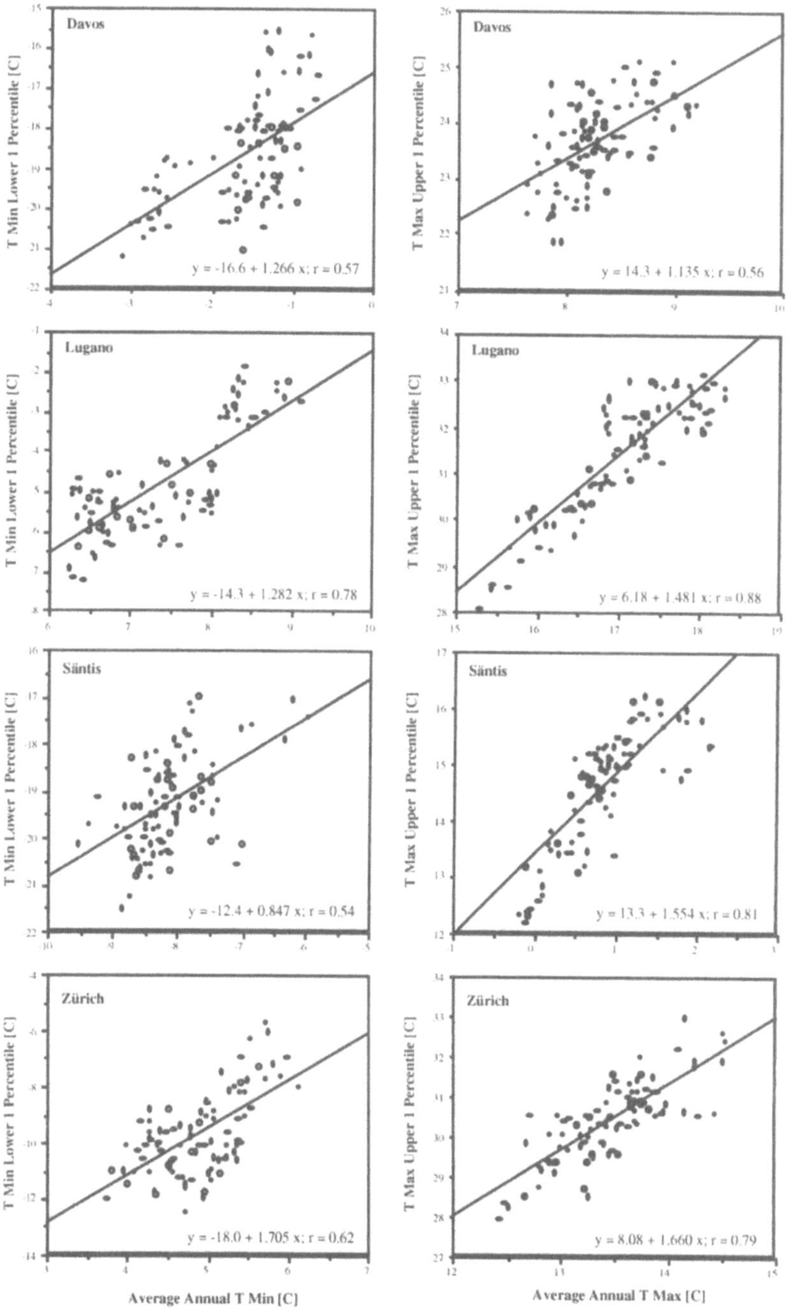

Figure 7.34: Linear regression relationships between mean annual minimum temperatures and the lower 1 percentile of the minima, and mean annual maximum temperatures and the upper 1 percentile of the maxima at all four sites, to illustrate the link between changes in means and the shifts in extremes

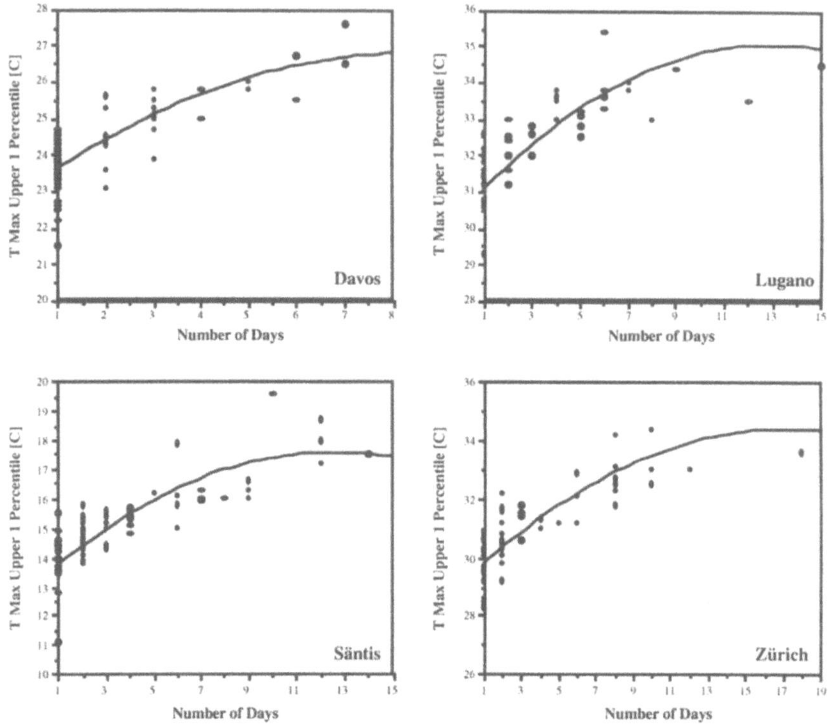

Figure 7.35: Non-linear regression (fits by polynomials of order 2) relationships between the persistence of days exceeding a given threshold and the upper 1 percentile of maximum temperature at all four sites

In the study of Mearns et al. (1984), the authors point out that extremes may be defined not only as independent extreme values (or percentiles as in the present study), but also by the number of consecutive days on which temperatures exceed a given threshold value. In terms of impacts, the duration of elevated temperatures may have a greater influence on certain environmental systems than a punctual extreme of temperature. One obvious example is the effect of temperature on snow melt: during a cold spell, a single day with temperatures rising to, say, 6 C, is less likely to influence snow-melt than a week where temperatures are not inferior to 2 C. Ecosystems are also frequently more sensitive to a run of very warm days than a single extreme temperature event. The number of consecutive days each year for which maximum temperatures have exceeded a particular threshold - here chosen as the 35 C sea-level temperature[1], where an adiabatic height correction for temperature has been applied to the data - have been computed for the four sites. The relation between the run of days where temperatures exceed the threshold level and the upper 1 percentile of temperature are illustrated in Figure 7.35; at all sites,

[1] The 35 C limit is frequently used in agricultural studies, because it represents a threshold beyond which corn and other types of cereal may undergo serious damage during sensitive phenological periods

the figure clearly shows that the longer the run of days exceeding the threshold, the higher the maximum temperature extremes.

The best fit is a non-linear one (regression through a second-order polynomial), which implies that there may be a limit to the temperature extremes beyond a certain number of consecutive warm days. While this may be an artifice of the statistical regression, there is often a physical justification to an upper temperature limit in that a series of very warm days in summer will ultimately end through convective precipitation events, which tend to reduce excessive temperatures, or through synoptic systems entering the region. The relationships between the length of warm weather episodes and temperature extremes given in Figure 7.35 are all highly significant (correlation coefficients all exceeding 0.8); they have an obvious physical meaning, i.e., that the longer the heat input, the larger the temperature response. The duration of periods of warm summer weather therefore govern the extremes which in turn influence the value of the annual means.

Changes in climate are likely to lead to shifts in the extremes of not only temperature but also precipitation and other climatological variables. Changes in extreme precipitation, i.e., the enhancement of dry or wet conditions, may indeed have impacts whose severity could be more significant than a change in temperature extremes alone. Taking into consideration the fact that in Switzerland, the temperature contrast between the 1990s and the earliest part of the century is more than 2 C on average in terms of minimum temperatures and well over 1 C in terms of mean annual temperatures, the possible links between this warming and changes in precipitation extremes has been investigated. A typical precipitation frequency distribution for the Alpine region takes the form of a decreasing exponential function, with roughly two-thirds of the values concentrated in the lowermost precipitation interval (the 0 mm per day of precipitation, i.e., no-precipitation conditions), rapidly diminishing to minor frequency values beyond the 100 mm/day range.

Figure 7.36 illustrates the differences (in %) of average daily precipitation totals at the Davos. Lugano, and Zürich sites between the current 1991-1994 warm period and the cold 1906-1910 period. Säntis records for this part of the study have been omitted because of the large uncertainties associated with precipitation measurements at high-elevation, exposed sites such as this one; in addition, the Säntis precipitation gauge has been displaced on several occasions, this century leading to large heterogeneities in the measurements.

It is seen that the warmer climate induces shifts towards both the dry and wet extremes of precipitation; in all cases, the number of dry days increases by 1 - 3% (i.e., 4 - 11 days per year), and the number of low-to-moderately moist days (<10 mm/day) diminishes by 0.5 - 3%. The number of events with extreme (up to 100 mm/day) to highly extreme (over 100 mm/day) precipitation is on the increase in the warmer period, particularly in Lugano to the south of the Alps, where strong convective activity is not uncommon from summer to early fall. While the increase in frequency is small (representing perhaps 1 - 2 occurences during a five-year period), it is notable that this shift does indeed occur in a warmer period; additionally, however minor an increase, any event exceeding 100 mm/day is generally accompanied in the Alpine regions of Switzerland by considerable damage to the environment and to human infrastructure, in the form of floods as well as mud and rock slides.

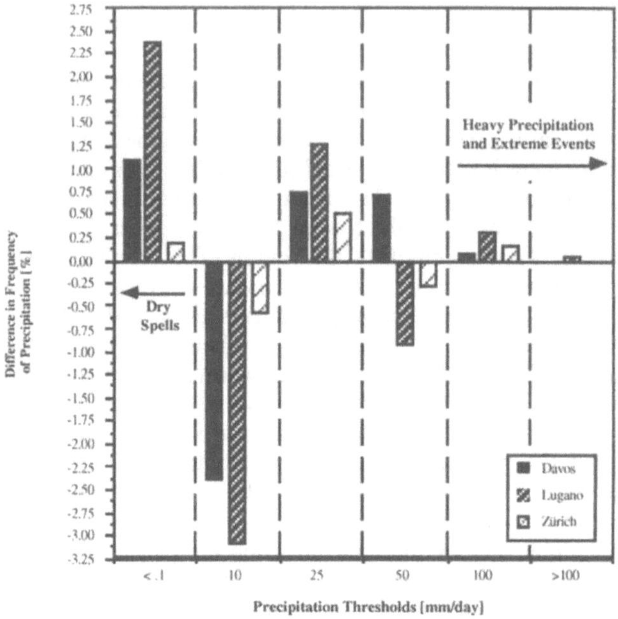

Figure 7.36: Shifts in selected ranges of precipitation frequency between the warmest and the coldest period of the century. Note the tendency of precipitation to be enhanced at the extremes (either dry or extremely high values) at all sites

Period	Davos	Lugano	Zürich
1901 - 1905	1041	1755	1068
1906 - 1910	1021	1558	1069
1911 - 1915	983	1776	1018
1916 - 1920	1008	1797	1019
1921 - 1925	920	1470	1018
1926 - 1930	982	1898	1073
1931 - 1935	1011	1863	1088
1936 - 1940	1024	1868	1225
1941 - 1945	981	1412	961
1946 - 1950	1019	1584	935
1951 - 1955	1084	1689	1143
1956 - 1960	991	2009	1180
1961 - 1965	964	1498	1099
1966 - 1970	1119	1596	1186
1971 - 1975	958	1560	1088
1976 - 1980	1073	1868	1110
1981 - 1985	971	1596	1059
1986 - 1990	986	1526	1194
1991 - 1994	1057	1545	1087

Table 7.14: Annual precipitation totals at the four climatological sites, averaged over 5-year periods

This observation is consistent with other studies (Rebetez et al., 1997) for the Alps which show that late summer extremes of precipitation have significantly increased in recent years; it is also consistent with the hypothesis of an enhancement of the hydrological cycle in a warmer climate (IPCC, 1996). Numerical studies by Lüthi et al. (1996) of Alpine precipitation under warmer global climatic conditions have concluded that increases in extreme precipitation events can be expected in coming decades, particularly south of the Alps, under warmer mean climatic conditions; in addition, the increase in means is essentially seen to be the result of enhanced extremes. It is interesting to note that while the precipitation extremes have shifted in the warmer period of the century, average values have shown no signifiant trend; average annual precipitation totals for the five-year sampling periods are given in Table 7.14. The difference in total annual precipitation between the warm and cold sampling periods ranges from - 0.1% in Lugano to + 7% in Säntis.

On an average annual basis, it is difficult to relate in a statisically meaningful manner the change in precipitation extremes to mean or extreme temperature changes, and there is no obvious statistical link to the upper 1 percentile of maximum temperatures. This is because extreme precipitation events can occur outside of periods of maximum annual temperatures; these latter generally occur towards the end of July, while many high precipitation episodes are observed later in the season. While the warmest periods of the year may often be accompanied by convective downpours, extreme precipitation events are related to specific synoptic situations (Schaer et al., 1996). In addition, Rebetez et al. (1997) have shown that flooding and mud or rock slides associated with such events are sometimes compounded by the melting of early snow which has fallen on the mountains prior to the event; significant amounts of snow are unlikely to fall before September, i.e., well after the peak in annual temperatures. Even though it is difficult to link mean annual temperatures to precipitation extremes, Rebetez (1996) has shown, however, that there are statistically-significant relationships between mean temperature and mean precipitation, especially on a seasonal basis, in the Alps. One notable conclusion of that study is that higher mean summer temperatures are related to lower mean summer precipitation.

More meaningful links between temperature and precipitation extremes need to be investigated on the basis of monthly or seasonal statistics and not solely on annual statistics, where the high intra-annual variabilty of precipitation is concealed. Even statistics at higher temporal resolution may fail to provide meaningful relationships with extreme precipitation events, however, since these are generally related to synoptic and not to local conditions, as already mentioned.

7.2.5.2 Comparisons with climate model results for current and future climates

It is of interest to assess the capability of a GCM such as ECHAM3, designed primarily for large-scale climate simulations, in reproducing certain regional climatic features. Much analysis has been carried out on the ECHAM3 simulations of the energy budget and other surface processes (Wild et al., 1995a; 1995b; 1995c; 1995d; 1996), but little work has been reported on the probability density functions of temperature in a particular region. Temperature data has been extracted from the GCM for the model grid-point cluster closest to (and most representative of) the geographical location of Zürich. Because the model surface at this location does not correspond to the true altitude of the Zürich observation site, an adiabatic correction has been applied to the model data to enable intercomparisons between the simulated results and the climatological data.

Figure 7.37 illustrates the frequency distribution of minimum temperature for current (1 x CO_2) climate as simulated by the GCM at the "Zürich model grid point"; the distribution of observed temperature minima for current climate (1990 - 1994) is also provided as an illustration of the reasonable performance of the GCM in reproducing the range and amplitude of the distribution. The extremes are fairly well captured in the model, though they tend to be overestimated; however the intermediate temperature range exhibits a larger and more peaked amplitude than the flatter profile of the observed distribution. The 5-year average in the GCM for minimum temperatures at this location is 6.9 C, while for the observations the corresponding figure is 6.1 C. Given the uncertainties associated with local or regional processes simulated by a global climate model, the fit between the observed and simulated curves can be considered to be a very encouraging result.

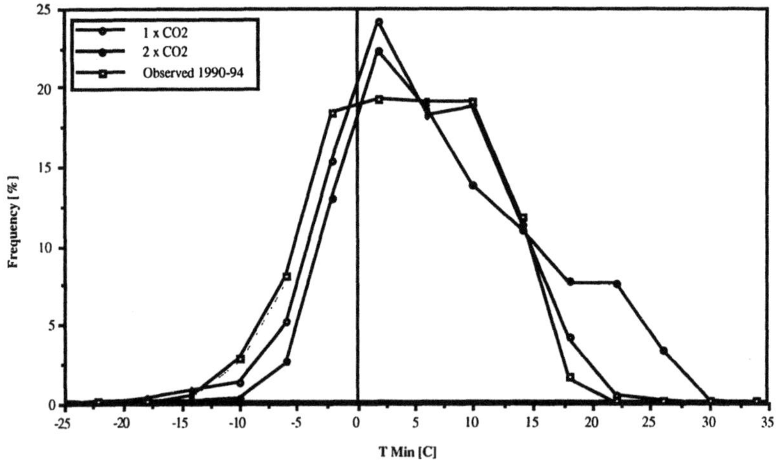

Figure 7.37: Comparison of GCM-simulated minimum temperature distribution for Zürich with the observed 1990-1994 distribution, and shifts of this distribution computed by the GCM for a 2 x CO_2 experiment

Also illustrated in this figure are the minimum temperature profiles for future (2 x CO_2) climate simulated by ECHAM3. The GCM simulation shows that the shift in the upper part of the extreme is far greater than the lower extreme; indeed, the peak frequency lies within the same temperature range but with a reduced amplitude. Tempertatures below -5 C, for example, occur 10.5% of the year for current climate (7.5% according to the GCM simulation), and these figures are reduced in the 2 x CO_2 climate to 5% in the GCM scenario simulation. In the upper range of temperatures, an arbitrarily-chosen minimum temperature of +18 C is exceeded 1.5% of the time in the current climate situation (4% in the GCM). In a warmer global climate, the GCM projects a 4% increase to 8% in the frequency for which this temperature threshold is exceeded, while the statistical approach estimates this increase to reach 14.5%.

In terms of maximum temperature distributions, an identical analysis has been undertaken and is illustrated in Figure 7.38. Despite obvious discrepancies, the GCM nevertheless captures the main features of the observed distribution, in particular the skewed distribution towards higher temperatures. The GCM-computed 5-year average is 12.5 C and the observed 1990 - 1994 average is 13.6 C. While this corresponds to a discrepancy of over 1 C between model and reality, the averages are given for reference purposes only, since the GCM is not necessarily simulating the observed 1990 - 1994 situation but a more generic "current climate" which lies within the period from 1979 to today.

The 2 x CO_2 climate estimations by the GCM emphasize the shifts of the upper extreme towards unprecedented temperatures. For current climate, maxima at or below 0 C represent 13.5% of the observed frequency (9% in the GCM), and those which exceed 30 C represent 2.2% and 2%, respectively. The 30 C temperature limit approximately represents a sea-level reduced temperature of 35 C, which as already mentioned is a threshold with significance for heat-stress damage to crops and certain ecosystems. Climatic conditions therefore currently account for about 7 - 8 days per year during which this threshold is exceeded. In the future climate projections, the GCM predicts an increase in the frequency of events exceeding the 30 C threshold in Zürich to 6.2%; this represents a substantial increase from 7 to 22 - 23 days per year where the limit is exceeded. Such a major increase would inevitably have consequences for the ecosystems concerned, and also for other sensitive systems at higher elevations (snow, permafrost, glaciers, and hydrological systems) where, presumably, the shifts in maxima and the corresponding extremes would be at least as large as in Zürich.

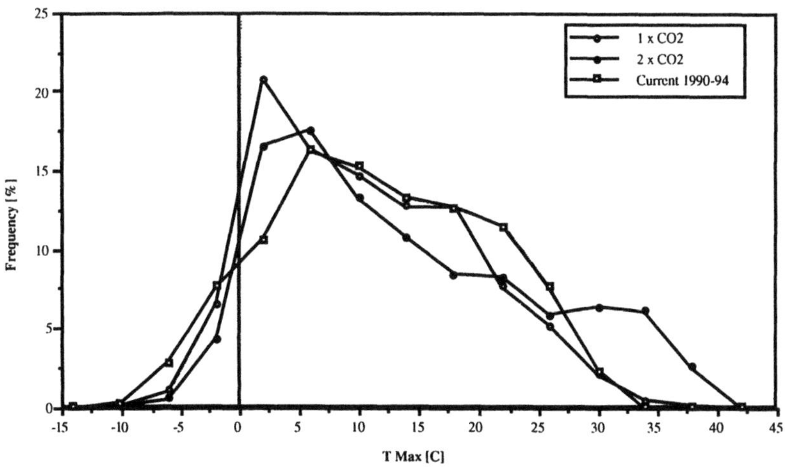

Figure 7.38: As Figure 7.37, except for maximum temperature distributions at Zürich

While the GCM simulations are satisfactory in this particular instance, the statistics of temperature may not always be as well captured in other areas, so that there is a potential value in a simple statistical approach to the problem. Firstly, a much simpler approach may attempt to bridge the gap between the inadequacies of current climate models in simulating extreme events in certain regions, and the requirements of regional climate scenarios in which extreme events are possibly the most important element for a range of impacts studies. They are also far more easily applicable to many end-users who may have little opportunity to access GCM-generated data. It may be erroneous, however, to assume that the change in mean temperature is the controlling factor for the change in extremes; it is quite likely, on the contrary, that the change in extremes is responsible for the changes in means, so that a statistical extrapolation is perhaps not as trivial as it initially appears.

The situation is even more difficult for precipitation extremes, where there is no obvious relationship on the annual average scale with temperature, so that no extrapolation is possible on the basis of the statistical method; it is also not particularly meaningful to use GCM precipitation data at one single model grid-point for a number of reasons, including the problem of representativity of the grid-point value in a region of complex topography. The increase in extreme precipitation events in a warmer mean climate is consistent with thoughts on future climatic change on the basis of modeling studies (IPCC, 1996), but the relationships are here too tenuous to be of any value in generating scenarios for precipitation statistics.

7.3 Conclusions pertaining to the observed regional climate in the Alps

7.3.1 Regional climate trends this century and links to large-scale processes

The study discussed in Section 7.2.1 has highlighted the fact that climate change this century is characterized by increases in minimum temperatures of about 2 K, a more modest increase in maximum temperatures (in some instances a decrease of maxima in the latter part of the record), little trend in the precipitation data, and a general decrease of sunshine duration through to the mid 1980s. The interannual variability is generally large, and filtering of the data to remove high-frequency noise shows that the regional climate undergoes a series of fluctuations of between 8 and 20 years' duration. The temperature change over this century is of greater magnitude than the global temperature changes published in the literature, reflecting an amplification of the global signal in the Apine region; warming has been most intense in the 1940s, followed by the 1980s; the cooling which intervened from the 1950s to the late 1970s was not sufficient to offset the warming in the middle of the century.

Pressure statistics have been compiled as a means of providing a link between the regional-scale climatological variables and the synoptic, supra-regional scale. These statistics show that pressure also exhibits a number of decadal-scale fluctuations, with the appearance of a new and anomalous behavior in the 1980s; in this decade, pressure reaches annual average values far higher than at other times this century. The pressure field is well correlated with the North Atlantic Oscillation (NAO) Index for distinct periods of the record (1931 - 1950 and 1971 - 1990) and is almost decorrelated from the NAO Index for the other decades of the century; this is indicative of transition from one climatic regime to another, dominated by zonal flow when the correlation with the NAO Index is high. In the 1980s, when zonal flow over the North Atlantic is strong, episodes of persistent, anomalously high pressures occur over Switzerland, particularly during the winter season. The difference between the zonal and non-zonal regimes is particularly marked between the decade of the 1950s and that of the 1980s.

The impact of this change between the 1950s and the 1980s on a number of climatological variables was investigated statistically in order to provide an illustration of the manner in which changes in synoptic regimes (i.e., 'climate change') impacts upon climate characteristics on a regional scale. The analysis shows that temperature, precipitation, snow depth, and sunshine duration are indeed sensitive to large-scale influences; not only were yearly mean changes quantified, but also seasonal and monthly fluctuations.

7.3.2 Altitudinal dependency of temperature anomalies

A series of anomalously cold and warm winters which occured in Switzerland during the 15-year period from 1979 to 1993 was analyzed in detail and reported in Section 7.2.2. The warm winters between 1988 - 1992 were particularly marked in the Alps, where lack of snow had severe consequences for the tourist-based economies of mountain communities.

Analyses of the departures of temperature minima from the 15-year means in warm and cold winters has shown that there is a very significant altitudinal dependency of the anomalies except at low elevations which are subject to fog or stratus conditions; the stratus tends to decouple the underlying stations from processes occuring at higher altitudes. It was also shown that there is a switch in the gradient of the temperature anomaly with height from cold to warm winters. For warm winters, the higher the elevation, the stronger will be the positive anomaly; the reverse is true for cold winters. The statistics for 88 observational sites provide a measure of the damping of the climate signal as an inverse function of height. The altitudinal dependency of temperature departures from the mean are the most important feature, followed by latitudinal effects (north and south of the Alps); continentality is not seen to be a major factor in determining the geogaphical distribution of temperature anomalies at this scale.

The investigation has also emphasized the fact that high elevation records can identify more readily significant interannual climatic fluctuations than at lower-elevation sites. This is also likely to be the case for longer-term climate change where possible greenhouse-gas warming would presumably be detected with more clarity. This type of study can help orientate future high-resolution climate model studies of climate change and in particular the assessment of model capability in reproducing a range of possible temperature anomalies and their altitudinal dependency.

7.3.3 Extremes and their sensitivity to changes in mean values

The investigation reported in Section 7.2.3 established that there are clear links between the extremes of temperature and their mean annual values in the Alpine region. This has been reported elsewhere in the literature; in this study, however, the density of data in both time and space has confirmed that these relationships are consistent, whatever the site characteristics or altitude of the observational site. Temperature extremes are seen to shift by a factor of over 1.5 in association with a unit change in means; there is additionally an asymmetry between the shifts in the lower (minimum) extreme and the upper (maximum) extreme, leading to a changed temperature distribution profile. Experiments with a high resolution global climate model have shown a reasonable agreement between observed temperature distributions, in the latter part of this century, and a 5-year current climate "time-window" simulation. Climate change simulations for the next century have highlighted shifts in extremes which are quite consistent with observed shifts in means and extremes between the coldest and warmest periods of this century (1906 - 1910 and 1991 - 1994, respectively).

On the basis of annual mean values, it was difficult to establish firm relations between the observed temperature changes this century and precipitation extremes, although there is a general bimodal shift in the warmer part of this century towards drier events, on the one hand, and more heavy precipitation events, on the other. More detailed analyses on seasonal, monthly, or case-by-case episodes would be required to assess whether any statistical basis could be found, which would enable extrapolation to different climatic conditions.

In terms of climate change studies, there is a crucial need by the climate impacts community for climate scenario data, particularly at high spatial resolution and for extreme values. Extrapolation of temperature extremes can be attempted, but with great caution, whereas scenarios of extreme precipitation events are unlikely to be achieved through statistical methods. This implies, therefore, that numerical modeling approaches, through improved physical parameterization schemes and ameliorated nesting of limited area models in general circulation climate models, or through climate models with variable grid meshes, urgently need to be pursued in order to overcome the inadequacies of statistically-based scenarios. Experiments with nested modeling techniques will be the focus of the following chapter.

8. Coupling the Macroscale to the Mesoscale: Regional Climate Simulations

8.1 Introduction

The necessity of obtaining regional scale climatological information is important for a number of reasons, which have been outlined in previous chapters of this text. One reason is obviously the requirements of the climatic impacts community and policy makers, who generally need information at scales much finer than those which even high-resolution GCMs are capable of delivering; another reason is that processes acting on local or regional scales often feed into the climate system, and therefore knowledge of the interactions between different atmospheric scales (cf. Figure 1.2, for example) is essential for furthering our understanding of the climate system in general.

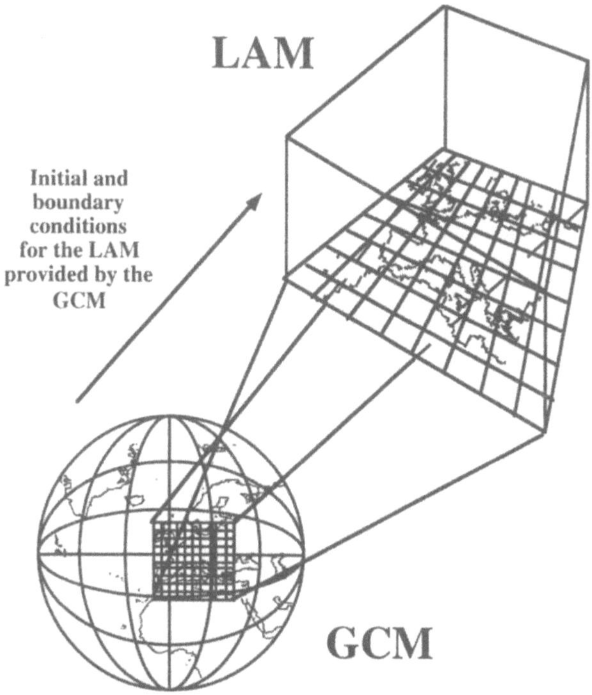

Figure 8.1: Schematic illustration of the coupling (or nesting) of a Limited Area Model (LAM) to a General Circulation Model (GCM)

A number of solutions exist which help improve the quality of climate data used in impacts assessments and economic decision making. Options include statistical techniques of downscaling from synoptic to local atmospheric scales (Gyalistras et al., 1994), use of paleoclimatic analogs (e.g., LaMarche, 1973; Webb et al., 1985; Davis, 1986; Schweingruber, 1988; Graumlich, 1993; Luckman, 1994), geographic analogs (Beniston and Price, 1991) and the coupling of mesoscale, or limited area, models (LAMs) to GCMs. It is this topic of coupled, or nested GCM/LAM modeling technique, which the present chapter will focus on.

The concept of nesting is relatively simple; results from a GCM for a particular region are used as initial and boundary conditions for a LAM, which operates at much higher resolution and, in many instances, with more detailed physical parameterizations. The LAM can theoretically be used to enhance the detail of regional climatology which would mostly be taking place at the subgrid scale of the GCM; a conceptual sketch of the nested technique is ilustrated in Figure 8.1.

8.2 Nested model simulations

The IPCC Second Assessment Report (IPCC, 1996) has used quite extensively the nested modeling technique in an attempt to improve regional climate information. The method, pioneered by Giorgi and co-workers (cf. Giorgi and Mearns, 1991, for example) at the National Center for Atmospheric Research (NCAR) in Boulder, Colorado, has been widely used in a number of case studies by different climate modeling groups worldwide.

Continuous monthly, season, or even multi-year simulations for current climate have for example been conducted by Bates et al. (1995), Chen et al. (1994), Copeland et al. (1994, 1995), Giorgi et al. (1993, 1994), and Leung and Ghan (1995) for North America; Hirakuchi and Giorgi (1995), Kanamitsu and Juang (1994), Lal et al (1995), Liu et al (1994), and Sasaki et al. (1995) for Asia, particularly Japan, China, and monsoonal India; Cress et al. (1994, 1995), Jones et al. (1995), Machenhauer et al. (1994), Marinucci et al. (1995), and Podzun et al. (1995) for various parts of Europe; McGregor and Walsh (1993, 1994) and Walsh and McGregor (1995) for the Australasian region; and Semazzi et al. (1993) for the Sahelian zone. Model grids for the regional component of the modeling system ranged from 15 - 125 km in the above experiments, and the length of simulations from 1 month to 10 years.

Fewer climate change experiments have been conducted with the nested modeling technique, though Hirakuchi and Giorgi (1995), Jones et al. (1995), McGregor and Walsh (1994) have attempted such experiments for selected parts of the world; the work of Beniston et al. (1995) and Rotach et al. (1997) over the Alpine region constitute the highest resolution modeling experiments for both current and future climate, and will be discussed in Sections 8.3 and 8.4 of this chapter.

Recent advances in nested modeling include continuous multi-year simulations (Marinucci and Giorgi, 1992; Giorgi et al., 1992) which consist in the equilibration of model climate with surface hydrology and the simulation of the full annual cycle for use in climatic impacts models.

When driven by analyses and observations, LAMs generally simulate a realistic structure and evolution of synoptic events. Model biases are in the range of a few tenths to a few degrees for temperature, and 10 - 40% for precipitation compared to observations, and these biases tend to decrease with increasing resolution or decreasing LAM domain size. GCM-driven LAMs tend to produce less convincing results, because of the poorer quality of the GCM data in terms of storm tracks and precipitation belts, for example. LAMs tend to produce more realistic regional detail of surface climatological parameters than GCMs, since the former models capture the regional detail of forcing elements such as orography or large lakes. Figure 8.2 shows the differences in GCM and LAM results for precipitation and temperature compiled by Giorgi et al. (1994) for a LAM operating at 50 km resolution nested into a GCM with a 300 km resolution.

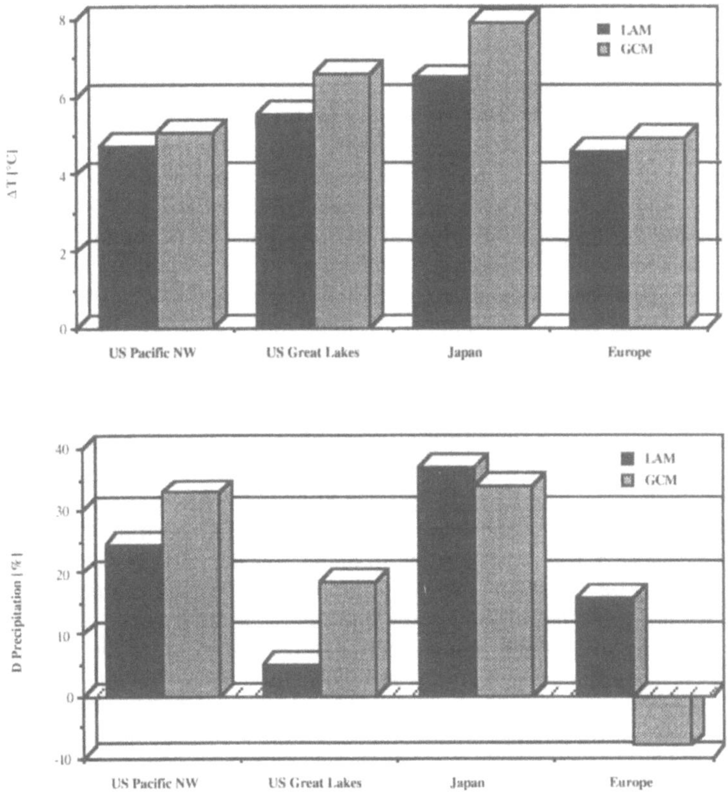

Figure 8.2: Comparison of GCM and LAM results for temperature and precipitation in selected regions

It is clearly seen that there are salient differences which result from the fact that regional detail is better captured by the LAM, and consequently so are the climatological processes which take place in the region. More detailed analyses have shown, however, that while a LAM can enhance climatological detail compared to the GCM simulation, they can also in some circumstances degrade the GCM results, especially if they are regionally averaged.

The nested models are generally seen to perform better at mid-latitudes than in the tropics, and that the LAM results improve with increasing resolution in both the driving GCM and the LAM itself. One major problem with validating regional climate simulations is the lack of climatological observations in many regions of the world. This problem is to a large extent overcome in the Alpine region, however as seen in the next sections.

8.3 Simulations of current climate over the Alpine region

8.3.1 Introduction

Climate over the central-western Alpine region is strongly influenced by the presence of topographical features which span a wide range of spatial scales (Beniston et al., 1994). Not only does the Alpine chain modify the structure of overlying synoptic systems through cyclogenetic processes (e.g., Mesinger and Pierrehumbert, 1986), but the effect of complex patterns of mountain ridges and valleys modulates the distribution of precipitation via orographic upslope lifting and the development of local mesoscale circulations (Wallen, 1970a; 1970b). As a first step towards the goal of improving simulations over the Alpine region, the RegCM2 (Regional Climate Model version 2) produced at NCAR was applied to this region and was nested within the high resolution version of the MPI GCM (ECHAM3). A number of test simulations for present day January and July climate conditions with the nested RegCM/ECHAM3 model were carried out to evaluate the performance of the modeling system before attempting any long term climate simulations. The RegCM was run at a grid point spacing of 20 km and was driven by output of a simulation with ECHAM3 at spectral T106 resolution, corresponding to a grid point spacing of about 120 km. Although a resolution of 20 km (the highest affordable for this work given the available computing resources) is still not sufficient to capture the fine details of the western Alpine topography, it captures some main topographic features lost at the T106 GCM resolution, and it is substantially higher than in any previously attempted experiments of this kind. Therefore, the present simulations offer the first opportunity to analyze the behavior (in climate mode) of a nested model at resolutions of a few tens of km. Moreover the model resolution is consistent with that of the gridded data used for the model validation.

In this part of Chapter 8, the analysis will focus on the simulated influence of the regional model representation of the Alpine topography on the precipitation and temperature climatology of the region. Both these quantities are in fact critical for assessment studies of the impact of climatic change on regional hydrology, cryosphere and ecosystems. The model domain consists of a 55 x 55 point grid centered over the Swiss Alps, stretching from the Mediterranean to Southern Scandinavia, and from the Atlantic coast to Central Europe. The availability of simulations at 20 km resolution allows to examine the performance, over an area of complex terrain, of model physics which have been mostly tested and used at coarser grid point spacings (50-80 km). Figure 8.3 illustrates the limits of the RegCM model domain and the central region for which results will be discussed.

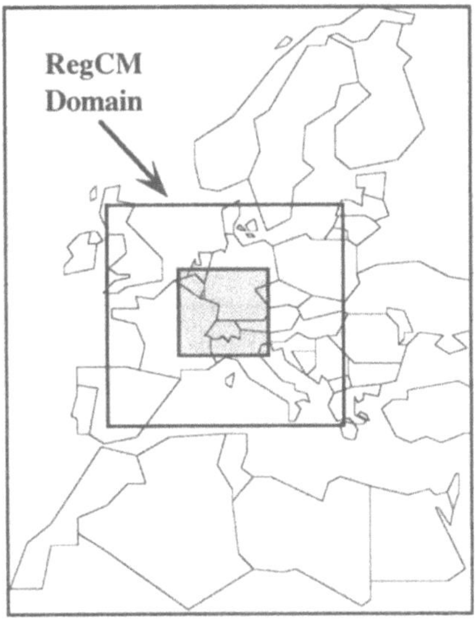

Figure 8.3: RegCM model domain of integration. The gray area represents the central domain for which results are discussed in the text

Moreover, since high resolution model data on a wide range of elevations are produced, aspects of the model precipitation climatology of relevance to impacts assessments, which have not been previously examined, can now be studied. These include the elevation dependency of precipitation and the precipitation intensity distribution.

In the validation analysis, the model results are compared with observations from a relatively dense station dataset covering the Swiss territory which was assembled for this study. The effects of different precipitation parameterizations and surface conditions are also investigated in a series of sensitivity experiments. The main goal of this work is to assess the degree of accuracy that can be expected from the present modeling system over the central-western Alpine region and the aspects of the model performance which are in greatest need of improvements before the model is used for the generation of climate change scenarios.

8.3.2 Results and model validation

In order to better understand the results of the nested model runs, it is useful to briefly describe the performance of the ECHAM3/T106 model simulation for January and July over the European region. Overall, the GCM reproduced reasonably well the location of the northeastern Atlantic jet, both in summer and in winter. Over northwestern Europe the simulated jet is stronger than observed in both seasons, while the Mediterranean branch of the jet is underpredicted, especially in July. These features of the simulated large scale flow fields can be explained by the model tropospheric temperature patterns. Mid-tropospheric temperatures are higher than observed by 0.5-2.5° between 35°-50°N, while they are lower than observed north of 60°N and south of 35°N in summer. This induces an enhancement of the meridional temperature gradient at mid-latitudes (50-60°N) and a reduction of the gradient over the Mediterranean, which in turn causes the simulated overprediction of the Atlantic jet and underprediction of the Mediterranean jet, respectively. The simulated tropospheric relative humidities are lower than in the ECMWF analysis over the continental regions, especially in summer and between 40-50°N, the area of RegCM nesting.

	Elevation [m]	**January**	**July**
Obs - LW		272.34	290.93
RegCM	886	274.99	292.68
ECHAM3	880	273.07#	292.17#
		273.11*	292.21*
Stations	1164	270.86	288.84
Obs - LW		270.53	288.15
RegCM	1514	271.45#	288.39#
		273.72	290.66
ECHAM3	1204	271.24#	289.52#
		271.51*	289.79*

Table 8.1: January and July average air temperatures.Top set: Observed data from Legates and Wilmott (LW), Reg CM, and ECHAM3 averaged over the RegCM inner domain. Bottom set: LW, RegCM and ECHAM3 data are first interpolated to the station locations and then compared to the Swiss station observations. Average model and station elevations are also presented. *: No lapse rate correction; #: Lapse rate correction

Throughout most continental areas, the model overestimates land temperatures by 1-4°C in winter and by up to 2-6°C in summer. The only exception is the central-eastern Alpine region, where the model temperatures are lower than observed. This is at least partially due to the fact that observing stations from which the Legates and Willmott (1990a; 1990b; hereafter referred to as LW) dataset is obtained tend to be located in valleys and at low elevations, which generally introduces a warm bias in observed datasets over mountainous regions unless suitable topographic corrections are introduced. January precipitation is close to observed, or slightly overestimated, in central and northern Europe, while it is substantially underestimated in the Mediterranean basin. In July, precipitation is underestimated throughout continental Europe. In January, the position of the Atlantic jet compares well with observations, and therefore the errors in surface temperatures and precipitation over central and northern Europe are small. Over the Mediterranean, however, where the jet is underpredicted, precipitation is less than observed and, because of the relatively low evaporation and high solar radiation, temperatures are too high. In July the predominant feature which ECHAM3 simulates is the strong underestimation of precipitation throughout the continent. Several factors contribute to this result: the underestimation of the eastern Mediterranean branch of the jet, the relatively low tropospheric relative humidity values, and a feedback process by which reduced rainfall causes a decrease in soil water and evaporation, which in turn tends to further inhibit precipitation. Wild et al. (1995a), who present an extended discussion of the model surface energy and water budgets, suggests that the one-layer soil water formulation of the ECHAM3 surface package tends to produce excessively rapid evaporation during springtime, which in turn generates excessively dry soil conditions at the beginning of summer.

In summary, the main discrepancies between model and observations found in the driving ECHAM3 monthly climatologies examined here are a slight underestimation of the Mediterranean branch of the jet and associated precipitation in January, and an underestimation of the Atlantic jet, moist convective activity and precipitation throughout the continent in July. It should be noted that the model run presented large interannual variability over the region for these months. For example, most of the January precipitation over the Alpine and central European region was produced during 3 of the 5 simulated months, and in July close to no precipitation was produced in the fifth year. More discussion on the model interannual variability will be presented in the next sections.

	Elevation [m]	January	July
Obs - LW		2.91	3.87
RegCM	886	4.10	2.54
ECHAM3	880	3.27	1.50
Stations	1164	3.50	3.70
Obs - LW		3.61	4.90
RegCM	1514	4.60	3.37
ECHAM3	1204	3.29	1.89

Table 8.2: January and July average precipitation [mm/day].Top set: Observed data from LW, RegCM, and ECHAM3 averaged over the RegCM inner domain. Bottom set: LW, RegCM and ECHAM3 data are first interpolated to the station locations and then compared to the Swiss station observations

Since the RegCM domain for these experiments is rather small, it cannot be expected that the large scale features in the driving ECHAM3 and nested RegCM domain be substantially different. Indeed, the upper tropospheric fields (e.g. wind, temperature, geopotential height) were very close in the ECHAM3 and RegCM simulations and are thus not discussed in this paper. Rather, the focus here is on surface air temperature, precipitation and low level circulation patterns, which can be substantially affected in the nested model simulations. In addition, we do not analyze the whole RegCM domain, but focus on an inner portion of it, which will be used in subsequent figures. This inner domain extends from 5°E to 12°E and from 45°N to 49°N, covering Switzerland and the western part of the Alps. A few short experiments showed that the results in the inner domain were not sensitive to relatively small modifications of the full domain employed for the simulations.

For an overall picture of the model performance, Tables 8.1 and 8.2 compare observed and model average surface air temperatures and precipitation for January and July. Two sets of data are presented. In the first (top portion of the table), the model data are averaged over the whole inner domain and are compared with observations averaged over the same domain from the LW dataset. In the second (bottom portion of the tables), the LW, RegCM, and ECHAM3 data are interpolated to the station locations and are compared with the Swiss station dataset. In the interpolation, it is assumed that the model and LW values at a station location are given by the values at the closest grid points. In previous studies, e.g by Giorgi and Marinucci 1991 and Giorgi and Bates 1989, the model precipitation and temperature values at a station location were bi-linearly interpolated from the surrounding four grid points. A few tests showed that the two interpolation procedures lead to similar regionally-averaged values. The averages shown in the table are taken over the whole station dataset.

Figure 8.4: Average January precipitation. Gray area: > 4 mm/day; black area > 6 mm/day. A: LW dataset; B: ECHAM3 simulation; C: RegCM simulation

Since the ECHAM3 and RegCM elevations are generally different from each other and the model elevations are different from the corresponding station elevations, temperature data are also presented in which a correction has been introduced to account for elevation differences. This correction is calculated by multiplying the elevation difference by a standard lapse rate of 6.5°C per km. It is to be noted that, when averaged over the whole inner domain, the RegCM and ECHAM3 elevations are about the same. When averaging elevations at the station locations (i.e. at the closest grid point), it can be seen that, while the station and ECHAM3 averages are about equal, the average elevation is about 400 m higher for RegCM than for the stations.

Both the comparison in the inner domain and at the station locations indicate that, after the lapse rate correction is accounted for, the models are too warm by 1 - 2.5°C. Wild et al. (1995a) indicate that one of the primary causes for this result is an overestimate of net radiation at the surface by the models. The average July temperatures in ECHAM3 and RegCM differ by a few tenths of a degree, while in January RegCM is about 2 degrees warmer than ECHAM3. The main reason for this resides in the treatment of radiative transfer in the presence of clouds in RegCM. Previous applications of the model have shown that the RegCM tends to produce relatively large amounts of low level clouds in the cold season (Giorgi et al., 1993b). Clouds in the model are treated as grey bodies with emissivity exponentially reaching unity as the liquid water path increases. In fact, in the lower troposphere the model clouds can be considered to be black bodies which substantially increase the downward longwave flux at the surface. Since in winter longwave radiation is the dominant component of the surface energy budget, large cloud amounts tend to significantly raise the surface temperatures. For these simulations, total January cloud amounts in the RegCM were in the range of 75 - 85 % over the western Alpine region, as opposed to values in ther range of 68 - 72 % estimated from surface observations (Warren et al., 1986).

Average precipitation data for models and observations are reported in Table 8.2. When comparing to both the LW long term climatological data and the station data, the ECHAM3/T106 simulation reproduces fairly well the average precipitation amount in January, but underestimates it by over a factor of 2 in July. This result is consistent with the fact that in January, the central-western Alpine region lies in a transition area where the ECHAM3 precipitation bias changes from positive (in central Europe) to negative (in the Mediterranean). In July, the central-western Alpine average data of Table 8.2 confirm the general precipitation underprediction computed by the GCM.

The RegCM produces more precipitation than the ECHAM3/T106 simulation, both in January and July, by factors of 1.3 to 1.8. As a result, precipitation is overpredicted by the RegCM in January, but still underpredicted in July (although less than by the ECHAM3). The difference in precipitation amounts produced by ECHAM3 and RegCM can be ascribed to a number of factors. First is the stronger topographical forcing resulting from the higher resolution of the regional model. Second is the use of different cumulus parameterizations, whose effect is relevant in July. The cumulus parameterizations in ECHAM3 and RegCM employ similar closure assumptions; however the ECHAM3 scheme includes two processes, lower level stabilization by the penetrative downdraft and evaporation of falling rain water, which are not accounted for in RegCM and tend to reduce precipitation (e.g. Giorgi et al., 1993b).

More insights into the effect of topography on precipitation production can be gained by examining the average precipitation spatial patterns. January and July precipitation for the LW observations, and the RegCM and ECHAM3 simulations are shown in Figures 8.4a-c and 8.5a-c, respectively. An analysis of the precipitation patterns of Figures 8.4b and c shows that both models tend to place wintertime precipitation maxima in the upslope regions of the mountain systems. ECHAM3 positions a primary precipitation maxima at about 47°N and 6°E, i.e., in the westward slopes of the model Alps. This is clearly associated with upslope orographic forcing within westerly winter storm systems. Secondary maxima also appear along the northern and southern model Alpine slopes.

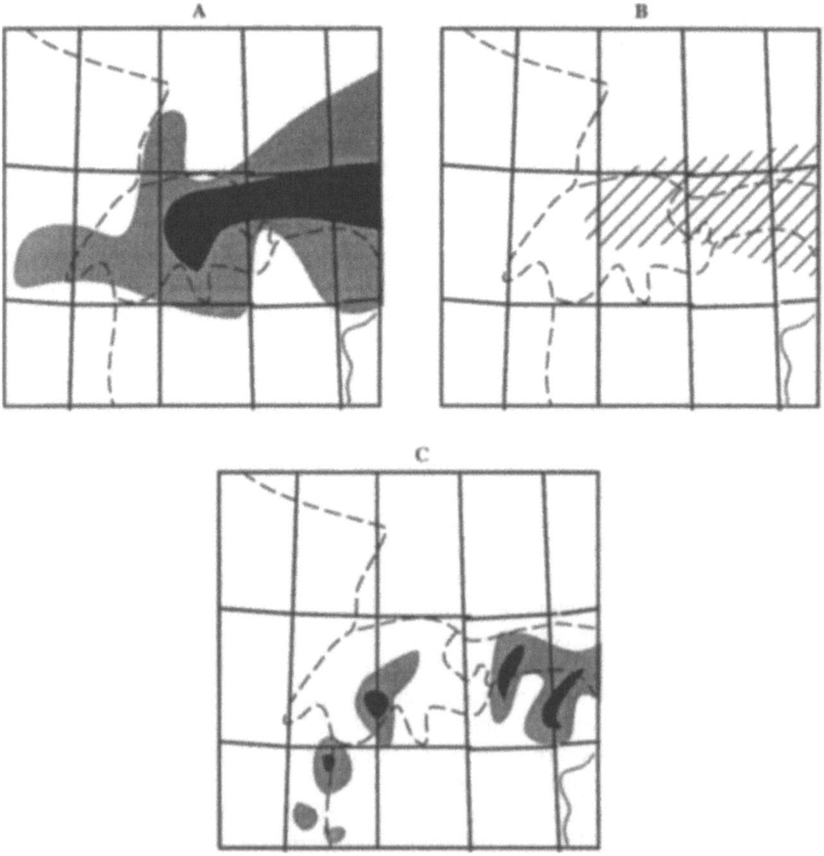

Figure 8.5: As Figure 8.4, except for average July precipitation. Hachured region in B indicates 2 mm/day (no area exceeding 4 mm/day)

Similarly, RegCM produces main maxima in the northern and southern model Alpine slopes, generally where the steepest topographical gradients are located. Other maxima can be found in the upslope regions of the Jura Mountains and southwestern Germany. Relative minima are produced in the regions east of the mountain peaks as a result of the topographic shadowing effect on eastward-moving winter storms. Comparison between Figures 8.4a and 8.4b indicates that the locations of most maxima in the northern Alpine upslope regions and in regions north of the Alps are realistically simulated.

The areas where the largest disagreement is found between observed and RegCM-simulated data are in the southern Alpine slopes. Two maxima are produced here in the RegCM runs, one (reaching about 10 mm/day) elongated in a northeastward direction and centered around 45.5°N and 8°E and the other further east, directed in an eastward direction, centered at 45.9°N and 9.5°E. These maxima, which for the purposes of the present discussion will be referred to as the south-western Alps and south-central Alps maxima respectively, are aligned with the southern slopes of the model Alps. Because such maxima are prominent features of the January simulation not shown by the LW dataset, a more detailed analysis of the possible mechanisms that led to these results was conducted.

For this purpose, the average precipitation over the inner domain for the individual simulated months over the region was investigated. These are shown in Table 8.3 along with the average temperatures for the individual months. As can be expected, ECHAM3 and RegCM runs show similar trends. The January temperature range is not large, about 3 K, while a marked variability in monthly values is found for the winter precipitation, the first three Januaries being much wetter than the last two.

	Temperature [K]		Precipitation [mm/day]	
	RegCM	ECHAM3	RegCM	ECHAM3
JAN 1	274.83	273.23	5.55	4.75
JAN 2	275.14	273.57	5.81	4.02
JAN 3	275.36	273.37	5.63	4.33
JAN 4	273.13	271.02	1.87	1.64
JAN 5	276.50	274.16	1.63	1.59
JUL 1	289.16	289.03	2.33	1.07
JUL 2	290.81	290.15	3.38	1.03
JUL 3	294.69	293.10	3.56	3.04
JUL 4	291.36	290.35	2.48	2.20
JUL 5	297.39	298.20	0.97	0.14

Table 8.3: RegCM and ECHAM3 average temperature and precipitation for individual simulated months (1 - 5 denote the year of the GCM simulations). Data are averaged over the RegCM inner domain given in Figure 8.3

From the analysis of individual months, two distinct regimes were identified, each responsible for one of the southern Alps maxima (a southerly flow and a southwesterly flow). The average precipitation for JAN1 and JAN2 are illustrated in Figures 8.6a and b, respectively. In the first regime, predominant for the months JAN1 and JAN5, a strong meridional component characterizes the flow over Europe. The winds are mostly directed from the southwest to the northeast, generating upslope conditions in the south-central Alps. As shown in Figure 8.6a, this southerly upslope flow regime is mostly responsible for the south-central Alps maximum of Figure 8.4c and produces only little precipitation north of the Alpine ridges. In this regime, a pronounced vortex activity is generated over the upper Po Valley by the steep Alpine barrier to the southerly flow.

The second regime was dominant during months JAN2, JAN3, and JAN4. The circulation here is stronger, more zonal and associated with perturbations moving over central and northern Europe. The westerly flow is split in two branches by the Alps. The northern branch is deflected in a northeastward direction to generate the upslope conditions which produce the precipitation maxima in the northern Alpine slopes (Figure 8.6b), while the southern branch veers in a southeastward direction towards the Adriatic Sea. When the westerly flow crosses the Alps, it is accelerated at high elevations, and as it descends the southern slopes, it encounters regions of lower wind speed, vortex activity, and relatively high moisture from the Gulf of Genoa, which generates areas of strong convergence and precipitation. Therefore, the southwestern Alps precipitation maximum of Figure 8.5c is associated with prevalent westerly rather than southwesterly flow.

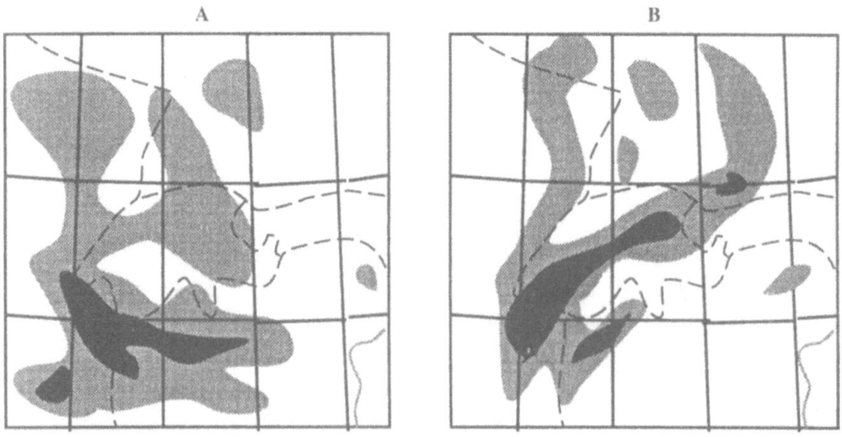

Figure 8.6: Precipitation in RegCM for the JAN1 (A) and JAN2 (B) simulations. Gray shading: > 5 mm/day; black shading: > 10 mm/day

As already mentioned, the winter precipitation maxima in the southern Alpine slopes simulated by the nested model do not appear realistic. Usually, the regions south of the western Alps experience their precipitation maximum during the intermediate seasons, autumn and spring (e.g. Air Ministry, 1962), under the influence of strong cyclogenetic events in the Gulf of Genoa which carry moist maritime air with southerly and southwesterly flow from the central Mediterranean. Often these events occur also in winter, but they are weaker than in other seasons due to the lower surface temperatures. Thus, the simulated southern Alpine maximum appears more typical of autumn or spring conditions, rather than winter conditions.

Referring back to the July precipitation results (Figures 8.5a-c), the LW dataset places the local maxima approximately at the same locations as in January, and in the observations the precipitation amounts are higher in July than in January. Conversely, in the model simulations, summer and winter maxima occur in different locations. Both in the ECHAM3 and in the RegCM runs, the primary precipitation maxima do not occur in the upslope regions, but rather close to the mountain tops. This is a consequence of the different mechanisms which control the simulated precipitation in the warm and the cold season. In the cold season, precipitation is mostly forced by orographically-induced upslope flow. In the warm season, the dominating mechanism is the instability induced by elevated surface warming near the mountain peaks. For example, as shown by previous work (Giorgi, 1991), the Kuo-type scheme used in RegCM for the present runs is especially sensitive to this process, since the vertical motions triggered by the surface heating induce moisture convergence, the primary controlling factor in the Kuo-type closure, and thus reinforce moist convection.

Table 8.3 indicates that also in July a wide range of month-by-month variability is found for the present runs. Average temperatures vary over a range of 8 - 9 K and precipitation varies by over a factor of 3. Overall, the RegCM results show less monthly variations than the ECHAM3 results. One possible contribution to this result is given by the different surface schemes in the two models for the present runs, and in particular by the treatment of soil water and evapotranspiration. For these simulations, the initial RegCM soil water was a function of surface type, and thus was the same for all years. Conversely, soil moisture in the ECHAM3 runs at the beginning of July depends on the previous climate and hydrology history over the region. This could contribute to the somewhat lower variability shown by the RegCM results compared to the ECHAM3 results. The lower variability in the RegCM results could also be an indication of the greater importance of local processes versus synoptic scale forcing at the higher resolution of the regional model. Finally, one of the July cases (JUL5) is extremely dry in both models, and because of the small sample size it strongly affects the overall average. It is worth noting that, when neglecting the JUL5 month, the July average precipitation would be in much closer agreement with observed precipitation amounts.

As mentioned, the availability of model and observed data on a wide range of elevations allows the study of the dependency of precipitation upon altitude. It has been observed that precipitation tends to increase with elevation (Sevruk, 1989) because of the orographic forcing mechanisms discussed above. It can be expected, however that at a certain height this increase would cease, and precipitation would start decreasing with elevation, since available water vapor would decrease at high altitudes. A clear identification of this precipitation maximum is however difficult, because it likely occurs at rather high elevations, where not many stations are located, and because at high elevations the gauged data become increasingly less reliable (Sevruk, 1989).

To study the relationship between precipitation and elevation in the model, and to compare it with the Swiss data, the area between 46 and 48°N and 6 to 10.5°E was examined in detail, i.e., the inner domain illustrated in Figure.8.3. Model and observed average precipitation for elevation intervals of 250 m were then calculated. The station averages were obtained from all stations with elevations within the selected intervals, while the model averages were calculated over all grid points with elevation within the same intervals. In January, the observations show a steep increase of precipitation with elevation up to about 2000 m and then a decreasing trend; accounting for the gauge precipitation correction, this would likely move this maximum somewhat to even higher elevations. The RegCM shows a similar trend, with precipitation increasing with height until about 1250 m and then decreasing at higher elevations. Even considering that the model maximum elevation is only 2500 m, the elevation of maximum precipitation appears clearly underestimated by the model. Indeed, from Figure 8.4b it appears that slope more than elevation controls the wintertime precipitation amounts in the model. This makes sense, since the orographic uplift process depends on the horizontal gradient of topography rather than elevation.

Two processes, however, contribute to the model-observation disagreement. First, in the upslope process the model precipitates water as soon as this condenses. In reality, water vapor will take some time to condense into cloud water and turn into precipitation. This delay time may move the precipitation maximum to higher elevations. Second, the model representation of the Alps provides a continuous barrier which strongly forces the upslope flow. The Alpine chain, however, is in fact constituted by complex patterns of troughs, ridges and valleys, so that much of the flow is found around the mountain barriers and the upslope orographic forcing is less intense. Note that in the ECHAM3 run, the precipitation maximum actually occurs at the lowest elevations and the trend is reversed from observed, with precipitation decreasing with height.

In July, the gradient of observed precipitation with height is less pronounced than in January, and the precipitation maximum occurs at higher elevations. The RegCM results reproduce both these features and show a better agreement with observations in July than in January, evidently a result of the decreased importance of upslope lifting with respect to local heating and destabilization effects. Also in the ECHAM3 run the maximum is shifted to higher elevations in July compared to January.

Another aspect of precipitation which has become a standard element of validation analysis in model testing is the frequency and intensity of individual precipitation events. In fact, for climate impact assessments, precipitation intensity distributions can be as important, or even more important, than average precipitation (Mearns, 1994). For this purpose, the model data for all grid points in the previously described Swiss sub-domain were used and the frequency of occurrence of daily precipitation events within given intervals of intensity were calculated (i.e. the number of precipitation events in a given interval divided by the total number of events). A similar procedure was used for the daily Swiss station data. Table 8.4 reports the total model and observed frequency of precipitation events, i.e., the percentage of the total number of days in which a daily precipitation amount in excess of 0.1 mm occurred at a particular station location.

It is obvious from Table 8.4 that both models tend to overestimate the frequency of occurrence of daily precipitation, especially in January. Even in July, when the average precipitation is less than observed, the precipitation frequencies are still higher than observed. This tendency of models to overpredict the number of rain events has been noted and extensively discussed in previous works (Mearns et al., 1990; 1994) and it has been attributed to two reasons. The first is related to model physics, i.e., especially in models which do not include explicit microphysics, the immediate precipitation of supersaturated water tends to produce an excessively large number of drizzle events, which tend to decrease in number when evaporation of falling rain is included. The second is related to model resolution: the observed data refer to point measurements, while the model data can be considered as averages over a grid volume. Assuming that precipitation effectively takes place only over a fraction of the total area of the grid box, the model precipitation frequencies should be rescaled accordingly. This latter process, which becomes more important as the model resolution decreases, would be especially significant in the summer and less in the winter. The fact that even the high resolution RegCM still substantially overestimates the total frequencies indicates that the model resolution cannot by itself explain the difference between observed and simulated data.

	January	**July**
Stations	46.2	42.4
RegCM	81.2	56.9
ECHAM3	77.1	45.9

Table 8.4: Total precipitation frequencies [%]

A much better agreement between models and observations is found for the intensity frequency distributions. Both models tend to overpredict light precipitation events and underpredict heavy events, but overall the differences with observed distributions are not large. It is interesting to compare the ECHAM3 and RegCM results for the two seasons. In January the RegCM produces more heavy events and less light events than ECHAM3. From the previous discussion, this can be expected in view of the higher resolution of the RegCM and the fact that most winter precipitation in the models is of resolvable scale nature. In July, however, the RegCM produces a greater percentage of heavy events than ECHAM3, but also a greater percentage of small events. The latter result is the opposite of what could be expected from model resolution alone. It is evidently due to the different convective parameterizations employed by the models. The Kuo scheme used in RegCM tends to produce higher frequency of light events than the Tiedke scheme of ECHAM3. One reason for this is due to the evaporation of falling rain, which is included in the ECHAM3 scheme and not in the RegCM Kuo scheme; another reason is the stabilizing influence of penetrating downdrafts in the Tiedke scheme. Both these processes would tend to inhibit the occurrence of light convective events.

Summarizing the results from the base runs discussed in this section, both ECHAM3 and RegCM reproduced some important aspects of the temperature and precipitation climatology of the western Alps. The temperature biases were generally small, as well as the January precipitation biases. The RegCM reproduced several features of the spatial precipitation patterns and the precipitation elevation dependency and intensity distributions. Among the main model biases: an underprediction of summer precipitation and of the elevation of winter maximum precipitation, and an overprediction of precipitation frequencies. In addition, the

RegCM produced significantly more rain than the driving GCM over the nesting region and simulated winter precipitation maxima in the southern Alpine slopes not present in the observation dataset. A set of sensitivity experiments aimed at investigating the influence of surface conditions and model parameterizations on the model results is described in the next sub-section.

8.3.3 Sensitivity experiments

In this section, three sets of sensitivity experiments are discussed, which are aimed at:

- examining the sensitivity of the results to the model cumulus parameterization
- evaluating the importance of the initialization of soil moisture
- isolating the effect of the model high resolution representation of topography

8.3.3.1 Sensitivity to model cumulus parameterization

A number of January and July simulations were conducted, using the cumulus parameterization developed by Grell (1994a; 1994b) in place of the Kuo-type scheme of the standard model. Briefly, in the Grell scheme clouds are pictured as two steady-state circulations caused by an updraft and a downdraft with no direct mixing between cloudy air and environmental air except at the top and bottom of the circulations. A simplistic cloud model is used which assumes that the mass flux in the updraft and downdraft is constant with height and the originating levels of updraft and downdraft are given by the levels of maximum and minimum ambient moist static energy, respectively.

	KUO		GRELL		SOILWAT	
	Temp	**Prec**	**Temp**	**Prec**	**Temp**	**Prec**
JAN 2	275.14	5.81	275.10	5.70		
JUL 1	289.16	2.33	289.35	2.20	288.85	2.36
JUL 2	290.81	3.38	291.07	1.99	290.48	3.22
JUL 5	297.39	0.97	297.06	0.98	297.32	1.04
JUL A	292.45	2.23	292.49	1.72	292.22	2.21

Table 8.5: Surface air temperature [K] and precipitation [mm/day] averaged over the RegCM inner domain given in Figure 8.3 for different sensitivity experiments. KUO: Kuo cumulus parameterization. GRELL: Grell cumulus parameterization. SOILWAT: ECHAM3 initial soil water content. JUL A represents the average over all the July experiments

The scheme is activated when a parcel lifted from the updraft originating level eventually attains moist convection, and condensation in the updraft is calculated by lifting a saturated parcel. The heating and moistening feedback to the large scale is determined by compensating mass fluxes and detrainment at cloud top and bottom, and the scheme includes the cooling effect of moist convective downdrafts (Grell 1994a). The scheme can adopt different closure assumptions to relate the mass flux at the bottom of the updraft to the large scale forcings. For the present experiments, it is assumed that the convective system removes the available buoyant energy with a time scale of 20 minutes, as described in Giorgi et al. (1993b). This closure is similar to that adopted by Fritsch and Chappell (1980).

Four simulations with the RegCM/Grell scheme were performed. Referring to Table 8.3, they are JAN2, JUL1, JUL2, and JUL5. Only one January experiment was chosen, that discussed in the previous sub-section (Figures 8.6a and b) since the effect of the cumulus scheme is minimum in the winter time. Of the summer simulations, three were selected for which a substantial variability in precipitation was found in the base runs. Average surface air temperature and precipitation for the Swiss sub-domain in the RegCM/Grell and corresponding RegCM/Kuo experiments are shown in Table 8.5. As expected, in January the effect of changing cumulus parameterization is small, both for temperature and precipitation.

In July, use of the Grell parameterization does not affect greatly the surface air temperatures. For the three months tested, precipitation is reduced by about 20 % when using the Grell scheme. It is to be noted, however, that most of this reduction occurs in one of the simulated months. This result is in the opposite direction of what was found by Giorgi et al. (1993a; 1993b) who carried out a similar comparison at the coarser resolution of 70 km. They found that the use of the Grell scheme increased precipitation over the European continent compared to that of the Kuo scheme. This increase in precipitation, however, was minimum over the Alpine region, mostly because the Grell scheme appears to be less sensitive to high elevation surface heating than the Kuo scheme. The main reason for this resides in the closure assumptions. Once convection is initiated in the Kuo scheme near a mountain top, strong local upward motion and convergence occurs, which tends to maintain the simulated convection. Conversely, since moisture convergence is not a driving criterion in the Grell scheme, this responds less strongly to the mountain top convergent flow. Also, in their experiments, Giorgi et al. (1993b) found that the topographical signal on precipitation was less pronounced when using the Grell scheme than the Kuo scheme.

8.3.3.2 Sensitivity to model soil water initialization

In the base runs, the soil water content was initialized as a function of vegetation type. Basically, the initial water content relative to saturation is about 15 % for grass and 30 % for forest, the two dominant types over the Alpine region. Since this initial soil water may be different from that which would have been obtained from a continuous simulation, the results of the present base runs may be affected by the soil water initialization during summertime, when local evaporation can significantly contribute to enhance precipitation (e.g. Giorgi and Marinucci, 1991).

To estimate the extent to which soil water initialization affected the model results, we conducted three sensitivity experiments, JUL1, JUL2 and JUL5, in which the initial soil moisture was directly interpolated from the ECHAM3 soil moisture at the beginning of the simulated months. The two models have different soil layer formulations. ECHAM3 employs a one layer soil scheme with 15 cm water capacity, while in BATS the soil is represented by three layers (a 10 cm top soil layer, a root zone layer of depth varying from 1 m for grass to 1.5 m for forest, and a 3 m deep soil layer) and the soil water capacity depends on the soil texture. For this reason, in the interpolation we first normalized the ECHAM3 soil water content by the total water capacity. The soil water content relative to saturation thus obtained was then interpolated to the RegCM grid and the RegCM soil water content was finally computed by multiplying this interpolated value by the local RegCM soil water capacity.

Table 8.5 compares average surface air temperature and precipitation results for the base and soil water sensitivity experiments. The comparison shows that the sensitivity of the present runs to the soil water initialization is minimal, less than a few percent, both for temperature and precipitation.

8.3.3.3 Sensitivity to model topography

Both in January and in July, the RegCM produced significantly more rain than the driving ECHAM3 over the Alpine region. To isolate the effect of topography from that of the different physics parameterizations employed by the two models, two additional simulations were conducted, one for JAN2 and one for JUL2, with the RegCM employing the ECHAM3/T106 topography. Figures 8.7a - d show the JAN2 and JUL2 precipitation over the inner domain for the ECHAM3 runs and the RegCM runs with ECHAM3 topography, respectively.

Comparison of Figures 8.7a and 8.7c shows that, when using the same topography, the two models produce generally similar patterns of precipitation in January. In both simulations, a broad maximum in excess of 6 mm/day is found in north-west Switzerland and a strong precipitation gradient occurs south of the model Alpine chain. The average precipitation and temperature over the inner domain for the JAN2 RegCM case with ECHAM3 topography are 3.11 mm/day and 274.33 K, respectively, which can be compared with 4.02 mm/day and 273.57 K in the ECHAM3 run and 5.81 mm/day and 275.14 K in the original RegCM run, respectively (see Table 8.3). In January, the high resolution topographic forcing thus accounted for about half of the total RegCM precipitation, while, with the same topographic forcing, the RegCM produced about 25 % less precipitation than ECHAM3. Hence the different topographical representation is the main cause for the differences between January ECHAM3 and RegCM precipitation amounts and spatial distribution. It should also be noted that, with decreased precipitation and cloudiness at the coarser resolution topography setting, the surface temperatures in the RegCM are much closer to those of the ECHAM3 run.

The situation is different in July, when local thermal and convective processes become dominant. Comparison of Figures 8.7b and 8.7d shows that, although the precipitation patterns in the JUL2 ECHAM3 simulation and RegCM simulation with ECHAM3 topography show some similarities (such as maxima in the north-western part of Switzerland and a marked southward gradient), the RegCM run produces a much more pronounced spatial precipitation variability. For example, a maximum is simulated over northeastern Italy, which is not found in the ECHAM3 run.

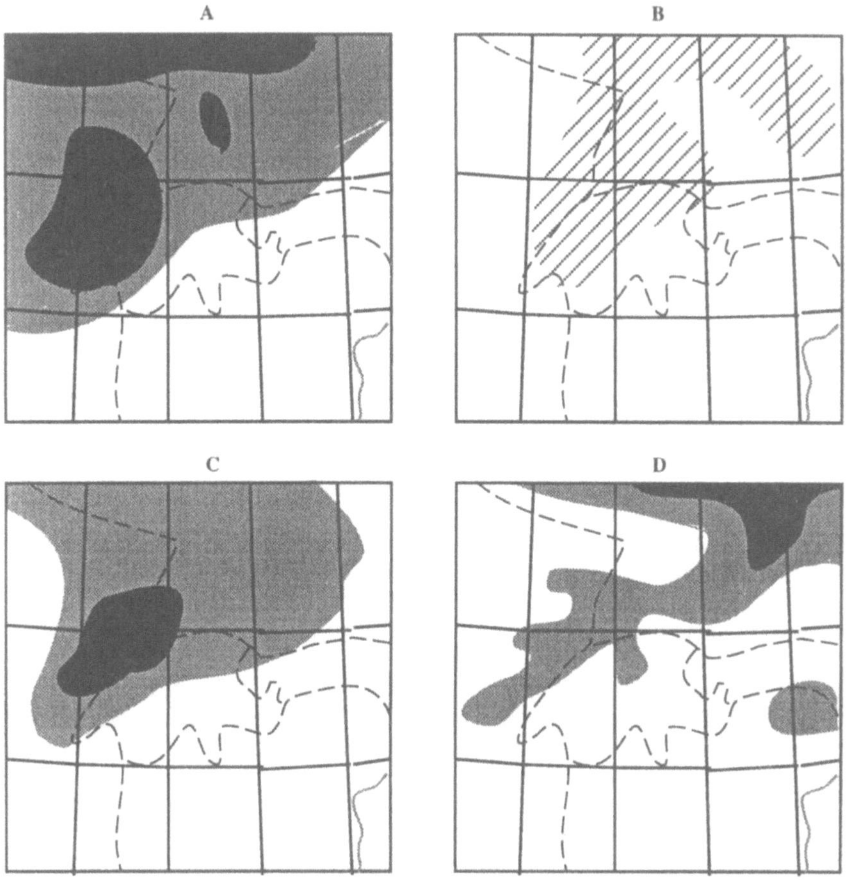

Figure 8.7: Average precipitation patterns; A: Simulation JAN2 with ECHAM3; B: simulation JUL2 with ECHAM3; C: Simulation JAN2 with RegCM; D: Simulation JUL2 with RegCM. Gray shading: > 4 mm/day; black shading: > 6 mm/day. In B, hachuring indicates > 2mm/day (no region exceeding 4 mm/day)

Similarly to the January case, the spatial precipitation distribution in the RegCM simulation with ECHAM3 topography is quite different from the precipitation obtained with the higher resolution topography (see Figure 8.6c). The average precipitation and temperature over the inner domain for the JUL2 RegCM case with ECHAM3 topography are 2.37 mm/day and 290.97 K, respectively, which can be compared with 1.03 mm/day and 290.15 K in the ECHAM3 run and 3.38 mm/day and 290.81 K in the original RegCM run, respectively (Table 8.3). Thus in July the effect of the different physical parameterization schemes appears to be of the same order of the effect of different topography representations in determining the RegCM and ECHAM3 precipitation amounts.

8.3.4 Summary and conclusions

As part of a joint effort to model the climate of the central-western Alpine region, this section has described a number of experiments with the NCAR RegCM LAM nested within a high resolution version of the MPI ECHAM3 global climate model. Two sets of simulations were conducted with RegCM for present day climate conditions, five for January and five for July, over a domain covering the Alpine region and adjacent areas with a grid resolution of 20 km. A number of additional experiments were conducted to examine the model sensitivity to cumulus parameterization, soil water content initialization and model topography. To analyze the model performance, the results from these simulations were compared with two observed climatological datasets, namely the half-degree resolution gridded data of LW and selected stations of the Swiss Climate Data Base.

Generally, the quality of the ECHAM3 driving fields was better over central Europe than over the Mediterranean Basin. The nested model reproduced several aspects of the central-western Alpine temperature and precipitation climatology, but also showed significant deficiencies which should be addressed before applying the model to climate change experiments. Most noticeable are the elevation dependency of winter precipitation, the precipitation frequencies and the simulation of unrealistic winter maxima over the southern Alpine slopes.

Overall, average model regional temperature errors were in the range of 2-3 °C, while average simulated precipitation was within 30% of observed values. These errors are of the same order of (or even smaller than) those found in previous experiments with the RegCM run at lower resolutions (e.g Giorgi and Marinucci, 1991, Marinucci and Giorgi, 1992). This indicates that even without any specific tuning the model performance was not significantly degraded by the relatively pronounced increase in resolution.

However, more tests are necessary to identify the best model configuration and parameterization options for the Alpine region. For example, a resolution even higher than 20 km would be required to capture certain narrow mountain and valley systems characteristic of Alpine topography. The 20 km grid point spacing is close to the limits of applicability of the Kuo and Grell-type parameterizations and may also require explicit cloud physics representations. Therefore, as computer capabilities begin to make regional climate simulations feasible at a resolution of a few kilometers, a major effort in model development should be devoted to the representation of convective processes in conditions where the scale separation assumption inherent in present cumulus parameterizations becomes less valid.

8.4 Simulations of future climate over the Alpine region

Nested simulations with the RegCM limited area model (LAM) were performed for conditions of doubled atmospheric concentrations of CO_2 (the '2xCO_2' scenarios, representing about the year 2050) in the same way as for the control simulation described in Section 8.3. Again, the months of January and July were chosen as representative for the winter and summer seasons, respectively. Thus a total of ten months (five Januaries and five Julies) were simulated with the nested modeling system. The simulations attempt to provide high-resolution regional climate parameters in a hypothetically warmer global atmosphere forced by the enhanced greenhouse effect. The high-resolution gridded data, on a 20-km mesh, can be of use to a range of potential end-users, particularly those working in the domain of climate change impacts on the natural and socio-economic environments.

8.4.1 Results of the nested model simulations

Table 8.6 compiles the area averaged 2m temperatures and total precipitation amounts for the control simulation and the 2 x CO_2 scenario. Additionally, the temporal averages over the five respective months is given. It is to be noted that the area averages are taken over the whole RegCM domain, i.e., about 1100x1100 km, rather than focusing on an 'inner domain' over the Alps as was presented in Section 8.3 (Tables 8.1 and 8.2). On average, the July temperature in the 2 x CO_2 scenario is increased by about 4 K as compared to the control run. Only one year of the control run exhibits a temperature that is larger than in the coldest year of the 2 x CO_2 scenario. In January, the temperature increase due to the doubling of CO_2 is only about 0.9 K on average. Here, the year-to-year range of the areally-averaged temperatures shows a considerable overlap between the control run and the 2 x CO_2 experiments. Overall, the year-to-year variability of both the January and July temperatures does not seem to be affected by the doubling of atmospheric CO_2 concentations. For the total precipitation amount the situation is less clear; there is virtually no difference in precipitation for the averaged winter months. In summer, precipitation is found to decrease by about 40% in the future climate. As a result of the smaller total precipitation amounts, the year-to-year range also appears to be somewhat reduced. However, it needs to be stated that the total number of five years (for each season) that were simulated with the LAM is too small to adequately capture the (observed) year-to-year variability in total precipitation.

The spatial distribution of the temperature and precipitation changes (2xCO_2 - 1xCO_2) is depicted in Figures 8.8 and 8.9, respectively. Focusing now on the alpine region, the following observations can be made:

- July 2m temperature increases by 2-4 K. Thus the least one can conclude from this is that the temperature changes over the whole domain (Western Europe) are not amplified over the sub-domain covering the Alps. Using the global model alone, a temperature increase of about 6 K was reported over the alpine region. Thus the higher resolution of the nested simulation seems to be able to reduce the erroneously high temperature change at least to a certain degree.
- In January, the 2 m temperature increases by 0.5-1.5 K, which is somewhat smaller than was predicted by the ECHAM3 simulation alone, where a temperature increase of about 2 K occurs over the Alps. In winter, the simulated temperature change is larger over the Alps than over the remaining part of western Europe.
- For July, the nested simulation predicts a decrease in total precipitation of up to 3.5 mm/day in certain regions on the one hand, particularly the French and Swiss Alps, while other regions experience significant precipitation increases.

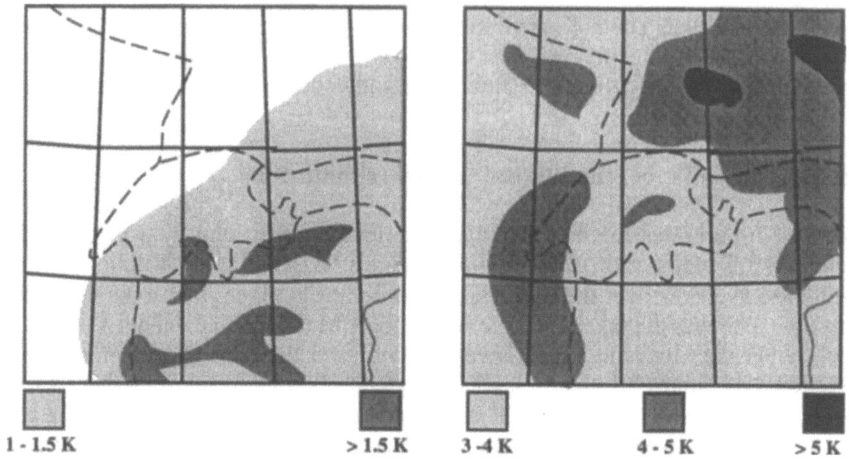

Figure 8.8: RegCM temperature changes in January (left) and July (right), between current and 2 x CO_2 climate

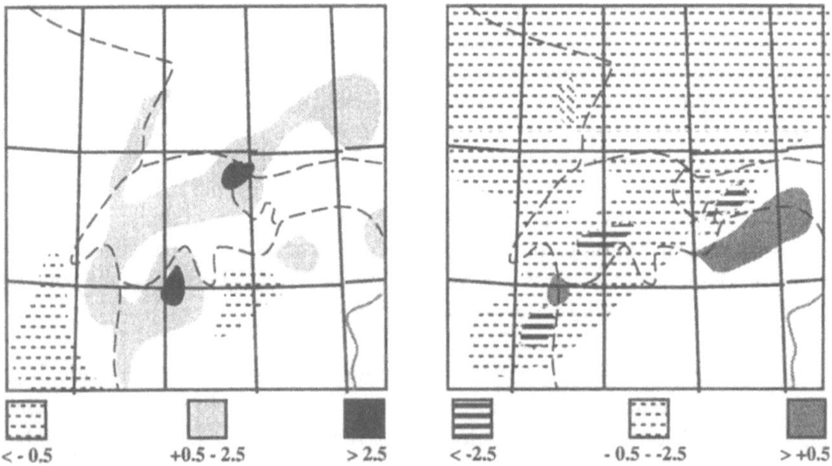

Figure 8.9: As Figure 8.8, except for average daily precipitation changes. Units: mm/day

Comparing these results to the ECHAM3 simulation, i.e., a decrease of about 1.2 mm/day over the alpine region, it can be seen that the RegCM is able to produce a much finer structure of precipitation patterns, which results in a high spatial variability over a region that is confined to the alpine chain. Over the remaining part of the model domain, the results of the driving GCM are not severely altered.

- For January, the precipitation results are similar to those for July, but not as pronounced. While ECHAM3 predicts an increase of up to 1.6 mm/day over the western alpine region, the nested simulation yields a much better resolved spatial structure with marked increases of precipitation, up to 3.5 mm/day over the

Mont Blanc region and increases of 0.5-1.5 mm/d over the remaining part of the alpine region. A zone of decreasing total precipitation amounts is predicted over France, where the driving GCM alone indicates an increase.

Season	**Year**	$T_{control}$ {K}	T_{2xCO} {K}	$P_{control}$ {mm/d}	P_{2xCO2} {mm/d}
Winter	01	277.1	278.6	2.7	1.4
	02	277.5	278.1	2.2	3.2
	03	277.5	277.2	2.0	0.78
	04	274.7	276.4	0.55	1.8
	05	278.3	279.2	0.76	1.2
	Average	277.0	277.9	1.6	1.7
Summer	01	291.6	301.1	0.78	0.37
	02	292.7	296.6	1.34	0.51
	03	296.0	300.4	1.09	0.23
	04	293.7	297.5	0.67	0.89
	05	298.2	295.9	0.22	0.65
	Average	294.4	298.3	0.86	0.53

Table 8.6: Average temperatures and total precipitation amounts for the five simulated years and their respective temporal averages. All the values represent area averages over the entire domain of the nested simulation

8.4.2 Interpretation of the results

At this point the question arises as to the confidence one can have in these predicted changes in 2 m temperature and total precipitation (ΔTCO_2, ΔPCO_2 for the $2xCO_2$ scenario as obtained with the limited area model. One approach is to compare the difference between the control simulation and the observations (ΔT_{obs}, ΔP_{obs}), i.e. the model biases to the predicted changes. It then becomes important to define a condition which allows a conclusion to be drawn to the effect that the predicted changes are significant. This will certainly be the case if the model bias for variable X (ΔX_{obs}) is smaller than the predicted change (ΔXCO_2) by, say, a factor of two or more, and at the same time, ΔX_{obs} is known not to be the result of compensating errors. On the other hand, no confidence can be attributed to the prediction if ΔX_{obs} is much larger than ΔXCO_2, in particular if the reason for a large value of ΔX_{obs} is not known . The reasons for the model biases must then be analysed in more detail in order to find out whether or not their nature is such that a prediction of regional climate change is likely to be severely affected. In any case, it is felt that this type of comparison can lead to both a deeper understanding of the model performance and of the conditions which are responsible for the model biases. Additionally, there are inevitable weaknesses and limitations in the observational record which need to be taken into consideration; the observations used in Section 8.3 are by no means perfect, even if they are the best available.

Focusing first on the summer results, the change in 2m temperature over the alpine region ($\Delta T_{CO_2} \approx$ 2-4 K) amounts to about the double of the model bias(ΔT_{obs}) in the area average (cf. Table 8.1). The predicted change in July precipitation is spatially highly variable (-3.5 to +3.5 mm/day), and the same is true for the difference between the control run and the climatological data. Averaged over the whole 'inner domain' of Section 8.3, RegCM underpredicts the July precipitation by 1.3 mm/day (about 35% of the observations). However, in the south-western and eastern Alps the overprediction of the nested simulation can be as large as 5 mm/day (or twice the amount observed, as seen in Figures 8.6a and 8.6c).

This 'local' overprediction can, at least partly, be attributed to the different spatial resolution of the modeling system and the climatological data. The most pronounced change in July precipitation is predicted for precisely those regions where the largest ΔP_{obs} is present. Thus for the alpine region as a whole, as well as for the south-western and eastern Alps, the changes in precipitation as predicted by the nested simulation are considerably smaller than the differences between observations and model results.

However, the interpretation of these findings is somewhat complicated by the fact that it is very difficult to distinguish between the error stemming from the driving GCM and that of the LAM itself. It is known that during the summer season, the ECHAM3 is too dry over Europe as a result of limitations in the surface parameterisation (cf. Wild et al., 1995a). Hence the evaporation becomes too small and, in turn, the temperature near the ground too high.

As a further consequence, the moisture content in the atmosphere is too low and the convective precipitation is underpredicted. For the scenario of a doubling of CO_2 concentrations, this means that a possibly increased ground temperature cannot lead to more evaporation, and hence possibly enhanced precipitation, since the soil is already too dry in the control simulation. Furthermore, the cooling due to the enhanced evaporation is absent, so that ΔT_{CO_2} as such is overpredicted. This fraction is smaller than what can be expected for the summer season in mid-latitudes (although it has to be noted that it is inherently difficult to determine, by measurements, the fraction of the total precipitation that stems from convective activity), thus confirming the situation described above.

Consequently, the change in convective precipitation ($\Delta P_{CO_2\text{-conv}}$) is similarly small and the results presented above for the change in 'total precipitation' reflect to a large extent the change in 'orographic precipitation' alone. As a preliminary conclusion for the July results, it may be stated that the predicted temperature change over the alpine region in the 2 x CO_2 experiment is probably too large and might well be of the order of, or smaller than, the model bias. For the assessment of the precipitation, a more detailed climatological data set, similar to the stations of the Swiss Climate Data Base mentioned in Chapter 7, but encompassing the entire alpine region, would be necessary. Furthermore, observations of the soil moisture should also be made available, in order to compare the model performance in the case of one of the precursor fields for precipitation.

Concerning the results for January, it is seen in Figure 8.8a that ΔT_{CO2} amounts to 0.5-1.5 K over the alpine region, i. e., somewhat smaller than the averaged ΔT_{obs} over the 'inner domain' (cf. Table 8.1). For precipitation, the changes range up to about 3.5 mm/day, as compared to an overprediction of RegCM over the inner domain of 1.2 mm/day (Table 8.2). In the temporal average, the largest changes are predicted in the south-western and, especially, south-central Alps, i. e., again exactly in those regions where the nested model was found to produce extremely high winter precipitation (cf. Figures 8.4a and 8.4c), which does not appear realistic. This indicates that the flow situation discussed in Section 8.3, which is leading to the overprediction of precipitation in the south-western and south-central Alps are more frequent under the 2 x CO_2 scenario. This fact might yield some further insight into the mechanisms producing the simulated large precipitation amounts. Overall, the modeled ΔT_{CO2} and ΔP_{CO2} for January are smaller in magnitude than the difference between the control run and the observational record.

The foregoing discussion leads to a major conclusion: in order to have reasonable confidence in the prediction of the possible change of climate variables such as temperature and precipitation under a 2 x CO_2 scenario, a more detailed analysis of the control simulation is necessary. In particular, the 'precursor fields' such as surface energy fluxes, cloud amount and surface properties, especially soil moisture, rather than the resulting temperature and precipitation, need to be studied in detail.

Apart from the possible refinement of some model components, the preparation of ameliorated climatological data, particularly for precipitation - in order to thoroughly interpret the maxima of winter precipitation over the south-western and south-central Alps - is believed to be of importance for reducing the uncertainty of the measured data themselves.

8.4.3 Concluding remarks

The changes in 2 m temperature and total precipitation over the Alpine region in a 2 x CO_2 world were calculated using the nested simulation technique. A first conclusion shows that the present nested modeling system (ECHAM3-RegCM) provides a tool capable of resolving a large part of the spatial variability over a region as complex as the Alps. In particular, topographic features could be treated with much more realistic definition. For the scenario of doubled atmospheric CO_2 concentrations, temperature is predicted to increase by 2 to 4 K in summer (July) and by 0.5 to 1.5 K in winter (January). Changes in total precipitation range from -3.5 mm/day to +3.5 mm/day (July), and from 0 to 3.5 mm/day (January). The nested simulation predicts smaller changes in temperature than the GCM alone, particularly in July. For total precipitation, the average changes (temporal average of the area-averaged precipitation) are small, with nevertheless a large spatial variability. In general, the LAM predicts changes in temperature and precipitation that are smaller than, or of the order of, the model bias, i.e., the difference between the simulation of the current climate and climatological data. Thus, in order to increase the confidence in the predicted changes of temperature and total precipitation amount, the model bias will have to be reduced. In view of possible future nested simulations with the refined version of the driving GCM (ECHAM4), care should be taken in the following areas:

- The LAM seems to produce too high precipitation amounts in certain regions of the Alps. However, climatological information at a high spatial resolution would be necessary to determine to what extent these large precipitation amounts are 'real' or overpredicted. Furthermore, the model components responsible for a possible overprediction need to be refined.
- The problem of soil moisture. Adequate climatological data are missing to initialise the LAM simulation, as well as for model validation purposes.
- The topography used in the present simulations is too coarse for the model resolution of 20 km. It would also be worthwhile to reduce the grid spacing to 15 or even 10 km, using appropriate topographical information.
- The analysis of the control run in terms of surface energy fluxes, cloud amount and other 'primary fields' would be desirable. However, as for other variables, the climatological data (at a resolution of 20 or even 10 km, and over a region as inhomogeneous as the Alps) are not generally available. A one-column mode could be a possible alternative.

8.5 Conclusions and outlook on the future of climate modeling in Switzerland

The ultimate objectives of the detailed modeling studies which have been presented in this Section are to further the basic understanding of climate processes in the Alpine region - both in present-day climate and in a changed global climate - and to provide a link to other members of the scientific community. The project funded by the Swiss National Science Foundation has enabled the use of advanced global and regional climate models for the first time in Switzerland.

High-resolution climatologies for January and July in both the 'present-day' and the '2 x CO_2' situations have been simulated over the Alps using the combined RegCM2 and ECHAM3 models. Based on the global information provided by the GCM for a five-year time window for both the present and future climates, RegCM2 has simulated detailed climatologies over the region of interest. A number of additional experiments were conducted to examine the model sensitivity to cumulus cloud parameterization, soil water content, and model topography. To analyze the model performances, the results from these simulations were compared with two observed climatological data sets, as well as a dataset from close to 100 observing stations distributed throughout Switzerland.

The nested model RegCM2 reproduced several aspects of the Alpine temperature and precipitation climatology but also showed significant deficiencies which are likely to be linked to failings in the initial and boundary conditions provided by ECHAM3. The model reproduces remarkably well the elevation-dependency of precipitation; however, the precipitation frequencies and the simulation of unrealistic winter maxima over the southern Alpine slopes are probably a result of insufficient water vapor advection from Mediterranean. Average model regional temperature errors in some cases exceeded 1 K, while average simulated precipitation was within 30% of observed values. These errors are, however, generally smaller than previous experiments with RegCM2 at lower spatial resolution, thereby confirming the value of increasing the spatial resolution of these models.

Indeed, the major benefit of high resolution resides in the more realistic simulation of precipitation.

In the 2 x CO_2 experiment reported in Section 8.4, generally higher temperatures in both summer and winter appear to be in accord with other international studies on global climate change. There are indications that temperature increases more at higher elevations than at lower altitudes, which is in the same sense as the observational studies of Chapter 7; precipitation is also higher and more intense. While the temperature and precipitation changes are within the range of model errors (when compared to observational data), comparing only model simulations for present and future climates reveals a trend of increasing temperatures and precipitation which is consistent with present-day hypotheses concerning greenhouse-gas forcing of the climate system.

Much effort has been invested in climate modeling since 1992, using the most up-to-date models available. This has been made possible thanks to an active collaboration with the Max-Planck-Institute for Meteorology (MPI) in Hamburg, Germany, the National Center for Atmospheric Research (NCAR) in Boulder, Colorado, and the technical support of the Swiss Scientific Computing Center (Centro Svizzero di Calcolo Scientifico or CSCS) in Lugano-Manno. Within the relatively short time period of two years (1992 - 1994), some of the longest high-resolution climate simulations worldwide have been undertaken, using both the T-106 version of the ECHAM3 General Circulation Climate Model (GCM) and the 20-km resolution RegCM2 Limited Area Model (LAM). In addition to the work carried out over the Alpine region, simulations have also taken place using this approach over Central Europe. This research took place within the context of a Swiss National Science Foundation initiative to encourage scientific contacts between Swiss research teams and scientists in Hungary and Poland.

Switzerland now has at its disposal very comprehensive and detailed global and regional datasets for both current and future climate, which can be made available to any interested end-user; the climatological information provided by the 20-km grid of the RegCM2 LAM is now at the spatial scale which begins to be of interest to scientists working in the field of impacts of climate change on different environmental and socio-economic systems.

On the basis of the experience gained in climate modeling, a number of scientific points have been raised which need to be addressed in the context of future global and regional climate simulations. The first major undertaking, based in part on past investigations, will has recently been to replace one of the principal modeling tools, namely ECHAM3, with the latest version of the GCM, ECHAM4. This updated model is currently recognized to be one of the most complete GCMs in existence today, and Switzerland is now in the priviliged position to undertake climate simulations in order to pursue independent research in Switzerland. While ECHAM3 was considered to be a tool to test the feasability of climate modeling in Switzerland, ECHAM4 is today the advanced GCM which will allow long-term climate research to be continued in this country.

8.5.1 Future research envisaged in the climate modeling domain

Because the debate over global climate change and its potential impacts is still very much at the forefront of environmental concerns, and because there is a recognition that uncertainties in climate model predictions remain very large, particularly on the regional scales, detailed modeling studies are expected to be a long-term goal in Switzerland. In the context of the United Nations Framework Convention on Climate Change, whose aim is to limit anthropogenic interference to the climate system and to allow climate changes to be sufficiently slow for ecosystems to adapt naturally, climatological information on the regional scale is of the utmost importance. Improved understanding of the regional-scale response to global climate forcing is one way of answering some of the numerous scientific questions raised by Articles the Climate Convention. Furthermore, the Intergovernmental Panel on Climate Change (IPCC), which has been instrumental in bringing the climate change debate to the forefront of public attention worldwide, has underlined that little progress in the understanding of future climate will be possible if regional climate processes, and the interactions between the global scale and the regional scale, are not better understood and modeled.

Future research will continue to focus on mountain regions as one of the many sensitive systems of the globe to climate change. Thorough investigations are being planned to evaluate the performance of present-day, state-of-the-art models in reproducing observed climate characteristics. The Alps will receive the most attention because of the density of research and data in the region, and in order to allow interaction with other groups working in the Swiss research community. However, other regions will be investigated in order to broaden current understanding of the climate system and to expand the Modeling Group's collaborations on the international level.

A number of inter-related research topics have been identified and are currently under way or will begin in the near future. The domains of research envisaged include the following:

- The intercomparison between observations and high-resolution global and regional climate model results, to evaluate model performances in reproducing some of the spatial and temporal variability of climate at different scales; this will include analyses of the new model simulations using the ECHAM4 GCM. The objective here is to determine major weaknesses in model parametric schemes which need to be addressed,
- The amelioration of specific model physical parameterization schemes to improve model performances in simulating present and future climate, with a particular emphasis on radiation, boundary-layer and surface processes in both the global and the regional models,
- Numerical and observational analyses of mechanisms controling temperature at high elevations; in particular, summer maxima seem to have an upper limit modulated by convective processes, while long-term trends indicate a stronger amplitude of warming with height. The causal processes will be investigated and related to the scale of the forcing mechanism; investigations will be made to determine whether these processes are also present in a warmer global climate as predicted for the 21st Century,
- A spin-off from this research will be the compilation of a high-resolution data base on alpine climatology for the specific use of research of this kind in future.

9. Concluding Remarks and Bibliography

9.1 Summary

The material contained in this work has discussed the results of a hierarchy of numerical models, from the small-scale Sommeria-Deardorf ABL model, to the DREAMS mesoscale model and on to the global-scale general circulation climate model, ECHAM4. In all cases, the attention has focused upon the role of atmospheric phenomena, which may be small in physical size relative to the characteristic model grid scale, in the evolution of atmospheric conditions. Many of these phenomena act on a subgrid scale in each type of model considered, but have been shown to have a substantial local, regional or global influence, i.e., the influence of the subgrid-scale processes ultimately feeds into the larger scales, sometimes fundamentally transforming the momentum, heat and moisture structure of the atmosphere. In order to enhance subgrid-scale processes in one model type, the nested model approach may be used whereby a finer grid model is used as a form of "numerical zoom" for a region and process of interest. However, the major drawback of the nested modeling technique is that there are no feedbacks from the smaller scales to the larger ones between models, since the forcing or controls are at the larger model scales only. This was shown to be the case for local atmospheric pollution embedded in mesoscale circulations, or regional atmospheric response to global climate change. While these examples have illustrated the importance of subgrid-scale processes *per se*; it is only because of limitations in computational resources and, to some extent, uncertainties in current understanding of the interactive mechanisms of processes at different scales, that fully-coupled models allowing two-way feedbacks are not yet available for in-depth studies.

The examples which have been chosen here are all relevant, directly or indirectly, to the issues of climate change. The anthropogenic forcing of climate through industry, agriculture and traffic begins at the very small scales and feeds through to the very large scales. The source of most pollution occurs over very small surfaces in the case of industry and traffic, i.e., chimneys and car exhausts, somewhat wider areas in the case of agriculture, and rather large areas in the case of biomass burning. Whaterver the scale of the source, the radiatively-active trace gases and aerosols which are released need to be transported from the surface to the free atmosphere through the ABL, thereby requiring knowledge of the nature of turbulence and stability which can enhance or inhibit pollutant transports. Theoretical models of ABL turbulence, and coupled mesoscale-air pollution models, can help provide a description of the mechanisms of pollution behavior from its source area to either the free atmosphere or back to the surface where it may accumulate, be recycled, or washed out. As the trace gases and aerosols become well mixed into the larger-scale atmosphere, they will force the system through complex radiative mechanisms. The presence of other elements of the climate system, either large-scale such as the oceans and the cryosphere, or much smaller scale such as clouds or portions of the surface, will act to produce a range of positive or negative feedbacks which can only be treated in a fully interactive manner in a three-dimensional GCM. The strength of the radiative forcing will be modulated by subgrid-scale features such as clouds, and a better understanding of the processes can be brought to light by detailed analyses of the type presented in Chapter 4 on the interactions between clouds and radiation. A changing climate will inevitably impact upon the regional scales, where an assessment of impacts requires a higher definition than that provided by a GCM. One option is the methodology described in Chapter 7, through the nesting of a LAM into a GCM. The LAM can itself be linked to smaller-scale modules which could, for example, evaluate the changes in pollution behavior in a particular region under different climatic conditions; this is in a way closing the circle of the hierarchy of models, where the smaller scale is used to determine how pollutants feed into the global system to force climate, and back to the small scales to see how a changed climate may modify the processes partially responsible for changing the climate.

This "full circle" has not really been attempted because of the enormous resources required to undertake numerical simulations from the small to the large scales and back to the small scales. However, the problem remains crucial in that the small to regional scales are those for which accurate climatological data is required for impacts studies and thereby for economic and political decision makers. The problem of climate change scenarios at the regional scale is a major source of concern (e.g., IPCC, 1990; 1996), and may be approached in manners other than numerical modeling. However, numerical modeling remains the only physically-based method of linking atmospheric processes at different scales.

9.2 Closing remarks: the necessity for an interdisciplinary approach

Predictions of future climates are carried out principally by General Circulation Models, which operate at present on spatial scales which are far too crude to be used directly for regional impact assessment studies. Because of the inevitable trade-off between spatial resolution and long time integrations imposed by computer resource limitations, GCMs will have difficulty in resolving the mountain areas of interest to the impact modelers. Indeed, in low-resolution spectral models (T-21 or T-42 scales), mountain chains such as the Alps are not even 'perceived' by such models and the physical influence of mountain barriers is taken into account through parameterizations such as gravity-wave drag. In the face of the scaling problem, a number of methods can be applied to extract information from global models and scale these down to the region of interest. Among the techniques available are mathematical procedures such as harmonic analysis and fractal geometry, statistical downscaling, or the coupling of different types of numerical model. None of the techniques will provide more than a crude scenario for a region of interest, but the refinement of the downscaling methods may reduce the margins of error in future.

Waiting for high-resolution GCM climate simulations in coming years as new generations of supercomputers come onto the market is probably not an optimal solution. Whereas higher resolution has led to marked improvements in forecasting skills for 5 - 10 days, a corresponding improvement for *climatological-scale* problems cannot be expected, because in the climate system scale is not the only dominant feature of the problem to be solved. The feedbacks from other components of climate, in particular the oceans, the cryosphere, and the biosphere, will also play a determining role in the evolution of climate forced by increased GHG concentrations. Despite improvements in recent years in the coupling of different systems to atmospheric GCMs, there are still many uncertainties and unknowns in the simulation of coupled systems. In addition, climate is an inherently chaotic system, so that a measure of unpredictabiltity will always need to be borne in mind when analysing GCM climate statistics, whatever the resolution of the climate model.

Regional climate predictions cannot be solved solely by increased computer power and higher resolution models. They will progress through a better integration of ocean, ice, and biological processes and feedback mechanisms with the atmosphere. This implies major investments in interdisciplinary research, which in turn requires breaking through the traditional barriers of the different disciplines, in order to achieve a convergent approach to research in climate and global change.

The role of the models will not only remain central to improving our undertanding of the control mechanisms on processes ranging from turbulence to climate change, they will inevitably gain from the ameliorations in truly interdisciplinary and integrated approaches to the science of climate change. This is one of the reasons that there have been digressions throughout the text from purely modeling issues to more interdisciplimary topics, in order to highlight the role of climate and atmospheric models in a broader environmental context.

Models will improve not only through better physics and faster computers, but also because the expertise brought by other disciplines may actually provide points of view other than those of atmospheric physicists and climatologists which would help in a better design of models and model-generated data sets for a range of potential end-users. Not only must the modeler face the challenges imposed by his own discipline, he must also accept to adapt his manner of thinking and working to incorporate the findings of other disciplines relevant to the problem at hand.

This is where numerical modeling of the atmosphere and climate will need to be directed in coming years.

9.3 Bibliography

Abegg, B., and Froesch, R., 1994: Climate change and Winter tourism: Impact on transport companies in the Swiss Canton of Graubünden. In Beniston, M. (Ed.): Mountain Environments in Changing Climates, Routledge Publishing Company, London and New York, 328 - 340.

Agee, E. M., 1976: Observational evidence of cell flatness as a function of convective depth and eddy anisotropy. J. Met. Soc. Japan 54, 68 -71.

Agee, E. M., T. S. Chen and K. E. Dowell, 1973: A review of mesoscale cellular-convection. Bull. Am. Meteorol. Soc.54,1004--1012.

Air Ministry, 1962: Weather in the Mediterranean; General Meteorology, 1, Meteorological Office, Her Majesty's Stationery Office, London U.K., 362 pp.

Albrecht, B., 1977: A time-dependent model of the trade-wind boundary layer. Ph.D. thesis, Colorado State University, Fort Collins.

Albrecht, B. A., 1981: Parameterization of Trade-Cumulus Cloud Amounts . J. .Atmos. Sci. 38, 97-105.

Ammann, B., and A. F Lotter, 1989: Late-glacial radiocarbon and palynostratigraphy on the Swiss Plateau , Boreas 18, 109-126.

Arakawa, A. and Schubert, W.H., 1974: Interaction of a cumulus cloud ensemble with the large scale environment, Part I. J. Atmos. Sci., 31: 674-701.

Arpe K, Bengtsson L, Dümenil L, Röckner E., 1994: The hydrological cycle in the ECHAM3 simulations of the atmospheric circulation. In: Desbois M, Desalmand F (eds) Global precipitation and climate change. NATO ASI Series, vol. I 26:361-377.

Asai, T. and I. Nagasuji, 1977: On the preferred mode of cumulus convection in a conditionally unstable atmosphere. J. Met. Soc. Japan 55,151 167.

Asselin, R., 1972. Frequency filter for time integrations. Mon. Weather Rev., 100: 487-490.

Auer, I., and Boehm, R., 1994: Combined temperature-precipitation variations in Austria during the instrumental record. Theor. and Appl. Clim., 49, 161 - 174

Augstein, E. and Wendel, M., 1981: Modelling of the Time-Dependent Atmospheric Tradewind soundary. Layer with Non-Precipilating Cumulus Clouds. Contr. Atm. Phys. 53, 509-538.

Baeriswyl, P.-A., and Rebetez, M., 1995: Répartition spatiale des modifications climatiques dans le domaine alpin. Swiss National Science Foundation, Intermediate Report, Project Nr. 31 - 038271.

Bakan, S., 1982: Open cellular structures during Kon l ur. Hamburger Geophys. Monogr., 57, 63 - 78.

Balling, R. C., 1992: The Heated Debate: Greenhouse predictions versus climate reality. Pacific Research Institute Press, San Francisco, 195 pp.

Bantle, H., 1989: Programmdokumentation Klima-Datenbank am RZ-ETH Zurich, Swiss Meteorological Institute, Zurich.

Barkstrom, B.R., 1984: The Earth Radiation Budget Experiment (ERBE), Bull. Amer. Meteor. Soc., 65, 1170-1185.

Barkstrom, B.R., E.F. Harrison, and R.B. Lee III, 1990: Earth Radiation Budget Experiment. EOS, 71, 297-305.

Barry, R. G., 1992: Mountain Weather and Climate, Second Edition., Routledge Publishing Company, London.

Barry, R.G., 1994: Past and potential future changes in mountain environments, Mountain Environments in Changing Climates, M. Beniston, (ed.), Routledge Publishing Company, London and New York, 3-33.

Bates, G. T., Hostettler, S. W., and Giorgi, F., 1995: Two-year simulation of the Great Lakes region with a coupled modeling system. Mon. Weather Rev., 123, 1505 - 1522

Beljaars ACM, Viterbo P., 1994: The sensitivity of winter evaporation to the formulation of aerodynamic resistance in the ECMWF model. Boundary Layer Meteorol 71: 135-150

Bénard, H., 1901: Les tourbillons cellulaires dans une nappe liquide transportant de la chaleur par convection en régime permanent. Ann. Chim. Phys. 23, 62-144.

Bengtsson, L., 1981: Numerical prediction of atmospheric blocking - a case study. Tellus 33, 19-42.

Bengtsson, L., M. Botzet, and M. Esch, 1995: Hurricane-type vortices in a general circulation model: Part 1. Tellus, 47A, 175-196.

Beniston M., 1978a: Aspects de la turbulence dans un modèle de couche limite en présence de cumulus. J. Rech. Atmos., 12, 203-207.

Beniston M., 1978b: Aspects of non-precipitating cumulus convection in a PLB model of the tropical boundary-layer. Proc. 11th Technical Conference on Hurricanes and Tropical Meteorology, Miami, USA, Dec., 11-16, 1983.

Beniston M., 1978c: Utilisation d'un modèle tridimensionnel pour étudier les cumulus des alizés et la paramétrisation de leurs effets sur le champ turbulent de la couche limite. Thèse de doctorat, Université Pierre et Marie Curie, Paris, France.

Beniston M., 1979: Modèles à petite et à moyenne échelle. Rapport sur les travaux français, 1975-1978. Publication of the International Union of Geodesy and Geophysics, pp. 227-233

Beniston M., 1980: Effets dynamiques et thermodynamiques d'un îlot de chaleur dans un modèle de couche limite à méso-échelle. J. Rech. Atmos., 14, 109-128.

Beniston M., 1982: Mesoscale modeling of cellular convection : proposals for a numerical model. Hamburg. Geophys. Monogr., 57, Appendix (15 pp.).

Beniston M., 1983a: Développement d'un modèle tridimensionnel pour l'étude de la convection cellulaire. La Météorologie, 29, 193-200.

Beniston M., 1983b: Mesoscale model for the study of three-dimensional convective phenomena. Model description and preliminary results from a KonTur simulation. Hamburg Geophys. Monogr., B2, 1-62.

Beniston M., 1984a: A numerical study of atmospheric cellular convection. Dyn. Atmos. Oceans, 8, 223-242.

Beniston M., 1984b: Verification of convective parameterization schemes using small-scale and mesoscale model results and in-situ observations. ECMWF Workshop Publications, European Centre for Medium-range Weather Forecasts, Reading, England, pp. 187-212

Beniston M., 1985 : Organization of convection in response to initial and lower boundary conditions in a mesoscale numerical model. Contr. Atm. Phys., 58, 31-52.

Beniston M., 1986 : The influence of a water surface on mesoscale dynamics as a function of atmospheric stability. Boundary Layer Meteorol., 36, 19-37

Beniston M., 1987a: A numerical study of atmospheric pollution over complex terrain in Switzerland. Boundary Layer Meteorol., 41, pp. 75-96

Beniston M., 1987b: Mesoscale numerical modeling of flow over topography in Switzerland : Application to pollutant transports. Abstract in Proc. IUGG-87 Conference in Vancouver, Canada, Aug. 1987.

Beniston M., 1987c: Overview and synthesis of the International Conference on Energy Transformations and Interactions with Atmospheric Processes (ICEA-87), Bull. Amer. Meteol. Soc., 68, pp. 1429-1430

Beniston M., 1988: Numerical simulations of atmospheric pollution dispersion in an Alpine valley. Supercomputing Rev., 1, pp. 2-3

Beniston M., 1989: Numerical modeling of regional-scale atmospheric flows, with applications to air pollution. Proc. 5th International Symposium on Numerical Methods in Engineering, Springer-Verlag, 1, 627 - 632.

Beniston M., 1990: Utilisation de modèles numériques détaillés pour l'analyse et la prévision locale et régionale de la qualité de l'air. In:"Les modèles d immissions", pp. 127-144, Edited by CerclAir (Société suisse des respondables de l hygiène de l air), Nov. 10, 1989, Ouchy, Switzerland

Beniston M., 1991a: Chapter 5.2 in "Bienne: Climat et Pollution Atmosphérique d'une Ville Suisse", H. Wanner (ed.), Haupt Verlag, Bern/Stuttgart, pp. 197 - 224

Beniston, M., 1993: Prévisions climatiques pour les Alpes: une revue des techniques de régionalisation, La Météorologie, 45, 38-44

Beniston, M. (Editor), 1994: Mountain Environments in Changing Climates. Routledge Publishing Co., London and New York. 492 pp

Beniston, M., 1997: Variations of snow depth and duration in the Swiss Alps over the last 50 years: links to changes in large-scale climatic forcings. Climatic Change, in press.

Beniston M., and G. Sommeria, 1981: Use of a detailed planetary boundary layer model for parametrization purposes. J. Atmos. Sci., 38, 780-797.

Beniston M. and R.A. Pielke (Editors), 1987: Interactions between Energy Transformations and Atmospheric Phenomena: A Survey of Recent Research. D. Reidel Publishing Co., Dordrecht/Boston, 426 pp.

Beniston M., D. Ruffieux, and J.A. Hertig, 1989: A combined numerical and wind-tunnel study of ventlation and air pollution episodes in a rural valley of Switzerland. Boundary Layer Meteorol., 48, 129-156.

Beniston M., and M. Price, 1991: Climate scenarios for the Alpine regions: A collaborative effort between the Swiss National Climate Program and the International Center for Alpine Environments. Env. Conserv., 18, 360 - 363

Beniston, M., and J. Perez-Sanchez, 1992: An example of climate-relevant processes unresolved by present-day General Circulation Models. Env. Conserv., 18, 165-169

Beniston, M., Rebetez, M., Giorgi, F., and Marinucci, M. R., 1994: An analysis of regional climate change in Switzerland. Theor. and Appl. Clim., 49, 135 - 159

Beniston, M., and Schmetz, J., 1995: Thermodynamic and radiative processes in the atmosphere. In: Gyr, A. and Rys, F. S. (eds.), Diffusion and Transport of Pollutants in Atmospheric Mesoscale Flow Fields, Kluwer Academic Publishers, Dordrecht and Boston (Chapter 5), pp. 57 - 88.

Beniston, M., and Rebetez, M., 1995: Regional behavior of minimum temperatures in Switzerland for the period 1979 - 1993. Theor. and Appl. Clim. In Press

Beniston, M., Ohmura, A., Rotach, M., Tschuck, P., Wild, M., and Marinucci, M. R., 1995: Simulation of climate trends over the Alpine Region: Development of a physically-based modeling system for application to regional studies of current and future climate. Final Scientific Report Nr. 4031 - 33250 to the Swiss National Science Foundation, Bern, Switzerland.

Beniston, M., Ohmura, A., Wild, M., Tschuck, P., Marinucci, M., Bengtsson, L., Schlese, U., Esch, M., Giorgi, F., and Bernasconi, A., 1993: Coupled simulations of global and regional climate in Switzerland. Supercomputing Switzerland, 80 - 86

Beniston, M., Fox, D. G., Adhikary, S., Andressen, R., Guisan, A., Holten, J., Innes, J., Maitima, J., Price, M., and Tessier, L., 1996: The Impacts of Climate Change on Mountain Regions. Second Assessment Report of the Intergovernmental Panel on Climate Change (IPCC), Chapter 5, Cambridge University Press (in press)

Berdnikov, V. S. and A. G. Kirdyashkin, 1978: Cellular convection in horizontal fluid layers for different boundary conditions. Izv. Atm. Oc. Phys. 15, 813-817.

Betts AK, Ball JH, Beljaars A.C.M., 1993: Comparison between the land surface response of the ECMWF model and the FIFE-1987 data. Q J R Meteorol Soc 119:975-1001

Betts, A. K., 1973: Non-precipitating cumulus convection and its parameterization. Quart. J. Roy. Meteor. Soc., 99, 178-196.

Betts, A. K., 1975: Parametric interpretation of Trade-Wind cumulus budget studies. J. Atmos. Sci., 32, 1934 - 1945

Betts, A. K., 1976: Modeling subcloud layer structure and interaction with a shallow cumulus layer. J. Atmos. Sci., 33, 2363-2382.

Blackmon, M. L., 1976: A climatological spectral study of the 500 mb geopotential height of the northern hemisphere. J. Atmos. Sci. 33, 1607-1623.

Blackmon, M. L., G. W. Branstator, and G. T. Bates, 1987: An analysis of equatorial pacific sea surface temperature anomaly experiments in general circulation models with and without mountains. J. Atmos. Sci. 44, 1828-1844.

Blondin C., 1991: Parameterization of land-surface processes in numerical weather prediction. In: land surface evaporation: Measurement and parameterization (ed. by TJ Schmugge and JC André). 31-54, Springer

Bluestein, H. B., 1993: Observations and theory of weather systems. Oxford UP, 594pp.

Boden, T. A., Kanciruk, P., and Farrell, M. P., 1990: Trends 90: A compendium of data on global change. Oak Ridge, Tennessee: Carbon Dioxide Information Analysis Center, Environmental Sciences Division, Oak Ridge National Laboratory, USA

Boer, G.J., 1993: Climate change and the regulation of the surface moisture and energy budgets. Climate Dynamics, 8, 225-239.

Bolin, B., 1950: On the influence of the earth's orography on the general character of the westerlies. Tellus 2, 184-195.

Bossert, J.E. Kao, C.E., Roads, J.O.,Xhen, S.C. and Ueyoshi, K.: 1992, 'Regional scale simulations of the western U.S. climate', Proc. Third Symp. on Global Change Studies, Atlanta, Amer. Meteor. Soc., 24-28.

Bradley, E.F., 1968: A Micrometeorological Study of Velocity Profiles and Surface Drag in Ihe Region Modified by a Change of Roughness. Quart. J. Roy. Meteorol. Soc. 94, 361-379.

Brassington, D.J., 1981: Sulfur dioxide absorption cross-section measurement from 290 nm to 317 nm, Appl. Opt., 20, 3774-3779.

Briegleb, B. P., 1992: Delta-Eddington approximation for solar radiation in the NCAR Community Climate Model. J. Geophys. Res., 97, 7603 - 7612

Broccoli, A. J. and S. Manabe, 1992: The effect of orography on midlatitude northern hemisphere dry climates. J. Climate 5, 1181-1201.

Brümmer, B., Fischer, T. and S. Zank, 1985: Aircraft observations of open cellular structures during KonTur. Beitr. Phys. Atmosph., to appear.

Brutsaert WH, 1982: Evaporation into the atmosphere. Reidel Publishing Company, Dordrecht, Holland

Budyko, M. I., 1962: Atlas of the heat balance of the earth. Akademiya Nauk SSSR, Presidium Mezhvedomstvennyi Geofizichesyi Komitet, Moscow.

Budyko, M. I., and Y. Izrael (Eds.), 1982: Anthropogenic Climate Change, Hydrometeoizdat, Leningrad (in Russian)

Burga, C., 1990, Vegetation history and paleoclimatology , Quart. Bull. of the Zürich Nat. Sci. Soc., 135, 17 - 30

Businger, J. A., J. C. Wyngaard, Y. Izumi and E. F. Bradley, 1971: Flux-profile relationships in the atmospheric surface layer. J. Atmos. Sci. 28,181—189.

Cermak, J., 1975: Applications of Fluid Mechanics of Wind Engineering , J. Fl. Eng. 97, 9-38.

Cess, R. D., and 31 co-authors, 1990: Intercomparison and interpretation of climate feedback mechanisms in 19 atmospheric general circulation models. J. Geophys. Res., 95 (D10), 16,601 - 16,615

Cess, R.D., M.H. Zhang, P. Minnis, L. Corsetti, E.G. Dutton, B.W. Forgan, D.P. Garber, W.L. Gates, J.J. Hack, E.F. Harrison, X. Jing, J. T. Kiehl, C.N. Long, J.-J. Mocrette, G.L. Potter, V. Ramanathan, B. Subasilar, C.H. Whitlock, D.F. Young, and Y. Zhou, 1995: Absorption of solar radiation by clouds: observations versus models.Science, 267, 496-499.

Chen, C. and Cotton, W. R., 1983: A One-Dimensional Simulation of the Stratocumulus-Capped Mixed Layer , Boundary-Layer Meteorol. 25, 289-321.

Chen, C.-T., and E. Roeckner, 1995: Validation of the earth radiation budget as simulated by the Max Planck Institute for Meteorology general circulation model ECHAM4 using satellite observations of the Earth Radiation Budget Experiment (ERBE). Max-Planck-Institut for Meteorology Report No. 166. Submitted to J. Geophys. Res.

Chen, C.-T., E. Roeckner, and B.J. Soden, 1995: A comparison of satellite observations and model simulations of column integrated moisture and upper tropospheric humidities. Max-Planck-Institut for Meteorology Report No. 155. Submitted to J. Climate.

Chen, S.-C., Roads, J., Juang, H. H.-M., and Kanamitsu, M., 1994: California precipitation simulation in the NMC nested spectral model. The January 1993 event. Proc. Symposium to Share Weather Pattern Knowledge, June 1994, Rocklin, California, USA.

Chopra, K.P., 1973: Atmospheric wake and oceanic flow problems introduced by islands. Adv. Geophys., 16, 297-421.

Claussen M, Lohmann U, Roeckner E, Schulzweida U , 1994: : A global data set of land-surface parameters. Max Planck Institute for Meteorology Report No. 135

Cliff, A. D., and J. K. Ord, 1981: Spatial Autocorrelation. Pion, London

Copeland, J. H., Chase, T., Baron, J., Kittel, T. G. F., and Pielke, R. A., 1994: Impacts of vegetation change on regional climate and downscaling of GCM output to the regional scale. Proc. 32nd Hanford Symposium on Health and the Environment, October 1993, Richland, Washington, USA

Copeland, J. H., Pielke. R. A., and Kittel, T. G. F., 1995: Potential climatic impacts on vegetation change: A regional modeling study. J. Geophys. Res., in press

Cress, A., Davies, H. C., Frei, C., Lüthi, D., and Schaer, C., 1994: Regional climate simulations in the Alpine region. LAPETH-32, report Atmospheric Sciences ETH, Zürich, Switzerland, 134 pp.

Cubasch, U., G.C. Hegerl, A. Hellbach, H. Höck, U. Mikolajewicz, B.D. Santer, and R. Voss, 1995b: A climate change simulation starting 1935 . Climate Dynamics, 11, 71-84.

Cubasch, U., Hasselmann, K., Hoeck, H., Meier–Reimer, E., Mikolajewicz, U., Santer, B.D. and Sausen, R.: 1992, 'Time–dependent greenhouse warming computations with a coupled ocean-atmosphere model', Climate Dynamics, 8, 55-69.

Cubasch, U., J. Waszkewitz, G.C. Hegerl, and J. Perlwitz, 1995: Regional climate changes as simulated in time-slice experiments. Climatic Change, 31, 273-304.

Davies, H.C. and Turner, R.E.: 1977, 'Updating prediction models by dynamical relaxation: An examination of the technique', Quart. J. Roy. Met. Soc., 103, 225–245.

Davies, R., Ridgway, W. L., and Kim, K.-E., 1984: Spectral Absorption of Solar Radiation in Cloudy Atmospheres: A 20 cm^{-1} Model , J. Atmos. Sci. 41, 2126-2137.

Deardorff, J. W., 1970: A three-dimensional numerical investigation of the idealized planetary boundary-layer. Geophys. Fluid Dyn., 1, 377-410.

Deardorff, J. W., 1972: Numerical investigation fof neutral and unstable planetary boundary layers. J. Atmos. Sci., 29, 91 - 115

Deardorff, J. W., 1973: Three-dimensional modeling of the planetary boundary layer. Workshop on Micrometeorology, A. Haugen, Ed., Amer. Meteor. Soc., 271-311.

Deardorff, J. W., 1980: Stratocumulus-Capped Mixed Layers Derived from a Three-Dimensional Model , Boundary-Layer Meteorol. 18, 495-527.

Deardorff, J. W., G. E. Willis and D. K. Lilly, 1969: Laboratory investigation of non-steady penetrative convection. J. Fluid Mech., 35, 7-32.

Dessens, J. and A. Bücher, 1994: Changes in minimum and maximum temperatures at the Pic du Midi in relation with humidity and cloudiness, 1882-1984. Atmos. Res. (in press)

Deutsches Klimarechenzentrum (DKRZ), 1992: The ECHAM3 atmospheric general circulation model. DKRZ Technical Report, 6; Edited by Modellbetreungsgruppe (DKRZ).

Diaz, H. F., Bradley, R. S., and Eischeid, J. K., 1989: Precipitation fluctuations over global land areas since the late 1800s. J. Geophys. Res., 94, 1195 - 1210

Diaz, H.F. and Markgraf, V. (Eds.), 1992: El Nino, Historical and Paleoclimatic Aspects of the Southern Oscillation. Cambridge University Press, Cambridge, UK., 476 pp.

Dickinson, R. E., A. Henderson-Sellers, and P. J. Kennedy, 1993: Biosphere-Atmosphere Transfer Scheme (BATS) versions 1E as coupled to the NCAR Community Climate Model. NCAR Technical Nore, NCAR/TN-387+STR, 72 pp.

Dickinson, R.E., Enrico, R.M. Giorgi, F. and Bates G.T.: 1989, 'A regional climate model for the western United States', Climatic Change, 15, 383-422.

DKRZ, 1992: The ECHAM3 atmospheric General Circulation Model. Deutsches Klimarechenzentrum DKRZ Technical Report Nr. 6

Dole, R. M. and N. D. Gordon, 1983: Persistent Anomalies of the Extratropical Northern Hemisphere Wintertime Circulation: Geographical Distribution and Regional Persistence Characteristics. Mon. Wea. Rev., 111, 1567-1586.

Dümenil, L., and E. Todini, 1992: A rainfall runoff scheme for use in the Hamburg Climate model. In: advances in numerical hydrology. A tribute of James Dooge (ed. by J.P. O'Kane). European Geophysical Society Series on Hydrological Sciences 1. Elsevier Science Publishers B.V.

Ehinger, J., and M. Beniston, 1993: APSIS Model Intercomparison Experiment. Part A: First results of atmospheric dynamics over the region of Athens with the DREAMS numerical modeling system. Env. Software, 8, 65 - 71

Estoque, M. A., 1961: A Theoretical Investigation of the Sea Breeze , Quart. J. Roy. Meteorol. Soc. 87, 136-146.

Fouquart, Y., and B. Bonnel, 1980: Computations of solar heating of the earth's atmosphere: a new parameterization. Beitr. Phys. Atmos., 53, 35-62.

Fouquart, Y., B. Bonnel, and V. Ramaswamy, 1991: Intercomparing shortwave radiation codes for climate studies. J. Geophys. Res., 96, 8929-8953.

Fraedrich, K., 1976: A mass budget of an ensemble of transient cumulus clouds determined from direct cloud observations. J. Atmos. Sci., 33, 262-268.

Fraedrich, K., and Mueller, K., 1992: Climate anomalies in Europe associated with ENSO events. J Climate, 12, 25 - 31

Fravalo, C., Fouquart, Y., and Rosset, R., 1981: The Sensitivity of a Model of Low Stratiform Clouds to Radiation , J. Atmos. Sci. 38, 1049-1062.

Frei, C.: 1995, 'An Alpine precipitation climatology based on high–resolution rain–gauge observations', MAP newnsletter, 3, 46-47.

Fritsch, J.M., and C.G. Chapell, 1980: Numerical prediction of convectively driven mesoscale pressure systems, Part I: Convective parameterization; J. Atm. Sci., 37, 1722 - 1733

Fürholz B, 1994: Die Berechnung der Wasserhaushaltsbilanz an einem voralpinen Einzugsgebiet mit einem Wasserhaushaltsmodell, Diploma thesis, Institut of Geography ETH Zurich, 63p

Gall H, Burck M, Klitsch S, Zachow B, 1994: Results of lysimeter measurements in Gross Lüsewitz with regard to plant production. Wasser und Boden 46/10:64-68

Garratt JR, Krummel PB, Kowalczyk EA, 1993: The surface energy balance at local and regional scales - a comparison of general circulation model results with observations. J Clim 6:1090-1109

Garratt, J. R., 1994: Incoming shortwave fluxes at the surface - a comparison of GCM results with observations. J. Climate, 7, 72-80.

Garratt, J. R., and A. J. Prata, 1996: Downwelling longwave fluxes at continental surfaces- a comparison with GCM simulations and implications for the global land-surface radiation budget. J. Climate, 7, 72-80.

Gates WL , 1992: AMIP: The atmospheric model intercomparison project. Bull Amer Meteorol Soc 73:1962-1970

Gates, W. L., 1992: AMIP: The atmospheric model intercomparison project. Bull. Amer. Meteor. Soc., 73, 1962-1970.

Geiger, R., 1965: The Climate Near the Ground, Harvard University Press, Cambridge, Massachusetts, 277.

Geleyn, J. F., and H. J. Preuss, 1983: A new data set of satellite-derived surface albedo values for operational use at ECMWF. Arch. Meteor. Geophys. Bioclim. Ser. A, 32, 353-359.

Giorgetta, M., and Wild, M., 1995: The water vapor continuum and its representation in ECHAM4. Max-Planck-Institute for Meteorology Publication Nr. 162, 38 pp.

Giorgi, F., 1990: On the simulation of regional climate using a limited area model nested in a general circulation model. J. Clim., 3, 941-963.

Giorgi, F., 1991: Sensitivity of simulated summertime precipitation over the western United States to different physics parameterizations. Mon. Wea. Rev., 119, 2870 - 2888.

Giorgi, F., and G. T. Bates, 1989: The climatological skill of a regional model over complex terrain. Mon. Wea. Rev., 117, 2325 - 2347

Giorgi, F., and L.O. Mearns, 1991: Approaches to the simulation of regional climate change. Rev. Geoph., 29, 191-216.

Giorgi, F., and M. R. Marinucci, 1991: Validation of a regional atmospheric model over Europe: sensitivity of wintertime and summertime simulations to selected physics parameterizations and lower boundary conditions. Quart. J. Royal Met. Soc., 117, 1171-1206

Giorgi, F., M. R. Marinucci, and G. Visconti, 1990: Application of a limited area model nested in a general circulation model to regional climate simulation over Europe J. Geophys. Res.,95, 18413-18431.

Giorgi, F., M. R. Marinucci, and G. Visconti, 1992: A 2 x CO_2 climate change scenario over Europe generated using a limited area model nested in a general circulation model. II: Climate change scenario. J. Geophys. Res., 97, 10,11 - 10,028.

Giorgi, F., M. R. Marinucci, and G. T. Bates, 1993a: Development of a second-generation regional climate model. Part I: Boundary-layer and radiative transfer processes. Mon. Wea. Rev., 121, 2794 - 2813.

Giorgi, F., M. R. Marinucci, G. T. Bates, and G. De Canio, 1993b: Development of a second-generation regional climate model. Part II: Convection processes and assimilation of lateral boundary conditions. Mon. Wea. Rev., 121, 2814 - 2832.

Giorgi, F., C.S. Brodeur, and G.T. Bates, 1994: Regional climate change scenarios over the United States produced with a nested regional climate model. J. Clim., 7, 375 - 399.

Glowienka-Hense, R., 1990: The North Atlantic Oscillation in the Atlantic-European SLP. Tellus, 42A, 497 - 507

Gregory, J.M. and Mitchell, J.F.B.: 1995, 'Simulation of daily variability of surface temperature and precipitation over Europe in the current 2xCO_2 climates using the UMKO climate model', Quart. J. Roy. Met. Soc., 121, 1451-1476.

Grell, G. A., 1994: A simple and highly efficient cumulus parameterization scheme with a choice of closure assumptions. Mon. Wea. Rev., in press

Grell, G.A., 1994b: A simple and highly efficient cumulus parameterization scheme with a choice of closure assumptions. In preparation.

Gribi, A., 1995: Klimatologische Auswirkungen der Nord-Atlantischen-Oszillation (NAO) auf den Alpenraum. Diploma Thesis, University of Bern, Switzerland

Gruber, R., K. Deckers, M. Beniston, M. Gengler, and S. Merazzi, 1991: Software development strategies for parallel computer architectures. Phys. Reps (Physics Letters), 207, 167 - 214

Gutowski, W. J., D. S. Gutzler, and W. C. Wang, 1991: Surface energy balances of three general circulation models: implications for simulating regional climate change. J. Climate, 4, 121-134.

Gyalistras, D., von Storch, H., Fischlin, A., and Beniston, M., 1994: Linking GCM-Simulated Climatic Changes to Ecosystem Models: Case Studies of Statistical Downscaling in the Alps. Clim. Res., 4, 167 - 189

Haltiner, G. C., and Williams, R. T., 1980: Numerical Predictions and Dynamic Meteorology. Wiley, New York.

Hansen, J. E., 1971: Multiple Scattering of Polarized Light in Planetary Atmosphere: 1. The Doubling Method. J. Atmos. Sci., 28, 120-125.

Hansen, A. R., J. P. Pandolfo, and A. Sutera, 1993: Midtropospheric flow regiomes and persistent wintertime anomalies of surface-layer pressure and temperature. J. Clim, 6, 2136 - 2143

Hasselmann, K., R. Sausen, E. Maier-Reimer, and R. Voss, 1993: On the cold start problem with coupled atmosphere-ocean models. Climate Dynamics, 9, 53-61.

Held, I. M., 1983: Stationary and quasi-stationary eddies in the extratropical troposphere: theory, pp. 127-168. Academic Press.

Helfand, H. M. and E. Kalnay, 1983: A model to determine open or closed cellular convection. J. Atmos. Sci. 40, 631-650.

Henderson-Sellers, A., and K. McGuffie, 1987: A Climate Modeling Primer. J. Wiley, 217 pp.

Henderson-Sellers A, Yang Z-L, Dickinson RE , 1993: The project for intercomparison of land-surface parameterization schemes. Bull Amer Meteor Soc 74, 1335-1349

Hense, A., Kerschgens, M., and Raschke, E., 1982: An Economical Method for Computing the Radiativc Energy Transfer in Circulation Models , Quart. J. Roy. Meteorol. Soc. 108, 231-252.

Hertig, J.-A. and Liska, P., 1983: Simulation of Regional Atmospheric Flows on Small-Scale Topographical Models. Sixth International Conference on Wind Engineering, J. Wind Eng. and Ind. Aerod. 15, 77-91.

Hill, G. E., 1974: Factors controlling the size and spacing of cumulus clouds as revealed by numerical experiments. J. Atmos. Sci., 31, 646-673.

Hinkelmann, K., 1959: A numerical experiment with the primitive equations. In Bolin, B. (ed.): The Atmosphere and the Sea in Motion, Rockefeller University Press, New York.

Hinkley, E.D., 1974: Laser Monitoring of the Atmosphere, Springer-Verlag, New York.

Hirakuchi, H., and Giorgi, F., 1995: Multi-year present-day and 2 x CO_2 simulations of monsoon climate over eastern Asia and Japan with a regional climate model nested in a general circulation model. J. Geophys. Res., in press.

Hoeber, H., 1982: KonTur Field Phase Report. Hamburg Geophys. Monogr., B1, 168 pp.

Holton, J. R., 1972: An Introduction to Dynamic Meteorology. Academic Press, New York

Holtslag, A.A.M., and Boville, B.A.: 1993, Local versus nonlocal boundary-layer diffusion in a global climate model, J. Climate, 10, 1825-1842.

Houghton, J. T., G. J. Jenkins, and J. J. Ephraums (Eds.), 1990: Climate change - The IPCC Scientific Assessment, Cambridge University Press, Cambridge, 365 pp.

Houghton, J. T., Callader, B., and Varney, S. (Eds.), 1992: Climate change - The IPCC Supplement Report. Cambridge University Press, Cambridge, 200 pp.

Hubert, L. F., 1966: Mesoscale cellular convection. Meteorological Satellite Laboratory Report Nr. 33, Washington, D.C., 68 pp.

Hurrell, J. W., 1995: Decadal trends in the North Atlantic Oscillation regional temperatures and precipitation. Science, 269, 676 - 679

Hurrell, J. W., and van Loon, H., 1996: Decadal variations in climate associated with the North Atlantic Oscillation. Climatic Change, submitted

Institut Suisse de Météorologie, 1988: Mesures SODAR, Chablais 8516 (dec. 1987 to jun. 1988), Rapport Interne, Payerne.

IPCC, 1990: Climate Change. The Intergovernmental Panel on Climate Change Scientific Assessment (see Houghton et al., 1990)

IPCC, 1990: Climate Change. The IPCC Scientific Assessment. Houghton et al. (Eds.), Cambridge University Press, Cambridge

IPCC, 1992: Climate Change. The Intergovernmental Panel on Climate Change Scientific Supplement (see Houghton et al., 1992)

IPCC, 1994: Climate Change 1994. The Intergovernmental Panel on Climate Change Report on the Radiative Forcing of Climate. Cambridge University Press.

IPCC, 1996: Climate Change. The IPCC Second Assessment Report. Cambridge University Press, Cambridge and New York (in press)

Ives, J.D., 1992: Preface, The State of the World's Mountains, P. Stone, (ed.), Zed Books, London, xiii-xvi.

Jacqmin, D. and R. S. Lindzen, 1985: The causation and sensitivity of the northern winter planetary waves. J. Atmos. Sci. 42, 724-745.

Jaeger, J., and W. W. Kellogg, 1983: Anomalies in temperature and rainfall during warm Arctic seasons , Climatic Change, 34-60

Jaeger, L., 1976: Monatskarten des Niederschlags fuer die ganze Erde. Ber. d. Dt. Wetterdienstes 139, Offenbach.

Jones, P. D., 1988: Hemispheric surface temperature variations: Recent trends and an update to 1987. Journal of Climate, 1, 654 - 660

Jones, P. D., and Kelly, P. M., 1983: The spatial and temporal characteristics of Northern Hemisphere surface air temperature variations. Journal of Climatology, 3, 243 - 252

Jones, P. D., and Wigley, T. M. L., 1990: Global warming trends. Scientific American, 263, 84 - 91

Jones, P. D., Wigley, T. M. L., and Wright, P. B., 1986: Global temperature variations between 1861 and 1984. Nature, 322, 430 - 434

Jones, P. D., Wigley, T. M. L., Folland, C. K., Parker, D. E., Angell, J. K., 1988: Evidence for global warming in the past decade. Nature, 332, 790

Jones, R.G., Murphy, J.M. and Nogeur, M.: 1995, Simualtion of climate change over Europe using a nested regional-climate model. I: Assessment of control climate, including sensitivity to location of lateral boundaries, Quart. J. Roy. Met. Soc., 121, 1413, 1450.

Kanamitsu, M., and Juang, H.-M. H., 1994: Simulation and analysis of an Indian Monsoon by the NMC nested regional spectral model. Proc. International Conference on Monsson Variability and Prediction, May 1994, Trieste, Italy.

Karl, T. R., Tarpley, J. D., Quayle, R. G., Diaz, H. F., Robinson, D. A., and Bradley, R. S., 1989: The recent climate record: What it can and cannot tell us. Revs. in Geophys., 27, 405 - 430

Karl, T. R., W.C. Wang, M. E. Schlesinger, R. W. Knight , and D. Portman, 1990: A method relating General Circulation Model simulated climate to the observed local climate, Part I: Seasonal statistics. J. Clim., 1053-1079

Karl, T. R., Jones, P. D., Knight, R. W., Kukla, G., Plummer, N., Razuvayev, V., Gallo, K. P., Lindseay, J., Charlson, R. J., and Peterson, T. C., 1993: Asymmetric trends of daily maximum and minimum temperature. Bull. American Meteorol. Soc., 74, 1007 - 1023

Kellogg, W. W., 1982: Precipitation trends on a warmer earth . In: Reck, R. A., and J. R. Hummel (Eds.) Interpretation of climate and photochemical models, ozone and temperature measurements. AIP conference proceedings, New York, 35-45

Kerschgens, M. U., E. Pilz, and E. Raschke, 1978: A modified two stream approximation for computations of the solar radiation budget in a cloudy atmosphere. Tellus, 30, 429-435.

Kessler, E., 1969: On the distribution and continuity of water substance in atmospheric circulations. Meteorol. Monographs, American Meteorological Society, Boston, 82 pp.

Kida, H., Koide, T., Sasaki, H. and Chiba, M.: 1991, A new approach for coupling a limited area model to a GCM for regional climate simulations, J. Meteor. Soc. Japan, 69, 723, 728.

Kim, J. W., J. T. Chang, N. L. Baker , D. S. Wilks, and W. L. Gates, 1984: The statistical problem of climate inversion: determination of the relationship between local and large-scale climate , Mon. Wea. Rev., 2069-2077

Kirchhofer, W., and Sevruk, B., 1992: Mean annual corrections of measured precipitation amounts, 1951 - 1980. In: Spreafico, M. and Weingartner, R. (Eds.): Hydrological Atlas of Switzerland. National Topographical Service, Bern

Klötzli, F., 1984: Neuere Erkenntnisse zur Buchengrenze in Mitteleuropa, Fukarek, Akad. Nauka um jetn. bosne Herc., rad. 72, Odj. Prir. Mat. Nauka 21, P. Festschr, (ed.), Sarajevo, 381-395

Klötzli, F., 1991: Longevity and stress, Modern Ecology: Basic and Applied Aspects, G. Esser, and D. Overdiek, (eds.), Elsevier, Amsterdam, 97-110

Klötzli, F., 1994: Vegetation als Spielball naturgegebener Bauherren, Phytocoenologia, 24, 667-675

Kölsch, H.J., P. Rairoux, J.P. Wolf, and L. Wöste, 1988: New perspectives in remote sensing using excimer-pumped dye lasers and BBO crystals, paper presented at the 14th ILRC, San-Candido, 484.

Kondratyev, K. Ya.: 1969, Radiation in the Atmosphere, Academic Press, New York, London, 912 pp. O Brien, J. J.: 1970, A Note on the Vertical Structure of the Eddy Exchange Coefficient in the Planetary Boundary Layer , J. Atmos. Sci. 27, 1213-1215.

Konzelmann, T., D.R. Cahoon, and C.H. Whitlock, 1996: Impact of biomass burning in Equatorial Africa on the downward surface shortwave irradiance: observations and calculations. Submitted to J. Geophys. Res.

Kraichnan, R., 1967: Inertial ranges in two-dimensional turbulence. Phys. Fluids, 10, 1417-1423.

Krishnamurti, R., 1975: On cellular cloud patterns. Parts 1, 2, 3. J. Atmos. Sci. 32,1353—1383.

Kunz, P., and Rey, J.-M., 1995: Intempéries du 23 septembre 1994 dans le Haut-Valais (Massif du Simplon). Bull. de la Murithienne, 1 - 13

Kutzbach, J. E., 1967: Empirical eigenvectors of sea-level pressure, surface temperature and precipitation complexes over North America. J. Appl. Met., 133, 791-802

Kutzbach, J. F., P. J. Guetter, W. F. Ruddiman, and W. L. Prell, 1989: Sensitivity of climate to late cenozoic uplift in southern asia and the american west: Numerical experiments. J. Geophys. Res. (Atmospheres) 94, 18393-18407.

Lal, M., Jacob, D,m Podzun, R., and Cubasch, U., 1995: Summer monsoon climatology simulated with a regional climate model nested in a general circulation model. Proc. International Workshop on Limited-Area and Variable-Resolution Models, October 1995, Beijing, China.

Lauer, W., 1979: La posicion de los páramos en la estructura del paisaje en los Andes tropicales, El Medio Ambiente Páramo - Actas del Seminario de Mérida, Venezuela, M. L. Salgado-Laboriau, (ed.), Centro de Estudios Avanzados, Caracas, 29-43.

Legates, D.R. and Willmott, C.J.: 1990a, Mean seasonal and spatial variability in gauge-corrected global precipitation, Int. J. Climatol., 10, 111, 127.

Legates, D.R. and Willmott, C.J.: 1990b, Mean seasonal and spatial variability in global surface air temperature, Theor. Appl. Climatol., 41, 11, 21.

Legg, B. J., and Raupach, M. R., 1982: Markov chain simulation of particle dispersion in inhomogeneous flows. Boundary-Layer Meteorol., 24, 3 - 13

Lejenas, H. and H. Økland, 1983: Characteristics of Northern Hemisphere blocking as determined from a long time series of observational data. Tellus 35A, 350-362.

Leung, R. L., and Ghan, S. J., 1995: A subgrid paramterization of orographic precipitation. Theor. and Appl. Clim., in press

Li, Z., and H. Barker, 1994: Solar energy disposition: Intercomparison between satellite estimation, GCM simulation, and surface observations. WCRP Report - IAMAS International Workshop on Clouds-Radiation Interactions and their Parameterization in Climate Models. Camp Springs, MD, Oct. 1993.

Li, Z., and H. G. Leighton, 1993: Global climatologies of solar radiation budgets at the surface and in the atmosphere from 5 years of ERBE data. J. Geophys. Res., 98, 4919-4930.

Li, Z., C. H. Whitlock, and T. P. Charlock, 1994: Assessment of the global monthly mean surface insolation estimated from satellite measurements using Global Energy Balance Archive data. J. Climate, 8, 315-328.

Li, Z., H. Barker, and L. Moreau, 1995: The variable effect of clouds on atmospheric absorption of solar radiation. Nature, 376, 486-490.

Lilly, D. K., 1967: The representation of small-scale turbulence in numerical simulation experiments. Proc. IBM Scientific Computation Symposium. IBM Form No. 320-1951, 195-210

Lilly, D. K., 1969: Numerical simulation of two-dimensionally isotropic turbulence. Phys. Fluids, Suppl II, 240-249.

Lindzen, R. S., 1986: Stationary planetary waves, blocking and interannual variability, Volume 29 of Advances in Geophysics, pp. 251-273. Academic Press.

Liu, Y. F., Giorgi, F., and Washington, W. M., 1994: Simulation of summer monsoon climate over eat Asia with an NCAR regional climate model. Mon. Weather Rev., 122, 2331 - 2348

Lopez, R. E., 1977: The lognormal distribution and cumulus cloud populations. Mon. Wea. Rev., 105, 865-872.

Lorenz, E. N., 1963: Deterministic non-periodic flow. J. Atmos. Sci., 20, 130 - 141

Lorenz, E. N., 1968: Climate determinism. Meteorol. Monogr., 8, AMS Publications, Boston

Lorenz, E. N., 1969: The predictability of a flow which possesses many scales of motion. Tellus, 21, 289 - 307.

Lorenz, E. N., 1982: Atmospheric predictability experiments with a large numerical model. Tellus, 34, 505 - 513.

Lough, J. M., T. M. L. Wigley, and J. P. Palutikof, 1984: Climate and climate impact scenarios for Europe in a warmer world , J. Clim. and Appl. Met., 1673-1684

Louis J.F., 1979: A parametric model of vertical eddy fluxes in the atmosphere. Boundary Layer Meteorol 17:187-202

Louis JF, Tiedke M, Geleyn JF, 1982: A short history of the PBL parameterization at ECMWF; Proceedings, ECMWF workshop on planetary boundary layer parameterization, Reading, 25-27 Nov. 81, 59-80

Lüthi, D., Cress, A., Davies, H. C., Frei, C., and Schaer, C., 1996: Interannual variability and regional climate simulations. Theor. and Appl. Clim., in press

Ma, Q. and R.H. Tipping, 1992: A far wing line shape theory and its application to the foreign-broadened water continuum absorption. J. Chem. Phys., 97, 818-828.

Machenhauer, B., Jacob, D., and Bozert, M., 1994: Using the MPI's nested limited area model for regional climate simulations. WMO/TO-592, 758 - 760

Mahfouf, J.F., D. Cariolle, J.-F. Royer, J.-F. Geleyn, and B. Timbal, 1994: Response of the Météo-France climate model to changes in CO_2 and sea surface temperature. Climate Dynamics, 9, 345-362.

Maier-Reimer, E., U. Mikolajewicz, and K. Hasselmann, 1993: Mean circulation of the Hamburg LSG OGCM and its sensitivity to the thermohaline surface forcing. J. Phys. Oceanography, 23, 731-757.

Manabe, S. and T. B. Terpstra, 1974: The effects of mountains on the general circulation of the atmosphere as identified by numerical experiments. J. Atmos. Sci. 31, 3-42.

Mandelbrot, B. B., 1983: The Fractal Geometry of Nature. W. H. Freeman and Co., New York

Marinucci, M. R.,and F. Giorgi, 1992: A 2 x CO_2 climate change scenario over Europe generated using a limited area model nested in a general circulation model. I: Present day climate simulation. J. Geophys. Res., 97, 9989-10009.

Marinucci, M. R., Giorgi, F., Beniston, M., Wild, M., Tschuck, P., and Bernasconi, A., 1995: High resolution simulations of January and July climate over the Western Alpine region with a nested regional modeling system. Theor. and Appl. Clim., 51, 119 - 138

Martin, C.L. and Pielke, R.A., 1983: The adequacy of the hydrostatic assumption in sea breeze modeling over flat terrain. J. Atmos. Sci., 40: 1472-1481.

May, W., 1994: On the intra-seasonal variability within the extratropics in the ECHAM3 general circulation model Technical Report 147, Max-Planck-Institut fur Meteorologie.

McBean, G. A., 1982: Priorities in Boundary-Layer Research - Thoughts from a Working Group of the WMO , Boundary-Layer Meteorol., 23, 125-128.

McClatchey, R. A., Bolle, H. J., and Kondratyev, K. Ya., 1980: A Cloudless Standard Atmosphere for Radiation Computations. Report of the Radiation Commission of the IAMAP.

McFarlane, N. A., 1987: The effect of orographically excited gravity wave drag on the general circulation of the lower stratosphere and troposphere. J. Atmos. Sci. 44, 1775-1800.

McGregor, J.L., and K. Walsh, 1993: Nested simulations of perpetual January climate over the Australian region. J. Geophys. Res., 98, 23283 - 23290

McGregor, J.L., and K. Walsh, 1994: Climate change simulations of Tasmanian precipitation using multiple nesting. J. Geophys. Res., 99, 20889 - 20905

McNider, R. T., 1981: Investigation of the impact of topographic circulations on the transport and dispersion of air pollutants. PhD Dissertation, University of Virginia, Charlotteville

Mearns, L.O., 1994: Methodological issues concerning the determination of the effetcs of changing climatic variability on crop yields. In: American Society of Agronomy Special Pubblication: Climate Change and International Impact, ASA, Madison, Wi. (submitted)

Mearns, L.O., Giorgi, F., McDaniel, L., and C.S. Brodeur, 1994: Analysis of daily variability of precipitation in a nested regional climate model: comparison with observations and doubled CO2 results. Global and Planetary Change, in press.

Mearns, L. O., Katz, R. W., and Schneider, S.H., 1984: Extreme high-temperature events: Changes in their probabilities with changes in mean temperature. J Clim. and Appl. Meteorol., 23, 1601 - 1613

Mearns, L. O., S. K. Schneider, S. L. Thompson, and L. R. McDaniel, 1990: Analysis of climate variability in general circulation models: comparison with observations and changes in variability in 2XCO2. J. Geophys. Res., 95, 20469 - 2-490.

Measures, R.M., 1988: Laser Remote Chemical Analysis, JohnWiley, New York.

Menyuk, N., and D. K. Killinger, 1981: Temporal correlation measurements of pulsed dual CO_2 lidar returns, Opt. Lett., 6, 301.

Menyuk, N., D. K Killinger, and C. R. Menyuk, 1982: Limitations of signal averaging due to temporal correlation in laser remote sensing measurements, Appl. Opt., 21, 3377.

Menyuk, N., D. K Killinger, and C. R. Menyuk, 1985: Error reduction in laser remote sensing: combined effects of cross-correlation and signal averaging, Appl. Opt. , 24, 118.

Menzel L, 1991: Water budget studies in the river Thur basin (Eastern Switzerland). Analysis of hydrological field measurements 1976 - 1985. Berichte und Skripten 46, Geography Institute ETH Zurich, Switzerland, 206p

Menzel L, 1995: Soil moisture measurements with time domain reflectometry. Berichte und Skripten 55, Geography Institute ETH Zurich, Switzerland, 68p

Mesinger, F., and Arakawa, A., 1976: Numerical methods used in atmospheric models. GARP Publications Series Number 17, 1, 1-64.

Mesinger, F., and R.T. Pierrehumbert, 1986: Alpine lee cyclogenesis: Numerical simulation and theory. In Scientific Results of the Alpine Experiment (ALPEX), Volume I, GARP Publication Series, No. 27, 141-163.

Mitchell, J. F. B., S. Manabe, V. Meleshko, and T. Tokioka, 1990: Equilibrium climate change - and its implications for the future , In: Houghton, J. T., G. J.Jenkins, and J. J. Ephraums (Eds.), Climate change - The IPCC Scientific Assessment, Cambridge University Press, Cambridge, pp. 139-173

Monasterio, M., 1980a: Los páramos andinos como region natural: caracteristicas biogeográficas generales y afinidades con otras regiones andinas, Estudios Ecologicos en los Páramos Andinos, M. Monasterio, (ed.), Ediciones de la Universidad de Los Andes, Mérida, Venezuela, 15-27.

Morcrette, J.J., 1991: Radiation and cloud radiative properties in the European centre for medium range weather forecasts forecasting system. J. Geophys. Res. , 96, 9121-9132.

Morcrette, J.J., L. Smith, and Y. Fouquart, 1986: Pressure and temperature dependence of the absorption in longwave radiation parameterizations, Beitr. Phys. Atmos., 59, 455-469.

Mullen, S. L., 1989: The impact of orography on blocking frequency in a general circulation model. J. Climate 2, 1554-1560.

Mullen, S. L., 1994: The impact of an envelope orography on low-frequency variability and blocking in a low-resolution general circulation model. J. Climate 7, 1815-1826.

Müller, H., and G. Kappenberger, 1991: Claridenfirn-Messungen, 1914 - 1984. Zürcher Geographische Schriften, 40, 79 pp.

Münch A., 1993: AKWA-M Programdocumentation version 2/93, Institute of Hydrology and Meteorology, TU Dresden, Germany

Nicolis, G., and Prigogine, I., 1989: Exploring Complexity. Freeman Co., New York

Nigam, S., I. M. Held, and S. W. Lyons, 1988: Linear simulations of the stationary eddies in a GCM. part II: The mountain model. J. Atmos. Sci. 45, 1433-1452.

Nordeng T.E., 1996: Extended versions of the convective parameterization scheme at ECMWF and their impact on the mean and transient activity of the model in the tropics. Q J R Meteorol Soc, in press.

O'Brien, J. J., 1970: A note on the vertical structure of the eddy exchange coefficient in the planetary boundary layer. J. Atmos. Sci. 27,1213-1215.

O'Neill, R. V., 1988: Hierarchy theory and global change , In Rosswall, T., R. G.Woodmansee, and P. G. Risser (Eds.), Scales and Global Change, J. Wiley and Sons, London

Ohmura A., Gilgen H.,1991: Global Energy Balance Archive GEBA. Report 2: The GEBA database: Interactive applications, retrieving data. Zürcher Geografische Schriften Nr. 44, Zürich, 60 p

Ohmura, A., and H. Gilgen, 1992: Global Energy Balance Archive (GEBA, WCP-water A7) and new aspects of the global radiation distribution on the earth's surface. In: IRS '92, Current problems in atmospheric radiation (ed. by Keevallik and Kärner). DEEPAK Publishing, 271-274.

Ohmura A, Gilgen H, and Wild M., 1989: Global Energy Balance Archive GEBA. Report 1: Introduction. Zürcher Geografische Schriften Nr. 34, Zürich, 62p

Ohmura A, Konzelmann T, Rotach M, Forrer J, Wild M, Abe-Ouchi A, Toritani H., 1994: Energy balance for the Greenland ice sheet by observation and model computation. In: Snow and Ice Covers: Interactions with the Atmosphere and Ecosystems. (ed. by HG Jones, TD Davies, A Ohmura and EM Morris). IAHS Publication No. 223

Ohmura A, Müller G, Schroff K, Konzelmann T., 1990: Radiation annual report ETH No.1 1987 and 1989. Zürcher Geografische Schriften No. 39, Zürich, 305p

Ohmura, A., M. Wild, and L. Bengtsson, 1996: A possible change in mass balance of Greenland and Antarctic ice sheets in the coming century. Accepted in J. Climate.

Oki, T., A. Mushiake, K. Masuda, and H. Matsuyama, 1993: Global runoff estimation by atmospheric water balance using ECMWF data-set. IAHS Publ. No. 214, 163-167.

Orlanski, J., 1975: A rational subdivision of scales for atmospheric processes. Bull. Amer. Meteorol. Soc., 56, 527 - 530

Ozenda, P., 1985: La Végétation de la Chaine Alpine dans l'Espace Montagnard Européen, Masson, Paris, 344.

Palmer, T. N., G. J. Shutts, and R. Swinbank, 1986: Alleviation of a systematic westerly bias in general circulation and numerical weather prediction models through an orographic gravity wave drag parametrization. Quart. J. R. Met. Soc. 112, 1001-1039.

Paltridge, G., 1973: Direct Measurements of Water Vapor Absorption of Solar Radiation in the Free Atmosphere. J. Atmos. Sci. 30, 156-160.

Palutikof, J. P., T. M. L. Wigley, and J. M. Lough, 1984: Seasonal climatic scenarios for Europe and North America in a high CO2, warmer world , Climatic Research Unit Publication, University of East Anglia, Norwich. NTIS, Springfield, 70pp.

Pasquill, F., 1962: Atmospheric Diffusion. Van Nostrand Rheinhold, New York

Perkey, D. J., and Kreitzberg, C. W., 1976: A time-dependent lateral boundary scheme for limited area primitive equation models. mon. Weather. Rev., 104, 744 - 755

Peterson, E. W., 1969: Modification of Mean Flow and Turbulent Energy by a Change in Surface Roughness Under Conditions of Neutral Stability. Quart. J. Roy. Meteorol. Soc., 95, 561-575.

Pfister, C., 1995: Scientific report to the Swiss National Science Foundation, National Program on Climate and Natural Hazards, Bern, Switzerland

Phillips, N. A., 1957: A coordinate system having some special advantages for numerical forecasting. J. Meteorol., 14, 184 - 185

Phillips, N. A., 1959: Numerical integration of the primitive equations on the hemisphere. Mon. Weather Rev., 87, 109 - 120

Pielke, R. A., 1974: A three-dimensional numerical model of the sea breezes over south Florida. Mon. Weather Rev., 102, 115 - 139

Pielke, R.A., 1982: Mesoscale atmospheric circulations related to the coast—a summary. Proc. First Int. Conf. Meteorol. and Air/Sea Interaction of the Coastal Zone, pp. 154-161.

Pielke, R. A., 1984: Mesoscale Meteorological Modeling. Academic Press, New York

Pielke, R. A., 1988: Evaluation of climate change using numerical models. In "Monitoring Climate for the Effects of Increasing Greenhouse Gas Concentrations". R. A. Pielke and T. Kittel (Eds.), CIRA Publications, Ft. Collins, Colorado, pp. 161-172

Pielke, R. A., 1990: Overlooked scientific issues in assessing hypothesized Greenhouse Gas warming. Electric Utility Business Environment Conference and Exhibition, March 1990, Denver, Colorado.

Pittock, A. B., and M. J. Salinger, 1982: Towards regional scenarios for a CO2-warmed earth , Climatic Change, 23-40

Plank, V. G., 1969: The size distribution of cumulus clouds in representative Florida populations. J. App/. Meteor., 8, 46-67.

Platt, C. M. R., 1976: Infrared Absorption and Liquid Water Content in Stratocumulus Clouds , Quart. J. Roy. Meteoro/. Soc. 102, 553-562.

Podzun, R., Cress, A., Majewski, D., and Renner, V., 1995: Simulation of European climate with a limited area model. Part II: AGCM boundary conditions. Contr. Atmos. Phys., submitted.

Poreh, M. and Kacherginsky. A., 1981: Simulation of Plume Rise using Small Wind-tunnel Models , J. Wind Eng. and Ind. Aerod. 7(1), 1-15.

Quezel, P., and J. Schevok, 1981: Essai de mise en place de la zonation altitudinale des structures forestières, Ecologia Medit., 8, 389-410.

Quezel, P., and M. Barbero, 1990: Les forêts méditerranéennes: problèmes posés par leur signification historique, écologique et leur conservation, Acta Botanica Malacitana, 15, 145-178.

Ramanathan, V., B. Subasilar, G. Zhang, W. Conant, R. Cess, J. Kiehl, H. Grassl, and L. Shi, 1994: Warm pool heat budget and shortwave cloud forcing: a missing physics? Science, 267, 499-503.

Rameau, J.C., D. Mansion, G. Dumé, A. Lecointe, J. Timbal, P. Dupont, and R. Keller, 1993: Flore Forestière Française, Guide Ecologique Illustré, Lavoisier TEC and DOC Diffusion, Paris, 2419.

Randall, D. A., R. D. Cess, J. P. Blanchet, G. J. Boer, D. A. Dazlich, A. D. Del Genio, M. Deque, V. Dymnikov, V. Galin, S. J. Ghan, A. A. Lacis, H. Le Treut, Z. X. Li, X. Z. Liang, B. J. McAveney, V. P. Meleshko, J. F. B. Mitchell, J. J. Mocrette, G. L. Potter, L. Riskus, E. Roeckner, J. F. Royer, U. Schlese, D. A. Sheinin, J. Slingo, A. P. Sokolov, K. E. Taylor, W. M. Washington, R. T. Wetherald, I. Yagai, and M. H. Zhang, 1992: Intercomparison and interpretation of surface energy fluxes in atmospheric general circulation models. J. Geophys. Res., 97, 3711-3725.

Rayleigh, O. M., 1916: On convection currents in a horizontal layer of fluid when the higher temperature is on the under side. Sci. Papers, Cambridge Univ. Press 6, 432-443.

Rebetez, M., 1988: Note concemant les resultats des mesures d immission SO_2, Broye 1987, Technical Report, Laboratoire Cantonal Vaudois, Epalinges.

Rebetez, M., 1994: Perception du temps et du climat: une analyse du climat de Suisse Romande sur la base des dictons populaires. PhD Thesis, University of Lausanne. Editions Stratus, Oron-la-Ville, 304 pp.

Rebetez, M., 1996: Seasonal relationships between temperature, precipitation and snow cover in Switzerland. Theor. and Appl. Clim., in press

Rebetez, M., Lugon, R., and Baeriswyl,. P.A., 1996: Climatic warming and debris flows in high mountain regions: The case study of Ritigrabe, Switzerland. Climatic Change, Submitted

Rex, D. F., 1950: Blocking action in the middle troposphere and its effect upon regional climate. Tellus, 2, 196-211.

Richardson, L. F., 1922: Weather prediction by numerical processes. Cambridge University Press, London.

Richtmeyer, K. D., and Morton, K. W., 1967: Difference Methods for Initial-Value Problems. Interscience Press, New York

Rind, D., Goldberg, R., and Ruedy, R., 1989: Change in climate variability in the 21st century. Climatic Change, 14, 5 - 37

Roberts, R.E., J.E.A. Selby, and L.M. Biberman, 1976: Infrared continuum absorption by atmospheric water vapor in the 8 - 12 μm Window. Appl. Opt., 15, 2085-2090.

Rockel, B., E. Raschke and B. Weyres, 1991: A parameterization of broad band radiative transfer properties of water, ice and mixed clouds. Beitr. Phys. Atmos., 64, 1-12.

Rodgers, C. D., 1967: The Use of Emissivity in Atmospheric Radiation Circulations. Quart. J. Roy. Meteorol. Soc., 93, 43-54.

Rodgers, J. C., 1984: The association between the North Atlantic Oscillation and the Southern Oscillation in the nothern hemisphere. Mon. Weather Rev., 112, 1999 - 2015

Rodgers, J. C., and van Loon, H., 1979: The seesaw in winter temperatures between Greenland and Northern Europe. Part II: Some oceanic and amospheric effects in middle and high latitudes. Mon. Weather Rev., 107, 509 - 519

Roeckner, E., M. Rieland, and E. Keup, 1991: Modeling of cloud and radiation in the ECHAM model: ECMWF/WCRP Workshop on "clouds, radiative transfer and the hydrological cycle". ECMWF, Reading, pp 199-222.

Roeckner, E., K. Arpe, L. Bengtsson, S. Brinkop, L. Dümenil, M. Esch, E. Kirk, F. Lunkeit, M. Ponater, B. Rockel, R. Sausen, U. Schlese, S. Schubert, and M. Windelband, 1992: Simulation of the present day climate with the ECHAM model: impact of model physics and resolution. Max Planck Institute for Meteorology Report No. 93, 171 pp.

Roeckner E, Arpe K, Bengtsson L, Christoph M, Claussen M, Dümenil L, Esch M, Giorgetta M, Schlese U, Schulzweida U., 1995: The Max Planck Institute for Meteorology fourth-generation atmospheric general circulation model ECHAM4: Model description and climatology, In preparation

Rohrer, M., 1992: Die Schneedecke im Schweizer Alpenraum und ihre Modellierung. Zürcher Geographische Schriften, 49, 178 pp.

Rossow, W. B. and L. C. Garder, 1993: Validation of ISCCP cloud detections. J. Climate, 6, 2370-2393.

Rossow, W.B., and Y.-C. Zhang, 1995: Calculation of surface and top of atmosphere radiative fluxes from physical quantities based on ISCCP data sets. Part II: Validation and first results. J. Geophys. Res., 100 (D1), 1167 - 1197.

Rotach, M., R. Marinucci, M. Wild, P. Tschuck, and M. Beniston, 1997: Nested regional simulation of climate change over the Alps for the scenario of a doubled greenhouse forcing. Theor. and Appl. Clim., in press.

Rothman, L.S., R.R. Gamache, A. Barbe, A. Goldman, J.R. Gillis, L.R. Brown, R.A. Toth, J.M. Flaud, and C. Camy-Peyret, 1983: AFGL atmospheric absorption line parameters compilation: 1982 Edition. Appl. Opt., 15, 2247-2256.

Ruddiman, W. F. and J. F. Kutzbach, 1989: Forcing of late cenozoic northern hemisphere climate by plateau uplift in southern asia and the american west. J. Geophys. Res. (Atmospheres) 94, 18409-18427.

Saltzmann, B., (Ed.), 1983: Theory of Climate. Advances in Geophysics, Vol. 25. Academic Press, New York, 505 pp.

Santer, B. D., T. M. L. Wigley, M. E. Schlesinger, and J. F. B. Mitchell, 1990: Developing climate scenarios from equilibrium GCM-results , Max-Planck-Institut für Meteorologie Report No. 47, Hamburg, 29 pp.

Sasaki, H., Kida, H., Koide, T., and Chiba, M., 1995: The performance of long-term integrations of a limited area model with the spectral boundary coupling method. J. Meteorol. Soc. Japan, in press.

Sausen, R., W. Konig, and F. Sielmann, 1995: Analysis of blocking events from observations and ECHAM model simulations. Tellus 47A, 421-438.

Schaer, C., Frei, C., Lüthi, C., and Davies, H.C., 1996: Surrogate climate change scenarios for regional climate models. Geophys. Res. Letters, in press

Schaller, E. and Kraus, H., 1981: The Role of Radiation in an Inversion-Capped Boundary Layer. Part 1: The Need for a Detailed Consideration of Radiative Processes , Boundary-Layer Meteorol. 20, 485-495.

Schmetz, J., 1981: Strahlungseigenschaften von Wolkenfeldern, Mitteilungen aus dem Institut fur Geophysik und Meteorologie, Universität Köln, Nr. 30.

Schmetz , J. and Raschke, E., 1981: An Approximate Computation of Infrared Radiative Fluxes in a Cloudy Atmosphere, Pure Appl. Geophys. 119, 248-258.

Schmetz J. and M. Beniston, 1986 : Relative effects of solar and infrared radiative forcing in a mesoscale model. Boundary Layer Meteorol., 34, 137-155

Schmetz, J., 1984: On the Parameterization of the Radiative Properties of Broken Clouds , Tellus 36A, 417-432.

Schmetz, J., Raschke, E., and Fimpel, H., 1981: Solar and Thermal Radiation in Maritime Stratocumulus Clouds. Contr. Atm. Phys., 54, 442-452.

Schorling, M., 1987: Lagrangian simulation of neutral and heavy gases in complex terrain and streets. Proc. International Conference on Energy Transformations and Interactions with Atmospheric Phenomena, Lausanne, Switzerland, March 6 - 10, 1987

Schüepp, M., and H. Schirmer, 1977, Climates of Central Europe , In: Wallén, C.C. (Ed.), Climates of Central and Southern Europe. World Survey of Climatology, Elsevier, Amsterdam-Oxford-New York, 3-74

Selby, I.E.A., and R.E. McClatchey, 1975: Atmospheric transmittance from 0.25 - 28.5 μm, computer code LOWTRAN 3, AFCLR-TR-75-0255.

Sellers PJ, Mintz Y, Sud YC, Dalcher A., 1986: A simple biosphere model (Sib) for use within general circulation models: J Atm Sci 43:505-531

Semazzi, F. H. M., Lin, N.-H., Lin, Y.-L., and Giorgi, F., 1994: A nested modeling study of the Sahelian climate response to sea-surface temperature anomalies. Geophys. Res. Letters, 20, 2897 - 2900.

Sevruk, B., 1975: Inaccuracy of precipitation measuraments. A serious problem of waterresources instrumentation. In: Proc. Second World Congress, International Water Resources Association, New Delhi, Vol. III, 429 - 440

Sevruk, B., 1989: Reliability of precipitation gradient estimates. In: Proc. XIV Internat. Conf. on Carpathian Meteorology, Sofia, 402 - 408.

Sevruk, B., Kirchhofer, W., Tihlarik, R., and Zahlavova, L., 1993: Precipitation corrections in Switzerland. Proc. Symp. on Precipitation and Evaporation, Bratislava, Slovakia, 155 - 156

Shettle, E.P., and R. Fenn, 1976: Models of the atmospheric aerosols and their optical properties. AGARD Conference Proceedings No. 183, AGARD-CP-183.

Sheu, P. J. and E. M. Agee, 1977: Kinematic analysis and air-sea heat flux associated with mesoscale cellular convection during AMTEX-75. J. Atmos. Sci. 34, 793-801.

Shuttleworth WJ, and Dickinson RE, 1989: Comments on "modelling tropical deforestation: a study of GCM land-surface parameterizations" by R.E. Dickinson and A. Henderson-Sellers. Q J R Meteorol Soc 115:1177-1179

Shutts, G. J., 1983: The propagation of eddies in diffluent jetstreams: eddy vorticity forcing of `blocking' flow fields. Quart. J. R. Met. Soc. 109, 737-761.

Slingo, A. and Schrecker, H. M., 1982: On the shortwave Radiative Properties of Stratiform Water Clouds , Quart. J. Roy. Meteorol., 108, 407-426.

Smagorinsky, J., 1962: A primitive equation model including condensation processes. Proc. International Symposium on Numerical Weather Prediction, Tokyo, November 1960. Meteorol. Soc. Japan.

Smagorinsky, J., 1967: Global atmospheric modeling and the numerical simulation of climate. In Hess, W. N. (Ed.): Weather and Climate Modification, J. Wiley, New York

Smith, R. B., 1979: The Influence of Mountains on the Atmosphere, Volume 21 of Advances in Geophysics, pp. 87-230. Academic Press.

Sommeria, G., 1976: Three-dimensional simulation of turbulent processes in an undisturbed Trade-Wind boundary layer. J. Atmos. Sci., 33, 216 - 241

Sommeria, G., and J. W. Deardorff, 1977: Subgrid-scale condensation in cloud models. J. Atmos. Sci., 34, 344-355.

Sommeria, G., and M. A. LeMone, 1978: Direct testing of a three-dimensional model of the planetary boundary layer against experimental data. J. Atmos. Sci., 35, 25-39.

Spencer, R.W., 1993: Global oceanic precipitation from MSU during 1979-1991 and comparisons to other climatologies. J. Climate, 6, 1301-1326.

Stearns, L. P., 1983: Infrared Cooling in Cloudy Atmospheres: Precision of Grid Section for Numerical Models. Monthly Weather Rev., 111, 1501-1505.

Stephens, G. L., 1978: Radiation profiles in extended water clouds: 2. parameterization schemes. J. Atm. Sci., 35, 2123-2132.

Stephens, G. L. and Webster, P. J.: 1979, Sensitivity of Radiative Forcing for Variable Cloud and Moisture , J. Atmos. Sci. 36, 1542-1556.

Stephens, G. L. and Wilson, K. J.: 1980, The Response of a Deep Cumulus Convection Model to Changes in Radiative Heating , J. Atmos. Sci. 37, 421-434.

Taylor, P. A., 1970: A Model of Air Flow Above Changes in Surface Heat Flux, Temperature, and Roughness for Neutral and Unstable Conditions , Boundary-Layer Meteorol. 1, 18-39. Townsend, A. A.: 1965, The Flow in a Turbulent Boundary Layer After a Change in Surface Roughness , J. Fluid Mech. 26, 255-266.

Tibaldi, S., P. Ruti, E. Tosi, and M. Maruca, 1995: Operational predictability of winter blocking at ECMWF: an update. Ann. Geophysicae 13, 305-317.

Tiedke, M., 1989: A comprehensive mass flux scheme for cumulus parameterization in large-scale models. Mon. Wea. Rev., 117, 1779 - 1800

Timbal, B., J.-F. Mahfouf, J.-F. Royer, and D. Cariolle, 1995: Sensitivity to prescribed changes in sea surface temperature and sea ice in doubled carbon dioxide experiments. Climate Dynamics, 12, 1-20

Tokioka, T. and A. Noda, 1986: Effects of large-scale orography on January atmospheric circulation: A numerical experiment. J. Meteor. Soc. Japan 64, 819-839.

Trenberth, K. E., 1990: Recent observed interdecadal climate changes in the Northern Hemisphere. Bull. Amer. Meteorol. Soc., 71, 988 - 993

Troll, C., 1968: The cordilleras of the tropical Americas. Aspects of climatic, phytogeographical and agrarian ecology, Geoecology of the Mountainous Regions of the Tropical Americas, C. Troll, (ed.), Ferd. Dümmlers Verlag, Bonn, 15-56.

Twomey, S., 1976: Computations of the Absorption of Solar Radiation by Clouds. J. Atmos. Sci. 33, 1087-1091.

Veyre, Ph., Sommeria, G., and Fouquart, Y., 1982: Attempts to Directly Simulate Cloud-Radiation Interaction in the Case of Small Cumuli , in E. M. Agee and T. Asai (eds.), Cloud Dynamics. D. Reidcl Publ. Co., Dordrecht, Holland, pp 135-147.

Viterbo P, Beljaars C. M., 1995: An improved land surface parameterization scheme in the ECMWF model and its validation. J Clim (in press)

Von Storch, H., E. Zorita, and U. Cubasch, 1991: Downscaling of global climate change estimates to regional scales: an application to Iberian rainfall in Wintertime , Max-Planck Institut für Meteorologie Report No. 64, Hamburg, 39pp.

Wallace, J. M., and Y. Zhang, 1993: Structure and seasonality of interannual and interdecadal variability of the geopotential height and temperature fields in the Northern Hemisphere troposphere. J. Clim., 6, 2063 - 2082

Wallace, J. M., S. Tibaldi, and A. J. Simmons, 1983: Reduction of systematic forecast errors in the ECMWF model through the introduction of an envelope orography. Quart. J. R. Met. Soc. 109, 683-717.

Wallen, C.C. (Ed.), 1970a: Climates of Northern and Western Europe; World Survey of Climatology, 5, Elsevier Scientific Publishing Company, New York, 253 pp.

Wallen, C.C. (Ed.), 1970b: Climates of Central and Southern Europe; World Survey of Climatology, 6, Elsevier Scientific Publishing Company, New York, 248 pp.

Walsh, K., and McGregor, J. L., 1995: January and July climate simulations over the Australian region using a limited area model. J. Clim., in press

Wanner, H., and J.A. Hertig, 1984: Studies of urban climates and air pollution in Switzerland. J. Cl. Appl. Met., 23, 1614-1625.

Warner, J., 1977: Time Variation of Updraft and Water-Content in Small Cumulus Clouds , J. Atmos. Sci. 34, 1306-1313.

Warren, S.G., C.J. Hahn, J. London, R.M. Chervin, and R. L. Jenne, 1986: Global Distribution of Total Cloud Cover and Cloud Type Amounts over Land. NCAR Technical Note, NCAR/TN-273+STR, Boulder, CO, 29 pp.

Warrick, R.A. and Barrow, E.M., 1991: Climate change scenarios for the U.K. Transactions. Institute of British Geographers vol 16, no 4, p.387-99.

WCRP, 1991: Radiation and Climate, Second workshop of the baseline surface radiation network, WCRP-64, Geneva.

Wegmüller, S., and A. F. Lotter, 1990: Palynostratigraphic investigations of the late-glacial and Holocene vegetation history of the northwestern calcareous Swiss Prealps , Botan. Helv., 100, 37 - 73.

Wigley, T.M.L., 1985: Impact of extreme events. Nature, 316, 106 - 107

Widmann, M., and Schaer, C., 1996: A principal component and long-term trend analysis of daily precipitation in Switzerland. Int. J. of Climatology, Submitted.

Wigley, T. M. L., and Jones, P. D., 1981: Detecting CO_2-induced climate change. Nature, 292, 205 - 208.

Wigley, T. M. L., and Raper, S. C. B., 1990: Natural variability of the climate system and detection of the greenhouse effect. Nature, 344, 324 - 327.

Wigley, T. M. L., P. D. Jones, and P. M. Kelly, 1980: Scenario for a warm, high CO2-world , Nature, 283, 17-21.

Wigley, T. M. L., P. D. Jones, K. R. Briffa, and G. Smith, 1990: Obtaining sub-grid scale information from coarse-resolution general circulation model output. J. Geophys. Res., 1943-1953.

Wild, M., A. Ohmura, H. Gilgen, and E. Roeckner, 1995a: Validation of General Circulation Model simulated surface radiation fluxes using surface observations. J. Clim., 8, 1309 - 1324.

Wild, M., A. Ohmura, H. Gilgen, and E. Roeckner, 1995b: Analysis of GCM simulated surface fluxes over Europe using ground-based observations. Clim. Dyn. In press.

Wild M et al., 1995c: Observational support for revised values of surface and atmospheric radiation budgets in GCMs, in preparation

Wild, M., L. Dümenil, and J.P. Schulz, 1995d: High resolution GCM simulations over Europe: surface processes. Max-Planck Institute for Meteorology Report Nr. 176.

Wild, M., Dümenil, L. and Schulz, J.P.: 1996a, Regional climate simulation with a high resolution GCM: surface hydrology, Climate Dynamics, in press.

Wild, M. , Roeckner, E., Ohmuara, A., and Gilgen, H., 1997: Observational support for revised surface and atmospheric radiation budgets in the ECHAM4 GCM, J. Clim., accepted.

Wilks, D. S., 1989: Statistical specification of local surface weather elements from large-scale information , Theor. Appl. Climatol., 119-134.

Williams, J., 1980, Anomalies in temperature and rainfall during warm arctic seasons as a guide to the formulation of climate scenarios , Climatic Change, 249-266.

Wilson, C. A. and J. F. B. Mitchell, 1987: Simulated climate and CO_2-induced climate change over Western Europe , Climatic Change, 10, 11-42.

Wolf, J.P., 1987: Applications de la spectroscopie laser à la pollution atmosphérique, thèse de doctorat 683, Ecole Polytechnique Fédérale de Lausanne, Switzerland.

Wolf, J.P., and L. Wöste, 1987: Détection sélective et à distance de la pollution atmosphérique par lidar, Helv. Phys. Acta, 60, 161.

Yanai, M., S. Esbensen and J. M. Chu, 1973: Determination of the bulk properties of tropical cloud clusters from large-scale heat and moisture budgets. J. Atmos. Sci. 30, 611-627.

Zdunkowski, W. G., R. M. Welch, and G. Korb, 1980: An investigation of the structure of typical two stream methods for the calculation of solar fluxes and heating rates in clouds. Beitr. Phys. Atm., 53, 147-166.

Appendix: List of Commonly Used Mathematical Symbols

Term	Definition	Units
A	Autoconversion term	s^{-1}
c_p	Specific heat at constant pressure	J/kg/K
c	Propagation velocity in CFL criterion	m/s
CB	Level of cloud base (LCL)	m
c_1 - c_5	Empirical constants used in precipitation scheme	
E	Raindrop evaporation	s^{-1}
e_s	Partial pressure of atmosphericwater vapor	Pa
f	Coriolis term	s^{-1}
F	Raindrop fallout term	s^{-1}
F_x	Turbulent flux divergence for any variable x	
g	Gravitational acceleration	m/s^2
G_x	Large-scale contribution of any variable x	
h	Height of surface layer	m
H	Height of the Atmospheric Boundary Layer	m
i, j, k	Loop indices for computations in x, y, z space	
K_M	Vertical diffusion coefficient for momentum	m^2/s
K_H	Vertical diffusion coefficient for temperature	m^2/s
K_{Horiz}	Horizontal diffusion coefficient	m^2/s
L	Monin-Obukhov stability parameter	m
p	Atmospheric pressure	Pa
p_0	Reference pressure (generally 10^5 Pa)	Pa
q	Specific humidity of water vapor	kg/kg
q_L	Specific humidity of cloud liquid droplets	kg/kg
q_R	Specific humidity of rain drops	kg/kg
q_s	Specific humidity at saturation	kg/kg
Q	Emission rate of pollutants	kg/s
R	Gas constant	J/kg/K
R_{ui}	Lagrangian autocorrelation factor	

Ri	Richardson number (stability parameter)	
t	Time	s
T_{Lui}	Lagrangian time-scale	s
T	Ambient air temperature	K
u	Zonal (east-west) velocity	m/s
U	Horizontal wind vector	m/s
u_*	Friction velocity along x	m/s
v	Meridional (north-south) velocity	m/s
v_*	Friction velocity along y	m/s
V	Horizontal wind vector	m/s
VF	Raindrop evaporation ventilation factor	s^{-1}
w	Vertical velocity	m/s
w_R	Raindrop fallout velocity	m/s
w_*	Convective vertical velocity	m/s
α, β	Constants used in detrainment factor	s^{-1}
β_{ui}	Ratio of Eulerian to Lagrangian time-scales	
δ	Cloud detrainment factor	m^{-1}
Δt	Computational time-step	s
Δx, Δy, Δz	Grid intervals in the x, y, z directions, repectively	m
$\varnothing_{M}$, $\varnothing_{H}$	Stability parameters for heat and moisture	
λ	Wavelength spectrum in Lagrangian turbulence	m^{-1}
ρ	Atmospheric density	kg/m^3
σ	Standard deviation of velocity fluctuations	m/s
θ	Potential temperature	K
θ_v	Virtual potential temperature	K
θ_*	"Friction" potential temperature	K
χ	Pollutant concentrations	kg/m^3

Index

Springer and the environment

MIX
Papier aus verantwortungsvollen Quellen
Paper from responsible sources
FSC® C105338

If you have any concerns about our products,
you can contact us on
ProductSafety@springernature.com

In case Publisher is established outside the EU,
the EU authorized representative is:
Springer Nature Customer Service Center GmbH
Europaplatz 3, 69115 Heidelberg, Germany

Printed by Libri Plureos GmbH
in Hamburg, Germany